Novel Approaches for the Delivery of Anti-HIV Drugs

Novel Approaches for the Delivery of Anti-HIV Drugs

Special Issue Editor

José das Neves

MDPI • Basel • Beijing • Wuhan • Barcelona • Belgrade

Special Issue Editor
José das Neves
University of Porto
Portugal

Editorial Office
MDPI
St. Alban-Anlage 66
4052 Basel, Switzerland

This is a reprint of articles from the Special Issue published online in the open access journal *Pharmaceutics* (ISSN 1999-4923) in 2019 (available at: https://www.mdpi.com/journal/pharmaceutics/special_issues/delivery_anti_HIV_drugs).

For citation purposes, cite each article independently as indicated on the article page online and as indicated below:

LastName, A.A.; LastName, B.B.; LastName, C.C. Article Title. *Journal Name* **Year**, *Article Number*, Page Range.

ISBN 978-3-03921-900-1 (Pbk)
ISBN 978-3-03921-901-8 (PDF)

Cover image courtesy of NIH/National Institute of Allergy and Infectious Diseases.

Contents

About the Special Issue Editor . . . . **vii**

José das Neves
Novel Approaches for the Delivery of Anti-HIV Drugs—What Is New?
Reprinted from: *Pharmaceutics* **2019**, *11*, 554, doi:10.3390/pharmaceutics11110554 . . . . **1**

Tetsuo Tsukamoto
Gene Therapy Approaches to Functional Cure and Protection of Hematopoietic Potential in HIV Infection
Reprinted from: *Pharmaceutics* **2019**, *11*, 114, doi:10.3390/pharmaceutics11030114 . . . . **5**

Nejat Düzgüneş and Krystyna Konopka
Eradication of Human Immunodeficiency Virus Type-1 (HIV-1)-Infected Cells
Reprinted from: *Pharmaceutics* **2019**, *11*, 255, doi:10.3390/pharmaceutics11060255 . . . . **23**

Fernando Notario-Pérez, Raúl Cazorla-Luna, Araceli Martín-Illana, Roberto Ruiz-Caro, Juan Peña and María-Dolores Veiga
Tenofovir Hot-Melt Granulation using Gelucire® to Develop Sustained-Release Vaginal Systems for Weekly Protection against Sexual Transmission of HIV
Reprinted from: *Pharmaceutics* **2019**, *11*, 137, doi:10.3390/pharmaceutics11030137 . . . . **36**

.Letícia Mesquita, Joana Galante, Rute Nunes, Bruno Sarmento and José das Neves
Pharmaceutical Vehicles for Vaginal and Rectal Administration of Anti-HIV Microbicide Nanosystems
Reprinted from: *Pharmaceutics* , *11*, 145, doi:10.3390/pharmaceutics11030145 . . . . **56**

Kevin M. Tyo, Farnaz Minooei, Keegan C. Curry, Sarah M. NeCamp, Danielle L. Graves, Joel R. Fried and Jill M. Steinbach-Rankins
Relating Advanced Electrospun Fiber Architectures to the Temporal Release of Active Agents to Meet the Needs of Next-Generation Intravaginal Delivery Applications
Reprinted from: *Pharmaceutics* **2019**, *11*, 160, doi:10.3390/pharmaceutics11040160 . . . . **76**

Ashana Puri, Sonalika A. Bhattacharjee, Wei Zhang, Meredith Clark, Onkar N. Singh, Gustavo F. Doncel and Ajay K. Banga
Development of a Transdermal Delivery System for Tenofovir Alafenamide, a Prodrug of Tenofovir with Potent Antiviral Activity Against HIV and HBV
Reprinted from: *Pharmaceutics* **2019**, *11*, 173, doi:10.3390/pharmaceutics11040173 . . . . **107**

Leah M. Johnson, Sai Archana Krovi, Linying Li, Natalie Girouard, Zach R. Demkovich, Daniel Myers, Ben Creelman and Ariane van der Straten
Characterization of a Reservoir-Style Implant for Sustained Release of Tenofovir Alafenamide (TAF) for HIV Pre-Exposure Prophylaxis (PrEP)
Reprinted from: *Pharmaceutics* **2019**, *11*, 315, doi:10.3390/pharmaceutics11070315 . . . . **136**

Haitao Yang, Jing Li, Sravan Kumar Patel, Kenneth E. Palmer, Brid Devlin and Lisa C. Rohan
Design of Poly(lactic-*co*-glycolic Acid) (PLGA) Nanoparticles for Vaginal Co-Delivery of Griffithsin and Dapivirine and Their Synergistic Effect for HIV Prophylaxis
Reprinted from: *Pharmaceutics* **2019**, *11*, 184, doi:10.3390/pharmaceutics11040184 . . . . **152**

About the Special Issue Editor

José das Neves graduated in Pharmaceutical Sciences and holds an MSc in Pharmaceutical Technology and a Ph.D. in Pharmaceutical Sciences from the University of Porto, Portugal. He is currently Assistant Researcher at i3S—Institute for Research and Innovation in Health as well as INEB—Institute of Biomedical Engineering, University of Porto. His broad research interests include nanomedicine, mucosal drug delivery, and biomaterials. Dr. das Neves' current research focuses specifically on the development of drug delivery strategies for the prevention or treatment of sexually transmitted infections and lower female genital tract diseases. He has contributed significantly to the development of nanotechnology-based microbicides for preventing sexual transmission of HIV. Dr. das Neves is the co-author of 93 articles in international peer-reviewed journals and co-Editor of 4 scientific books. He is currently Associate Editor of *Frontiers in Pharmacology* (Frontiers Media) and Editorial Board member of *PLoS ONE* (Public Library of Science), *Nanomaterials* (MDPI), *Drug Delivery Letters* (Bentham Science), and *4Open* (EDP Sciences). He is also an active collaborator of the Global Burden of Disease Study.

pharmaceutics

MDPI

Editorial

Novel Approaches for the Delivery of Anti-HIV Drugs—What Is New?

José das Neves [1,2,3]

1 i3S—Instituto de Investigação e Inovação em Saúde, Universidade do Porto, 4200-135 Porto, Portugal; j.dasneves@ineb.up.pt; Tel.: +351-220-408-800
2 INEB—Instituto de Engenharia Biomédica, Universidade do Porto, 4200-135 Porto, Portugal
3 CESPU, Instituto de Investigação e Formação Avançada em Ciências e Tecnologias da Saúde, 4585-116 Gandra, Portugal

Received: 21 October 2019; Accepted: 24 October 2019; Published: 28 October 2019

HIV/AIDS continues to be one of the most challenging individual and public health concerns of our days. According to the latest UNAIDS data, in 2018, roughly 37.9 million individuals were infected with HIV globally, while around 770,000 people died of AIDS-related illness [1]. During that same year, an estimated 1.7 million new infections occurred, mainly due to unprotected sexual intercourse. Investment in the field has been considerable, but a cure to the infection remains elusive. Nonetheless, tremendous advances have been made over the last 36 years since HIV-1 was identified, namely in prevention, diagnostics, and treatment. The development of antiretroviral drugs and the introduction of highly active antiretroviral therapy (HAART) in the mid-1990s—currently referred to as combination antiretroviral therapy (cART)—led to a dramatic shift of AIDS from a fatal disease into a chronic and often stable medical condition [2]. In fact, cART contributed decisively to a steady decrease in the number of HIV-related deaths since the first years of the new millennium [3]. Antiretroviral drugs have also been found useful in the prevention field, particularly in post-exposure prophylaxis or mother-to-child transmission. Treatment as prevention and pre-exposure prophylaxis (PrEP) have further contributed to the reduction of sexually transmitted HIV infections. Long-lasting injectable products and antiretroviral-based microbicides that are currently in late stages of clinical development or regulatory approval may soon provide new options for prevention [4]. Gene therapy and the use of broadly neutralizing antibodies are also attracting a great deal of interest as possible approaches to HIV/AIDS management [5–7].

Still, different challenges remain in anti-HIV drug therapy/prophylaxis, and these include the following, among others: (i) the onset of severe adverse effects leading to the discontinuation or interruption of therapy or even prophylaxis [8,9]; (ii) sub-optimal biodistribution and pharmacokinetics, particularly in reservoir sites or mucosae involved in sexual transmission [10,11]; (iii) the occurrence of viral resistance [12]; (iv) troublesome regimens and/or drug delivery routes that lead to poor adherence by patients/users [13,14]; (v) low stability and reduced shelf-life of active molecules, which may be particularly challenging in tropical climates and low-resource regions lacking adequate refrigerated distribution channels and storage [15]; (vi) lack of suitable dosage forms for particular populations (e.g., children and women) [16,17]; (vii) costly drug products that are often inaccessible to populations in need of therapy/prophylaxis [15]; and (viii) social and legal constraints resulting in poor access to and the discontinuation of anti-HIV therapy/prophylaxis [18,19]. The response from the scientific community could not be more affirmative, and novel ideas and concepts have been emerging throughout the last decade or so. More important, innovative products are now under development, holding great promise for mitigating many of the challenges identified above.

This Special Issue presents an exciting series of reviews and original research articles from eminent scientists in academia and different nonprofit organizations involved in the development of antiretroviral drug products, focusing mainly on novel strategies for the formulation and delivery of anti-HIV compounds. Innovative approaches towards improved gene therapy and immunotherapy are also addressed. The presented reports provide not only interesting overviews and opinion on recent developments in the broad field of antiretroviral therapy/prophylaxis and drug delivery, but also describe the development of new products that are currently tracked for clinical testing.

The Special Issue starts with an interesting review by Tsukamoto at Kindai University, Japan, on strategies explored for curing HIV infection using a combination of gene therapy and host immunization [20]. In particular, the author emphasizes the possible role of anti-HIV intracellular immunization using gene silencing, among other approaches, in the protection of bone marrow hematopoietic stem/progenitor cells. Still in the same field, Düzgüneş and Konopka at the University of the Pacific, USA, contributed a stimulating review on a potential strategy for the eradication of cellular reservoirs of HIV [21]. This thought-provoking piece explores how such an objective could be achieved by using suicide gene therapy for killing HIV-infected cells, excision of chromosome-integrated viral DNA, and cytotoxic liposomes targeted to latency-reversed HIV-infected cells.

In the first original research study included in the Special Issue, the group of Veiga at the Complutense University of Madrid, Spain, provides details on the development of mucoadhesive tablets for the vaginal delivery of tenofovir, in the context of topical PrEP [22]. The combination of drug-loaded hydrophobic granules obtained by hot-melt granulation and hydrophilic matrices not only allowed the adhesive behavior of tablets to be increased, but also provided sustained drug release. This new formulation could be potentially beneficial in providing longer protective time windows against male-to-female transmission of HIV. The Special Issue continues with a review article on topical nano-microbicides, this time from my research team [23]. We provide an overview on useful vaginal and rectal platforms for the delivery of anti-HIV microbicide nanosystems. Critical topics and relevant studies concerning the development and testing of vehicles such as aqueous suspensions, gels, thermosensitive systems, films and fiber mats, among others, are detailed. Steinbach-Rankins and colleagues at the University of Louisville, USA, contributed an excellent revision of their own work, as well as the work of others, concerning the development and potential of electrospun fibers for vaginal drug delivery [24]. They particularly focus on the formulation of anti-HIV compounds, and how suitable material selection and the engineering of fibers can contribute to the modulation of the time required for complete drug release, ranging from a few minutes to over one week.

Still in the area of prophylaxis, the team led by Banga at Mercer University and CONRAD, USA, propose a new transdermal delivery system for tenofovir alafenamide, a nucleotide reverse transcriptase inhibitor [25]. The silicone-based patch was shown to be able to provide in vitro sustained drug release that may potentially allow weekly cutaneous applications for the purpose of systemic PrEP. Another exciting alternative for the delivery of tenofovir alafenamide was reported by Johnson et al. at RTI International and PATH, USA [26]. These researchers provide details on the manufacturing and in vitro evaluation of a subcutaneous reservoir-style implant for long-term delivery of the drug. In particular, sustained release was achieved for an impressive period of 180 days, representing an important step towards the development of a putative long-acting product for systemic PrEP or even therapy.

Rohan and colleagues at the University of Pittsburgh, Magee-Womens Research Institute, University of Louisville and International Partnership for Microbicides, USA, contributed an interesting study that further endorses the potential of nanotechnology-based microbicides [27]. In their study, poly(lactic-*co*-glycolic acid)-based nanoparticles were developed as carriers for the co-delivery of griffithsin and dapivirine, two potent candidate microbicide compounds. Studies in vitro showed that the proposed formulation not only featured interesting technological properties (including biphasic drug release), but also allowed a synergistic antiretroviral effect to be obtained. In another paper pertaining to the application of nanotechnology against HIV infection, Grande et al. (University of Calabria, Italy) reviewed the literature for nanocarriers of reverse transcriptase inhibitors [28]. In this interesting article, the authors provide a critical analysis on how nanosystems such as liposomes, niosomes, and solid lipid nanoparticles can help with overcoming technological and pharmacokinetic problems of this important class of antiretroviral drugs.

I hope that researchers involved in the fields of antiretroviral drug delivery and anti-HIV therapy/prophylaxis may find useful and stimulating information here that can be translated into their own ongoing and future work. A final word of appreciation is due to all the contributing authors, anonymous reviewers, and editorial staff at MDPI for making the publication of this Special Issue of *Pharmaceutics* possible.

Conflicts of Interest: The author declares no conflict of interest.

References

1. UNAIDS. *UNAIDS Data 2019*; UNAIDS: Geneva, Switzerland, 2019; Available online: https://www.unaids.org/en/resources/documents/2019/2019-UNAIDS-data (accessed on 16 October 2019).
2. Cihlar, T.; Fordyce, M. Current status and prospects of HIV treatment. *Curr. Opin. Virol.* **2016**, *18*, 50–56. [CrossRef] [PubMed]
3. GBD HIV Collaborators. Global, regional, and national incidence, prevalence, and mortality of HIV, 1980–2017, and forecasts to 2030, for 195 countries and territories: A systematic analysis for the Global Burden of Diseases, Injuries, and Risk Factors Study 2017. *Lancet HIV* **2019**. [CrossRef]
4. Piot, P.; Abdool Karim, S.S.; Hecht, R.; Legido-Quigley, H.; Buse, K.; Stover, J.; Resch, S.; Ryckman, T.; Møgedal, S.; Dybul, M.; et al. UNAIDS-Lancet Commission, Defeating AIDS–advancing global health. *Lancet* **2015**, *386*, 171–218. [CrossRef]
5. Pernet, O.; Yadav, S.S.; An, D.S. Stem cell based therapy for HIV/AIDS. *Adv. Drug Deliv. Rev.* **2016**, *103*, 187–201. [CrossRef]
6. Hua, C.; Ackerman, M.E. Engineering broadly neutralizing antibodies for HIV prevention and therapy. *Adv. Drug Deliv. Rev.* **2016**, *103*, 157–173. [CrossRef]
7. Swamy, M.N.; Wu, H.; Shankar, P. Recent advances in RNAi-mediated therapy and prevention of HIV-1/AIDS. *Adv. Drug Deliv. Rev.* **2016**, *103*, 174–186. [CrossRef]
8. Hawkins, T. Understanding and managing the adverse effects of antiretroviral therapy. *Antivir. Res.* **2010**, *85*, 201–209. [CrossRef]
9. Riddell, J.T.; Amico, K.R.; Mayer, K.H. HIV preexposure prophylaxis: A review. *JAMA* **2018**, *319*, 1261–1268. [CrossRef]
10. Cory, T.J.; Schacker, T.W.; Stevenson, M.; Fletcher, C.V. Overcoming pharmacologic sanctuaries. *Curr. Opin. HIV AIDS* **2013**, *8*, 190–195. [CrossRef]
11. Else, L.J.; Taylor, S.; Back, D.J.; Khoo, S.H. Pharmacokinetics of antiretroviral drugs in anatomical sanctuary sites: The male and female genital tract. *Antivir. Ther.* **2011**, *16*, 1149–1167. [CrossRef]
12. Iyidogan, P.; Anderson, K.S. Current perspectives on HIV-1 antiretroviral drug resistance. *Viruses* **2014**, *6*, 4095–4139. [CrossRef] [PubMed]
13. Chen, Y.; Chen, K.; Kalichman, S.C. Barriers to HIV medication adherence as a function of regimen simplification. *Ann. Behav. Med.* **2017**, *51*, 67–78. [CrossRef] [PubMed]
14. Woodsong, C.; MacQueen, K.; Amico, K.R.; Friedland, B.; Gafos, M.; Mansoor, L.; Tolley, E.; McCormack, S. Microbicide clinical trial adherence: Insights for introduction. *J. Int. AIDS Soc.* **2013**, *16*, 18505. [CrossRef] [PubMed]
15. Crawford, K.W.; Ripin, D.H.; Levin, A.D.; Campbell, J.R.; Flexner, C. Participants of Conference on Antiretroviral Drug Optimization. Optimising the manufacture, formulation, and dose of antiretroviral drugs for more cost-efficient delivery in resource-limited settings: A consensus statement. *Lancet Infect. Dis.* **2012**, *12*, 550–560. [CrossRef]
16. Dubrocq, G.; Rakhmanina, N.; Phelps, B.R. Challenges and opportunities in the development of HIV medications in pediatric patients. *Paediatr. Drugs* **2017**, *19*, 91–98. [CrossRef]
17. Woodsong, C.; Holt, J.D. Acceptability and preferences for vaginal dosage forms intended for prevention of HIV or HIV and pregnancy. *Adv. Drug Deliv. Rev.* **2015**, *15*, 146–154. [CrossRef]

18. Elopre, L.; Kudroff, K.; Westfall, A.O.; Overton, E.T.; Mugavero, M.J. The right people, right places, and right practices: Disparities in PrEP access among African American men, women, and MSM in the deep south. *J. Acquir. Immune Defic. Syndr.* **2017**, *74*, 56–59. [CrossRef]
19. Vella, S.; Schwartlander, B.; Sow, S.P.; Eholie, S.P.; Murphy, R.L. The history of antiretroviral therapy and of its implementation in resource-limited areas of the world. *AIDS* **2012**, *26*, 1231–1241. [CrossRef]
20. Tsukamoto, T. Gene therapy approaches to functional cure and protection of hematopoietic potential in HIV infection. *Pharmaceutics* **2019**, *11*, 114. [CrossRef]
21. Düzgüneş, N.; Konopka, K. Eradication of human immunodeficiency virus type-1 (HIV-1)-infected cells. *Pharmaceutics* **2019**, *11*, 255.
22. Notario-Pérez, F.; Cazorla-Luna, R.; Martín-Illana, A.; Ruiz-Caro, R.; Peña, J.; Veiga, M.D. Tenofovir hot-melt granulation using Gelucire((R)) to develop sustained-release vaginal systems for weekly protection against sexual transmission of HIV. *Pharmaceutics* **2019**, *11*, 137.
23. Mesquita, L.; Galante, J.; Nunes, R.; Sarmento, B.; das Neves, J. Pharmaceutical vehicles for vaginal and rectal administration of anti-HIV microbicide nanosystems. *Pharmaceutics* **2019**, *11*, 145. [CrossRef] [PubMed]
24. Tyo, K.M.; Minooei, F.; Curry, K.C.; NeCamp, S.M.; Graves, D.L.; Fried, J.R.; Steinbach-Rankins, J.M. Relating advanced electrospun fiber architectures to the temporal release of active agents to meet the needs of next-generation intravaginal delivery applications. *Pharmaceutics* **2019**, *11*, 160. [CrossRef] [PubMed]
25. Puri, A.; Bhattaccharjee, S.A.; Zhang, W.; Clark, M.; Singh, O.; Doncel, G.F.; Banga, A.K. Development of a transdermal delivery system for tenofovir alafenamide, a prodrug of tenofovir with potent antiviral activity against HIV and HBV. *Pharmaceutics* **2019**, *11*, 173. [CrossRef] [PubMed]
26. Johnson, L.M.; Krovi, S.A.; Li, L.; Girouard, N.; Demkovich, Z.R.; Myers, D.; Creelman, B.; van der Straten, A. Characterization of a reservoir-style implant for sustained release of tenofovir alafenamide (TAF) for HIV pre-exposure prophylaxis (PrEP). *Pharmaceutics* **2019**, *11*, 315. [CrossRef]
27. Yang, H.; Li, J.; Patel, S.K.; Palmer, K.E.; Devlin, B.; Rohan, L.C. Design of poly(lactic-*co*-glycolic acid) (PLGA) nanoparticles for vaginal co-delivery of griffithsin and dapivirine and their synergistic effect for HIV prophylaxis. *Pharmaceutics* **2019**, *11*, 184. [CrossRef]
28. Grande, F.; Ioele, G.; Occhiuzzi, M.A.; De Luca, M.; Mazzotta, E.; Ragno, G.; Garofalo, A.; Muzzalupo, R. Reverse transcriptase inhibitors nanosystems designed for drug stability and controlled delivery. *Pharmaceutics* **2019**, *11*, 197. [CrossRef]

pharmaceutics

MDPI

Review

Gene Therapy Approaches to Functional Cure and Protection of Hematopoietic Potential in HIV Infection

Tetsuo Tsukamoto

Department of Immunology, Kindai University Faculty of Medicine, Osaka 5898511, Japan; ttsukamoto@med.kindai.ac.jp; Tel.: +81-723-66-0221

Received: 7 February 2019; Accepted: 6 March 2019; Published: 11 March 2019

Abstract: Although current antiretroviral drug therapy can suppress the replication of human immunodeficiency virus (HIV), a lifelong prescription is necessary to avoid viral rebound. The problem of persistent and ineradicable viral reservoirs in HIV-infected people continues to be a global threat. In addition, some HIV-infected patients do not experience sufficient T-cell immune restoration despite being aviremic during treatment. This is likely due to altered hematopoietic potential. To achieve the global eradication of HIV disease, a cure is needed. To this end, tremendous efforts have been made in the field of anti-HIV gene therapy. This review will discuss the concepts of HIV cure and relative viral attenuation and provide an overview of various gene therapy approaches aimed at a complete or functional HIV cure and protection of hematopoietic functions.

Keywords: human immunodeficiency virus; acquired immunodeficiency syndrome; hematopoietic stem/progenitor cells; gene therapy

1. Introduction

Human immunodeficiency virus (HIV) infects CD4$^+$ T cells and causes acquired immunodeficiency syndrome (AIDS). AIDS remains as a global threat due to multifactorial reasons, including the difficulty in developing an effective vaccine [1]. According to The Joint United Nations Programme on HIV/AIDS (UNAIDS), in 2017, about 36.9 million people were living with AIDS while only 21.7 million patients were receiving antiretroviral therapy (ART), resulting in about 1.8 million newly HIV-infected people per year [2]. Although ART can limit the size and distribution of HIV reservoirs depending on the earliness of its initiation, it cannot eliminate latent HIV infections from the host and thus, a lifelong prescription is required for suppressing viral rebound from the reservoirs [3]. Therefore, further exploration is vital to discover new treatment options and effective vaccines [4].

The depletion of memory CD4$^+$ T cells preceding AIDS manifestation may be mainly due to the infection of these cells. However, HIV may also reduce the production of naïve T cells by infecting CD4$^+$ thymocytes [5–8]. On the other hand, although the dynamics of hematopoietic stem/progenitor cells (HSPCs) in HIV-infected settings is still unclear, it is well established that HIV infections are associated with hematological changes, such as anemia and pancytopenia [9]. These hematological changes are likely due to the modified HSPCs and hematopoietic potential of the host. Therefore, a cure for HIV disease should consider not only the absence of newly HIV-infected CD4$^+$ cells but also the normal production rates of CD4$^+$ T cells and other hematopoietic cells. To achieve an HIV cure in its strict sense, the protection of bone marrow hematopoietic functions is essential (Figure 1).

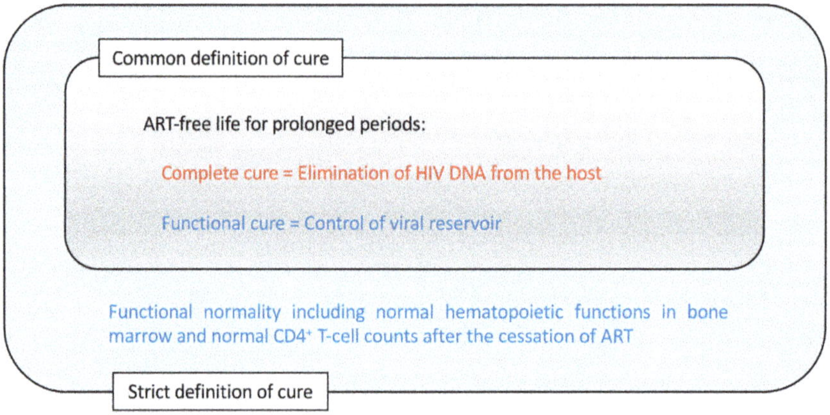

Figure 1. The concepts of human immunodeficiency virus (HIV) cure. A cure for the HIV disease is commonly interpreted as antiretroviral drug therapy (ART)-free life without viral rebound for prolonged periods. In addition, the cure for bone marrow dysfunctionalities observed in HIV-infected patients could be included in a stricter definition of a HIV cure.

This review first describes the current evidence of modified bone marrow hematopoietic potential in HIV infection, leading to the strict definition of an HIV cure. It then explains how anti-HIV gene therapy methods applied to HSPCs can support the preservation of hematopoietic potential and a functional cure. This will be followed by an overview of different potential gene therapy methods for achieving this goal.

2. Evidence of Modified CD34$^+$ Cell Dynamics and Functions in HIV Infection

HIV-1 may cause the loss of primitive hematopoietic progenitors without directly infecting these cells [10]. However, HIV infection does not cause the complete loss of CD34$^+$ stem cells and therefore, it is possible to harvest stem cells from HIV-infected patients suffering from lymphoma [11] albeit with reduced efficiencies in relation to the reduction of peripheral CD4$^+$ T-cell counts [12] or reduced in vitro lymphopoiesis capacities [13]. The recovery of CD4$^+$ T-cell counts after successful antiretroviral drug therapy treatment may depend on the recovery of CD34$^+$ cell counts [14].

A number of potential mechanisms have been suggested for the changes of CD34$^+$ cells in the presence of HIV, such as reduced expression of the proto-oncogene *c-mpl* on CD34$^+$ cells [15] and elevated plasma stromal cell-derived factor 1 (SDF-1) levels [16]. HIV-1 infection causes the upregulation of inflammatory cytokine production that may affect the dynamics and functions [17] or induce Fas-mediated apoptosis [18] of bone marrow CD34$^+$ cells. On the other hand, HSPCs themselves may contribute to inflammation and allergies [19]. This may be partly due to the fact that inflammatory signals are involved in HSPC development [20]. Recent evidence has suggested that CD34$^+$CD226(DNAM-1)^{bright}CXCR4$^+$ cells may represent a subset of common lymphoid progenitors associated with chronic HIV infection and inflammation, reflecting the altered dynamics of natural killer cells and α/β T cells [21].

Humanized mouse models are useful for analyzing bone marrow CD34$^+$ loss or changes after the HIV-1 challenge. In studies with humanized mice infected with CXCR4-tropic HIV-1$_{NL4-3}$, CD34$^+$ hematopoietic progenitor cells were depleted and showed impaired ex vivo myeloid/erythroid colony forming capacities after the challenge [22,23]. A reduction of bone marrow CD34$^+$ cell counts after CCR5-tropic HIV-1 infection was also detected in another study [24]. Interestingly, the depletion of bone marrow CD34$^+$ cells following CCR5-tropic HIV infection has been reported to depend on plasmacytoid dendritic cells [25] or to be associated with the expression of CXCR4 [26]. The latter implicates a potential role of the SDF-1/CXCR4 axis in the loss of CD34$^+$ cells. Another recent in vitro study suggested that CD34$^+$CD7$^+$CXCR4$^+$ lymphoid progenitor cells may be depleted in the presence of CXCR4-tropic HIV-1 in the coculture of HIV-infected cord-derived CD34$^+$ cells with mouse stromal OP9-DL1 cells, which allow the differentiation of T cells [27].

3. The Idea of Intracellular Immunization of HSPCs to Replace the Whole Hematopoietic System

After this, it is important to consider how we could deal with hematopoietic changes in HIV infection. A potential solution is gene therapy. In 1988, David Baltimore presented his idea of intracellular immunization by gene therapy [28] and his concepts are still valid today. First, he suggested expressing inhibitory molecules against HIV in target cells. Second, he proposed using retroviral vectors to transduce cells although lentiviral vectors are widely used today. Third, he conceived the use of gene-modified HSPCs to replace the immune system of the hosts with an HIV-resistant one. These concepts may be summarized as intracellular artificial immune systems designed against HIV and working independently from HIV-specific CD4$^+$ helper T cells, which are the most vulnerable HIV targets [29]. Since his work, a number of candidate gene therapies have been proposed and tested and are described later in this article.

4. The Protection of Bone Marrow CD34$^+$ Cells by an Anti-HIV Gene Therapy Demonstrated In Vivo

However, there have been few reports so far that have tested the protection of CD34$^+$ cells after HIV infection by gene therapy. This may be because viral suppression and CD4$^+$ counts have been widely accepted as measures for the effect of gene therapies against HIV. However, the true goal for any gene therapy against HIV should be the protection of hematopoietic potential because this is another arm of the definition of AIDS, i.e., the loss of cellular immunity (Figure 1).

Regarding this, we have recently reported that a transcriptional gene silencing (TGS) approach using a short hairpin (sh) RNA, which is called shPromA (Figure 2), resulted in limited CXCR4-associated depletion of bone marrow CD34$^+$ cells following CCR5-tropic HIV infection in humanized mice (Figure 3). This suggests that anti-HIV gene therapy can support the preservation of the hematopoietic potential of the hosts [26]. Further characteristics of shPromA and previous studies testing its efficacy as a functional cure gene therapy method is discussed in Section 8.

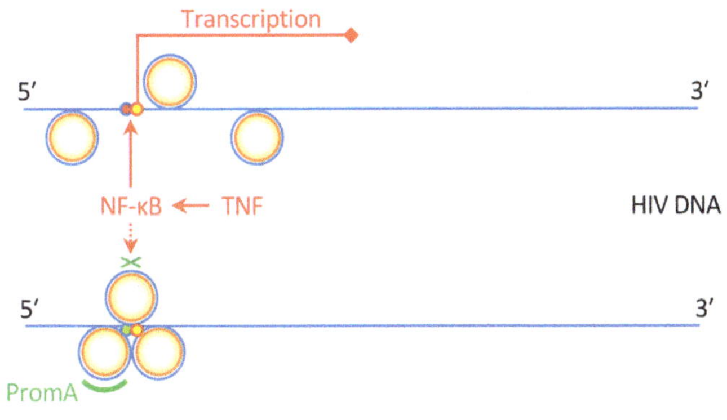

Figure 2. A schematic overview of PromA. PromA induces chromatin compaction in the human immunodeficiency virus (HIV)-1 promoter. This prevents HIV-1 DNA from reactivation, such as NF-κB-mediated reactivation by tissue necrosis factor (TNF). For details on the molecular mechanisms involved in transcriptional gene silencing induced by PromA, see Klemm et al., 2016 [30] and Mendez et al., 2018 [31].

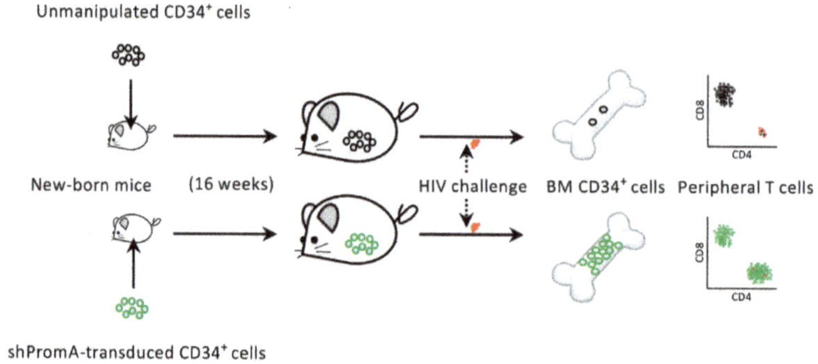

Figure 3. Summary of the humanized mouse study to test the efficacy of shRNA PromA (shPromA) [26]. Newborn NOD/SCID/Jak3^{null} mice were intrahepatically transfused with unmanipulated cord-derived CD34$^+$ cells or CD34$^+$ cells lentivirally transduced with shPromA. Those mice showing engraftment of human cells were challenged with CCR5-tropic HIV-1$_{JRFL}$. Two weeks after the challenge, the mice were sacrificed and their bone marrow (BM) CD34$^+$ cells and peripheral T cells were analyzed. Interestingly, mice transplanted with unmanipulated CD34$^+$ cells showed unexpectedly low BM CD34$^+$ cell counts 2 weeks after HIV infection, with concomitant depletion of peripheral CD4$^+$ T cells. On the other hand, mice engrafted with shPromA-expressing CD34$^+$ cells showed preserved BM CD34$^+$ cell and peripheral CD4$^+$ T-cell populations at 2 weeks post challenge.

5. Target Cells for Anti-HIV Gene Therapies

Recent studies, including the above shPromA study, indicate that ideal anti-HIV gene therapy targets should be hematopoietic stem cells rather than more differentiated cells, such as peripheral CD4$^+$ T cells, because the transduced cells could engraft the host bone marrow and act as a lifelong source of HIV-resistant CD4$^+$ cells [26,32,33]. Potential gene therapies using CD34$^+$ cells have been investigated in vitro using cell culture experiments [26,34–36] or in vivo using humanized mice [26,35,37–40]. Furthermore, the transplantation of macaques with gene-modified autologous CD34$^+$ cells followed by an infection with SIV has also been tested [41,42] although strategies may differ between gene therapies [33]. Based on such basic study results, the clinical trials using the transplantation of retrovirally or lentivirally gene-modified CD34$^+$ cells in HIV-positive patients have been carried out [43–45]. Gene therapies of CD34$^+$ cells have been considered as a cure for monogenic immune diseases. For example, the patients with adenosine deaminase deficiency [46], Wiskott–Aldrich syndrome (WAS) [47] and X-linked severe combined immunodeficiency [48,49] were successfully treated in clinical trials by transplantation of autologous CD34$^+$ cells retrovirally or lentivirally transduced with the wild-type gene. Lentiviral vectors may be more efficient in gene transfer into resting stem cells at the G0/G1 phase compared with murine retroviral vectors [50]. If applied to the gene therapy of HSPCs, both retroviral and lentiviral vectors could have adverse effects, including the deregulation of gene expression [51] and the triggering of the p53 protein [52]. However, lentiviral vectors may be safer than retroviral vectors because the latter may occasionally cause insertional mutagenesis near the active start regions of genes, which could possibly lead to oncogenesis and cancers, such as leukemias [48]. Self-inactivating retroviral or lentiviral vectors lacking the U3 region of 3′ LTRs have further safety advantages [53]. Moreover, recent evidence has shown that the transplantation of WAS patients with autologous CD34$^+$ cells transduced with lentiviral vectors encoding WAS protein results in the long-term survival of genetically engineered hematopoietic stem cells and lymphoid-committed progenitors [54]. Thus, this provides hope for lifelong protection from HIV.

Induced pluripotent stem cells (iPSCs) may also be candidates for anti-HIV gene transfer. iPSCs can be generated from the somatic cells of patients, which can differentiate to any cells in vitro and are expected to be utilized for the treatment of a broad range of genetic diseases [55–58]. Although CD34$^+$ cells can engraft in the bone marrow following transplantation and differentiate to hematopoietic cells in vivo, iPSCs may be more convenient for in vitro hematopoiesis compared to CD34$^+$ cells because of their ease of culture [59]. Interestingly, the impact of shPromA-transduced iPSCs on the suppression of viral replication in vitro has recently been demonstrated, suggesting that the large-scale production of gene-modified monocytes or lymphocytes in vitro for adoptive therapy could be a future option [60]. Additionally, the generation of iPSCs from HIV epitope-specific CD8$^+$ cytotoxic T cells followed by their redifferentiation into the identical epitope-specific CD8$^+$ T cells for adoptive transfer could be an effective immunotherapy [61].

6. Complete Cure vs. Functional Cure for HIV Infection

Before describing individual anti-HIV gene therapy methods, this review looks back on Figure 1 to summarize two major strategies for the treatment of HIV infection. One is to eliminate all the HIV DNA copies within the host, which is termed a complete cure (Figure 1). In pursuing the feasibility of this goal, tremendous efforts have been made to (1) find a method to detect all the latently infected HIV DNAs in viral reservoirs and to (2) eliminate all the detected HIV DNAs so that the host would become sterile in terms of HIV infection [62]. Among the methods to achieve this, the so-called "shock and kill" method, in which the reactivation of the viral reservoir is attempted with a shock-inducing agent followed by the immune-mediated killing of the reactivated cells, has been widely investigated [63–67]. These efforts have been partly successful [62,68]. However, the difficulty of viral eradication in vivo is not limited to HIV but include other viruses that induce long-lasting latent infections, such as herpes simplex viruses, varicella–zoster virus, cytomegalovirus and Epstein–Barr virus, making them ineradicable [69]. HIV may differ from other latently infecting viruses as the viral replication from the latent reservoir can resume quickly even if the host is not considered to be immunocompromised [70]. Moreover, even in the case of the Berlin patient who exhibited no sign of HIV existence following allogeneic transplantation with CCR5-Δ32/Δ32 hematopoietic stem cells, a complete cure was assumed rather than being fully demonstrated [71,72].

Alternatively, some potential gene therapy methods aim at a functional cure that is evidenced by the control of HIV replication below the limit of detection and the immune system being functionally normal despite residual cells harboring HIV proviral DNAs in the host (Figure 1) [68,73]. This approach might be more practical than the complete cure approach, given that many successful vaccines for chronic viral infections so far exert a functional cure rather than achieving the elimination of the targeted viruses [74]. In light of this, it could be stated that for those pathogens where an effective vaccine has not been developed to date, researchers could instead develop gene therapies aimed at a functional cure. In this way, there is an overlap between the concept of functional-cure gene therapy and the concept of vaccines against chronic pathogens [75]. In the next paragraph, the relevance of this is better clarified by looking at a similarity between live-attenuated vaccines and functional-cure gene therapy.

7. Connection between Functional-Cure Gene Therapies and Live-Attenuated Vaccine Approaches

Anti-HIV gene therapy might be compared to some of the vaccine candidates tested so far in order to better foresee its future direction. Live-attenuated vaccines have been tested in macaque AIDS models using simian immunodeficiency virus (SIV) strains [76–82]. After the infection of a host with a live-attenuated SIV or simian-human immunodeficiency virus (SHIV), the vaccine strain is controlled by T-cell response but remains slowly replicating in the infected host. This results in further immunization of the host to prepare for the subsequent superinfections of wild-type SIV or SHIV. Therefore, even if live-attenuated vaccines are powerful, they provide a functional but not a complete cure. This means that there is a scientific connection between live-attenuated vaccines and gene therapy approaches for a functional cure because the latter confer viral attenuation indirectly by rendering the host cells HIV-resistant (Figure 4a). The two distinct strategies can be collectively interpreted as the relative attenuation of the infected virus to the unmanipulated/gene-modified host cells (Figure 4b). Thus, relative viral attenuation might help the host immunity to control the virus [83].

(a)

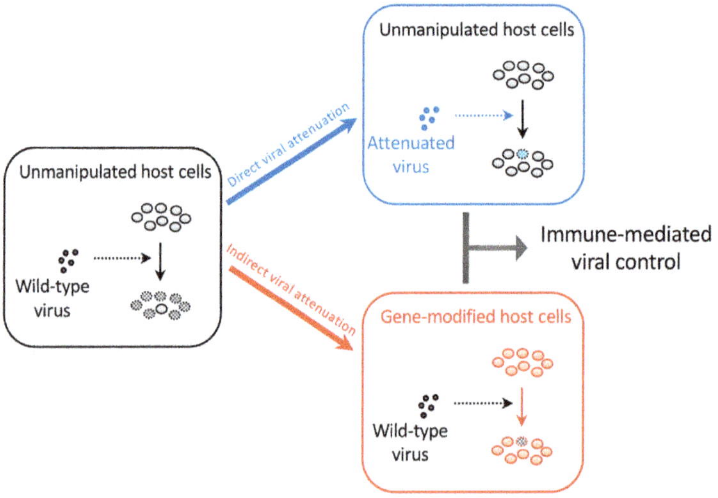

(b)

Given the following parameters:

v_0 ... Virulence of the wild type virus
v_1 ... Virulence of the live-attenuated virus
r_0 ... Viral resistance of the unmanipulated host cells
r_1 ... Viral resistance of the gene-modified host cells

"Relative viral attenuation $f(t_1)$" can be defined as follows, using the direct viral attenuation index (v_0/v_1) and the indirect viral attenuation index (r_1/r_0).

$$f(t_1) = \frac{v_0}{v_1} \times \frac{r_1}{r_0}$$

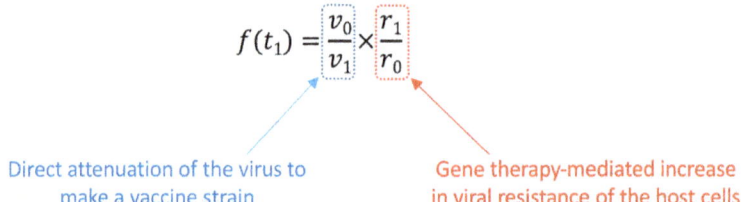

Direct attenuation of the virus to make a vaccine strain

Gene therapy-mediated increase in viral resistance of the host cells

Figure 4. The concept of relative viral attenuation. (**a**) A schema describing direct and indirect viral attenuation. HIV usually infects host CD4$^+$ cells efficiently and replicates rapidly. As a result, the host immune system fails to control viral replication (left). However, accumulating evidence in macaque AIDS models suggests that a live-attenuated virus, which infects and replicates slowly because of partial defects in the viral genome, can be controlled by the immune system and helps in further immunization against a potential superinfection with immunodeficiency virus strains that are homologous to the vaccine strain (upper right). The live attenuation method cannot be directly applied to HIV infection in humans because of safety concerns. However, indirect viral attenuation can be achieved by rendering the host cells HIV-resistant by an "intracellular immunization" gene therapy (lower right); (**b**) The definition of relative viral attenuation. This idea can connect live-attenuated vaccine studies and gene therapy approaches to achieve a functional cure. While live-attenuated HIV vaccines are not testable in humans for safety concerns, functional-cure gene therapy could be an alternative.

8. Gene Therapy Strategies against HIV

The Berlin patient, an HIV-positive male United States citizen who was diagnosed with HIV while attending university in Berlin and later suffered from acute myelogenous leukemia, received a transplantation of allogeneic hematopoietic stem cells homozygous for CCR5-Δ32. This resulted in a subsequent functional HIV cure [71,72]. Because CCR5 is critical in HIV infection and transmission, as observed with CCR5-Δ32 homozygous cells resistant to HIV infection [84], the manipulation of CCR5 expression on HIV target cells has been intensively investigated and is considered to be effective [34,35,37,39,40,85–102]. CCR5 can be targeted by zinc finger nucleases [103,104], ribozymes [105], CRISPR/Cas9 methods [106], transcription activator-like effector nucleases [106] and shRNAs [86,107,108]. Among these, several gene therapy methods, including one using lentiviral vector LVsh5/C46 that expresses shRNA against CCR5 and HIV-1 entry inhibitor C46 [109] have been tested in clinical trials [93]. While CCR5 is involved in numerous pathologic states, including inflammatory and infectious diseases [110], a complete knockout of CCR5 can be related to an increased sensitivity to some viral infections [111,112]. Therefore, CCR5 gene editing should only be considered for an HIV cure [110].

The targeting of HIV RNA sequences by ribozymes or RNAs [44,99,113–123] and HIV DNA sequences by the CRISPR/Cas9 system [124] has been investigated and is also considered a major strategy [100,125]. The latter method has recently been of great interest, which is primarily because of its potential for targeting and disrupting integrated HIV DNA sequences to achieve a complete cure. A recent study targeted and inactivated the HIV-1 long terminal repeat (LTR) U3 region in vitro by Cas9 and guide RNAs (gRNAs), with no off-target gene editing to the host cells being detected [126]. Another study also successfully targeted the HIV-1 LTR U3 region using the CRISPR/Cas9 system. However, this study also detected the emergence of escape variant viruses mediated by the error-prone non-homologous end joining (NHEJ) DNA repair following the CRISPR/Cas9 targeting in the host cells [127]. The mutagenesis problem with CRISPR/Cas9 has also been observed in the treatments of other diseases [128] but can be a serious problem when targeting the HIV DNAs because the strategy might require sustained expression of Cas9 and gRNA in the potential HIV target cells, which means a sustained risk of mutagenesis [129]. Therefore, an improved method for disrupting HIV DNA while prohibiting the emergence of replication-competent escape variants caused by the NHEJ repair system might be necessary. Nevertheless, radical approaches can still be tested in cultured cells and animals. For example, a recent study demonstrated that the in vivo gene delivery of multiplex single-gRNAs and *Staphylococcus aureus* Cas9 to transgenic mice bearing HIV DNA using an adeno-associated virus (AAV) vector resulted in an efficient excision of HIV DNA in various tissues and organs [130]. If safety concerns are met, such a gene delivery method can be a powerful tool to achieve the systemic elimination of latent viral reservoirs in hematopoietic cells and nonhematopoietic cells, such as astrocytes [131]. Wang et al. (2018) have written a thorough review of topics regarding the targeting of HIV DNA by CRISPR/Cas9 [124].

An alternative to the CRISPR/Cas9 strategy is the silencing approach, which aims to reduce the production rate of HIV viral particles per integrated HIV DNA copy [68]. Lentiviral gene delivery enables RNA-based gene silencing, including the previously characterized small interfering RNA (siRNA) called PromA [30,132]. PromA is a short RNA sequence specific for the two NF-κB binding sites in the HIV LTR U3 region. While specific mRNA cleavage by post-transcriptional gene silencing is the best-known mechanism for siRNAs, PromA triggers TGS, which is mediated by epigenetic changes, such as DNA methylation and heterochromatin formation [116,133]. In fact, PromA has been shown to induce chromatin compaction in the HIV-1 promoter region [133]. This means that in contrast to methods attempting to eradicate HIV DNA, PromA locks and stabilizes the latently infecting HIV provirus and prevents the reactivation of viral reservoirs from stimuli, such as tissue necrosis factor (Figure 2) [31,134,135]. The efficacy of PromA in suppressing HIV-1 replication in vivo was first demonstrated by an HIV challenge study using humanized NOD/SCID/JAK3^{null} (NOJ) mice transplanted with human peripheral mononuclear cells expressing shPromA [135]. Our recent study to extend the results using NOJ mice engrafted with shPromA-transduced CD34$^+$ cells and their derivatives further demonstrated that PromA could be an effective gene therapy for protecting bone marrow CD34$^+$ cells and the hematopoietic potential of the host from HIV infection (Figure 3) [26].

Other potential gene therapy methods include the secretion of soluble HIV entry inhibitors [38]; the rescue of hematopoiesis, including myelopoiesis, erythropoiesis and megakaryopoiesis using *c-mpl* [136]; the expression of a chimeric human-simian TRIM5α [137]; the expression of p68 kinase [138]; and the expression of HIV Gag mutants [36].

9. Application of Gene Therapy Methods to Immunotherapies

This section sheds lights on a different application of gene therapy methods to fight against HIV. Immunotherapy approaches based on gene therapy methods have been extensively investigated. Chimeric antigen receptor (CAR) T cells are engineered T cells expressing CARs for the recognition and killing of target cells [139,140]. Most typical CARs are engineered to recognize an antigen with a monoclonal antibody-derived extracellular domain that is conjugated to T-cell receptor-derived transmembrane and intracellular domains. Therefore, despite the use of T-cell signaling pathways, such CAR T-cell therapies might be regarded as an enhancement of antibody-based therapies [141,142]. To date, most successful CAR T-cell therapies have been against cancers [143]. For example, the high efficacy of the adoptive transfer of CAR T cells recognizing CD19 has been demonstrated for the treatment of patients with B-cell acute lymphoblastic leukemia [144] and diffuse large B-cell lymphoma [145]. In contrast, CAR T-cell therapies may require the manufacture of autologous CAR T cells for each patient and thus are not yet widely available [145]. Several broadly neutralizing antibodies have been considered for generating CAR T cells against HIV infection [146–148]. Despite the shared concern of escape mutations with antibody-based therapies, CAR T cells are MHC-independent and more potent than the administration of neutralizing antibodies so better outcomes can be expected. A further improvement of HIV CAR gene therapy has been tested to make CAR T cells HIV-resistant by the insertion of the HIV CAR gene expression cassette into the CCR5 locus, which results in the disruption of CCR5 [148]. Finally, an adoptive transfer therapy using autologous CD34$^+$ cells transduced with lentivirus expressing a CD4-based CAR that is able to bind the HIV envelope protein has been tested in humanized mice infected with HIV-1 [149] and pigtail macaques infected with a SHIV [150].

Another exciting gene transfer-based immunotherapy involves programming the production of specific anti-HIV antibodies [151,152]. Compared with vaccination, passive immunization using a set of broadly neutralizing antibodies is customizable, MHC-independent and provides instant and reliable protection against HIV [153,154]. However, neutralizing antibodies need repeated administration to provide prolonged protection [154]. Thus, the concept of antibody gene transfer is to overcome the limitation of passive immunization, which is only transiently effective [155]. It was demonstrated in a humanized mouse model that antibody gene transfer by intramuscular inoculation of an AAV vector encoding a full-length antibody was able to induce the production of the antibody by muscle cells and confer protection against intravenous HIV-1 challenge [156]. In another study, an adenovirus serotype 5 (Ad5) vector encoding an HIV-1-specific broadly neutralizing antibody PGT121 (Ad5.PGT121) afforded a more rapid and robust antibody response than an AAV encoding PGT121 (AAV1.PGT121) in HIV-1-infected bone marrow-liver-thymus humanized mice [157].

10. Biosafety and Bioethics Concerns Regarding the Application of Anti-HIV Gene Therapies to Human Germline Cells for Pregnancy

In the last part of the review, I would like to comment on the recent issue raised against anti-HIV gene therapy. In late 2018, it was reported that a Chinese researcher used the CRISPR/Cas9 technology to create twins bearing CCR5 double knockouts to confer HIV resistance [158]. However, the inheritable gene modification of human germline cells culminating in human pregnancy is currently unacceptable [159]. If applied to germline cells, CRISPR/Cas9 could cause additional inheritable mutations in the host genome DNA [128,160–162] and the influence of this is not entirely predictable at this moment. Therefore, such an investigation on human germline cells should be limited to nonclinical (i.e., in vitro) studies. Regarding the targeted gene, it should be emphasized that CCR5 knockout has not been proven to be safe. Even if a small population of people, mostly of Caucasian origin, is living without functional CCR5 alleles, this does not mean that the loss of CCR5 is universally harmless. This is partly because CCR5 has been reported to play important roles in some viral infections [111,112]. Moreover, there is an unexcluded possibility that the lack of CCR5 function among those carrying the CCR5-Δ32/Δ32 double mutations could be compensated by accompanying genetic variations that do not exist in the majority of human populations with the wild-type CCR5 alleles. Regarding the protection from HIV-1 infection, CCR5-knockout individuals are still susceptible to the infection of CXCR4-tropic HIV-1 strains despite these risks. We refer to the statement published in 2017 by an American Society of Human Genetics workgroup regarding human germline genome editing [163].

11. Conclusions

The evidence suggests that the HIV infection alters the bone marrow hematopoietic potential of the host. This can lead to impaired CD4$^+$ T-cell generation and contributes to the loss of peripheral CD4$^+$ T cells and the manifestation of AIDS. Further investigations on the topics discussed in this review will collectively enhance our understanding of the important role that HIV gene therapy can contribute toward an HIV cure. Intracellular immunization gene therapies, including silencing approaches, are expected to confer relative viral attenuation without interfering with the HIV genome and assist cellular immunity to kill HIV-infected cells. This will ultimately lead to better viral control and a functional cure. In light of this, the long-term preservation of bone marrow CD34$^+$ cells and hematopoietic potential as well as aviremic states and restored peripheral CD4$^+$ T-cell counts may be an appropriate endpoint of future anti-HIV gene therapies. With this regard, our recent in vivo study using humanized mice transplanted with shRNA (shPromA)-transduced CD34$^+$ cells and challenged with HIV-1 has been described to show that the anti-HIV intracellular immunization gene therapy can indeed protect bone marrow HSPCs. Although the efficacy and safety of those novel gene therapy strategies need further improvements for the future applications to autologous HSPC transplantation of HIV-infected patients, a combination of a gene therapy and host immunization based on preserved bone marrow hematopoietic potential and relative viral attenuation will give a hope for a functional cure. Furthermore, the recent advances in gene therapy-based immunotherapy approaches against HIV have also been described in this review.

Funding: This work received no external funding.

Acknowledgments: I thank Enago (www.enago.jp) for the English language review.

Conflicts of Interest: The author declares no conflict of interest.

References

1. McMichael, A.J. Is a Human CD8 T-Cell Vaccine Possible, and if So, What Would It Take? Could a CD8(+) T-Cell Vaccine Prevent Persistent HIV Infection? *Cold Spring Harb. Perspect. Biol.* **2018**, *10*. [CrossRef] [PubMed]
2. UNAIDS. *UNAIDS Data 2018*; UNAIDS: Geneva, Switzerland, 2018.
3. Hong, F.F.; Mellors, J.W. Changes in HIV reservoirs during long-term antiretroviral therapy. *Curr. Opin. HIV AIDS* **2015**, *10*, 43–48. [CrossRef] [PubMed]
4. Ghosn, J.; Taiwo, B.; Seedat, S.; Autran, B.; Katlama, C. HIV. *Lancet* **2018**, *392*, 685–697. [CrossRef]
5. Zou, W.; Denton, P.W.; Watkins, R.L.; Krisko, J.F.; Nochi, T.; Foster, J.L.; Garcia, J.V. Nef functions in BLT mice to enhance HIV-1 replication and deplete CD4+CD8+ thymocytes. *Retrovirology* **2012**, *9*, 44. [CrossRef] [PubMed]
6. Meissner, E.G.; Duus, K.M.; Loomis, R.; D'Agostin, R.; Su, L. HIV-1 replication and pathogenesis in the human thymus. *Curr. HIV Res.* **2003**, *1*, 275–285. [CrossRef] [PubMed]

7. Zaitseva, M.B.; Lee, S.; Rabin, R.L.; Tiffany, H.L.; Farber, J.M.; Peden, K.W.; Murphy, P.M.; Golding, H. CXCR4 and CCR5 on human thymocytes: Biological function and role in HIV-1 infection. *J. Immunol.* **1998**, *161*, 3103–3113. [PubMed]
8. Taylor, J.R., Jr.; Kimbrell, K.C.; Scoggins, R.; Delaney, M.; Wu, L.; Camerini, D. Expression and function of chemokine receptors on human thymocytes: Implications for infection by human immunodeficiency virus type 1. *J. Virol.* **2001**, *75*, 8752–8760. [CrossRef] [PubMed]
9. Parinitha, S.; Kulkarni, M. Haematological changes in HIV infection with correlation to CD4 cell count. *Australas. Med. J.* **2012**, *5*, 157–162. [CrossRef] [PubMed]
10. Marandin, A.; Katz, A.; Oksenhendler, E.; Tulliez, M.; Picard, F.; Vainchenker, W.; Louache, F. Loss of primitive hematopoietic progenitors in patients with human immunodeficiency virus infection. *Blood* **1996**, *88*, 4568–4578. [PubMed]
11. Re, A.; Cattaneo, C.; Skert, C.; Balsalobre, P.; Michieli, M.; Bower, M.; Ferreri, A.J.; Hentrich, M.; Ribera, J.M.; Allione, B.; et al. Stem cell mobilization in HIV seropositive patients with lymphoma. *Haematologica* **2013**, *98*, 1762–1768. [CrossRef] [PubMed]
12. Schooley, R.T.; Mladenovic, J.; Sevin, A.; Chiu, S.; Miles, S.A.; Pomerantz, R.J.; Campbell, T.B.; Bell, D.; Ambruso, D.; Wong, R.; et al. Reduced mobilization of CD34+ stem cells in advanced human immunodeficiency virus type 1 disease. *J. Infect. Dis.* **2000**, *181*, 148–157. [CrossRef] [PubMed]
13. Nielsen, S.D.; Clark, D.R.; Hutchings, M.; Dam-Larsen, S.; Repping, S.; Nielsen, J.O.; Mathiesen, L.; Miedema, F.; Hansen, J.E. Treatment with granulocyte colony-stimulating factor decreases the capacity of hematopoietic progenitor cells for generation of lymphocytes in human immunodeficiency virus-infected persons. *J. Infect. Dis.* **1999**, *180*, 1819–1826. [CrossRef] [PubMed]
14. Sauce, D.; Larsen, M.; Fastenackels, S.; Pauchard, M.; Ait-Mohand, H.; Schneider, L.; Guihot, A.; Boufassa, F.; Zaunders, J.; Iguertsira, M.; et al. HIV disease progression despite suppression of viral replication is associated with exhaustion of lymphopoiesis. *Blood* **2011**, *117*, 5142–5151. [CrossRef] [PubMed]
15. Koka, P.S.; Kitchen, C.M.; Reddy, S.T. Targeting c-Mpl for revival of human immunodeficiency virus type 1-induced hematopoietic inhibition when CD34+ progenitor cells are re-engrafted into a fresh stromal microenvironment in vivo. *J. Virol.* **2004**, *78*, 11385–11392. [CrossRef] [PubMed]
16. Ikegawa, M.; Yuan, J.; Matsumoto, K.; Herrmann, S.; Iwamoto, A.; Nakamura, T.; Matsushita, S.; Kimura, T.; Honjo, T.; Tashiro, K. Elevated plasma stromal cell-derived factor 1 protein level in the progression of HIV type 1 infection/AIDS. *AIDS Res. Hum. Retroviruses* **2001**, *17*, 587–595. [CrossRef] [PubMed]
17. Bordoni, V.; Bibas, M.; Viola, D.; Sacchi, A.; Cimini, E.; Tumino, N.; Casetti, R.; Amendola, A.; Ammassari, A.; Agrati, C.; et al. Bone Marrow CD34(+) Progenitor Cells from HIV-Infected Patients Show an Impaired T Cell Differentiation Potential Related to Proinflammatory Cytokines. *AIDS Res. Hum. Retroviruses* **2017**, *33*, 590–596. [CrossRef] [PubMed]
18. Isgro, A.; Mezzaroma, I.; Aiuti, A.; Fantauzzi, A.; Pinti, M.; Cossarizza, A.; Aiuti, F. Decreased apoptosis of bone marrow progenitor cells in HIV-1-infected patients during highly active antiretroviral therapy. *AIDS* **2004**, *18*, 1335–1337. [CrossRef] [PubMed]
19. Fischer, K.D.; Agrawal, D.K. Hematopoietic stem and progenitor cells in inflammation and allergy. *Front. Immunol.* **2013**, *4*, 428. [CrossRef] [PubMed]
20. Luis, T.C.; Tremblay, C.S.; Manz, M.G.; North, T.E.; King, K.Y.; Challen, G.A. Inflammatory signals in HSPC development and homeostasis: Too much of a good thing? *Exp. Hematol.* **2016**, *44*, 908–912. [CrossRef] [PubMed]
21. Bozzano, F.; Marras, F.; Ascierto, M.L.; Cantoni, C.; Cenderello, G.; Dentone, C.; Di Biagio, A.; Orofino, G.; Mantia, E.; Boni, S.; et al. 'Emergency exit' of bone-marrow-resident CD34(+)DNAM-1(bright)CXCR4(+)-committed lymphoid precursors during chronic infection and inflammation. *Nat. Commun.* **2015**, *6*, 8109. [CrossRef] [PubMed]
22. Koka, P.S.; Fraser, J.K.; Bryson, Y.; Bristol, G.C.; Aldrovandi, G.M.; Daar, E.S.; Zack, J.A. Human immunodeficiency virus inhibits multilineage hematopoiesis in vivo. *J. Virol.* **1998**, *72*, 5121–5127. [PubMed]
23. Jenkins, M.; Hanley, M.B.; Moreno, M.B.; Wieder, E.; McCune, J.M. Human immunodeficiency virus-1 infection interrupts thymopoiesis and multilineage hematopoiesis in vivo. *Blood* **1998**, *91*, 2672–2678. [PubMed]
24. Arainga, M.; Su, H.; Poluektova, L.Y.; Gorantla, S.; Gendelman, H.E. HIV-1 cellular and tissue replication patterns in infected humanized mice. *Sci. Rep.* **2016**, *6*, 23513. [CrossRef] [PubMed]

25. Li, G.; Zhao, J.; Cheng, L.; Jiang, Q.; Kan, S.; Qin, E.; Tu, B.; Zhang, X.; Zhang, L.; Su, L.; et al. HIV-1 infection depletes human CD34+CD38- hematopoietic progenitor cells via pDC-dependent mechanisms. *PLoS Pathog.* **2017**, *13*, e1006505. [CrossRef] [PubMed]
26. Tsukamoto, T. Transcriptional gene silencing limits CXCR4-associated depletion of bone marrow CD34+ cells in HIV-1 infection. *AIDS* **2018**, *32*, 1737–1747; Erratum in **2018**, *32*, 2857–2858. [CrossRef] [PubMed]
27. Tsukamoto, T. HIV Impacts CD34+ Progenitors Involved in T-Cell Differentiation During Coculture With Mouse Stromal OP9-DL1 Cells. *Front. Immunol.* **2019**, *10*, 81. [CrossRef] [PubMed]
28. Baltimore, D. Gene therapy. Intracellular immunization. *Nature* **1988**, *335*, 395–396. [CrossRef] [PubMed]
29. Tsukamoto, T.; Yamamoto, H.; Okada, S.; Matano, T. Recursion-based depletion of human immunodeficiency virus-specific naive CD4(+) T cells may facilitate persistent viral replication and chronic viraemia leading to acquired immunodeficiency syndrome. *Med. Hypotheses* **2016**, *94*, 81–85. [CrossRef] [PubMed]
30. Klemm, V.; Mitchell, J.; Cortez-Jugo, C.; Cavalieri, F.; Symonds, G.; Caruso, F.; Kelleher, A.D.; Ahlenstiel, C. Achieving HIV-1 Control through RNA-Directed Gene Regulation. *Genes* **2016**, *7*, 119. [CrossRef] [PubMed]
31. Mendez, C.; Ledger, S.; Petoumenos, K.; Ahlenstiel, C.; Kelleher, A.D. RNA-induced epigenetic silencing inhibits HIV-1 reactivation from latency. *Retrovirology* **2018**, *15*, 67. [CrossRef] [PubMed]
32. Savkovic, B.; Nichols, J.; Birkett, D.; Applegate, T.; Ledger, S.; Symonds, G.; Murray, J.M. A quantitative comparison of anti-HIV gene therapy delivered to hematopoietic stem cells versus CD4+ T cells. *PLoS Comput. Biol.* **2014**, *10*, e1003681. [CrossRef] [PubMed]
33. Kitchen, S.G.; Shimizu, S.; An, D.S. Stem cell-based anti-HIV gene therapy. *Virology* **2011**, *411*, 260–272. [CrossRef] [PubMed]
34. Chattong, S.; Chaikomon, K.; Chaiya, T.; Tangkosakul, T.; Palavutitotai, N.; Anusornvongchai, T.; Manotham, K. Efficient ZFN-Mediated Stop Codon Integration into the CCR5 Locus in Hematopoietic Stem Cells: A Possible Source for Intrabone Marrow Cell Transplantation. *AIDS Res. Hum. Retroviruses* **2018**. [CrossRef] [PubMed]
35. Li, L.; Krymskaya, L.; Wang, J.; Henley, J.; Rao, A.; Cao, L.F.; Tran, C.A.; Torres-Coronado, M.; Gardner, A.; Gonzalez, N.; et al. Genomic editing of the HIV-1 coreceptor CCR5 in adult hematopoietic stem and progenitor cells using zinc finger nucleases. *Mol. Ther.* **2013**, *21*, 1259–1269. [CrossRef] [PubMed]
36. Joshi, A.; Garg, H.; Ablan, S.; Freed, E.O.; Nagashima, K.; Manjunath, N.; Shankar, P. Targeting the HIV entry, assembly and release pathways for anti-HIV gene therapy. *Virology* **2011**, *415*, 95–106. [CrossRef] [PubMed]
37. Khamaikawin, W.; Shimizu, S.; Kamata, M.; Cortado, R.; Jung, Y.; Lam, J.; Wen, J.; Kim, P.; Xie, Y.; Kim, S.; et al. Modeling Anti-HIV-1 HSPC-Based Gene Therapy in Humanized Mice Previously Infected with HIV-1. *Mol. Ther. Methods Clin. Dev.* **2018**, *9*, 23–32. [CrossRef] [PubMed]
38. Falkenhagen, A.; Singh, J.; Asad, S.; Leontyev, D.; Read, S.; Zuniga-Pflucker, J.C.; Joshi, S. Control of HIV Infection In Vivo Using Gene Therapy with a Secreted Entry Inhibitor. *Mol. Ther. Nucleic Acids* **2017**, *9*, 132–144. [CrossRef] [PubMed]
39. Petit, N.Y.; Baillou, C.; Burlion, A.; Dorgham, K.; Levacher, B.; Amiel, C.; Schneider, V.; Lemoine, F.M.; Gorochov, G.; Marodon, G. Gene transfer of two entry inhibitors protects CD4(+) T cell from HIV-1 infection in humanized mice. *Gene Ther.* **2016**, *23*, 144–150. [CrossRef] [PubMed]
40. Myburgh, R.; Ivic, S.; Pepper, M.S.; Gers-Huber, G.; Li, D.; Audige, A.; Rochat, M.A.; Jaquet, V.; Regenass, S.; Manz, M.G.; et al. Lentivector Knockdown of CCR5 in Hematopoietic Stem and Progenitor Cells Confers Functional and Persistent HIV-1 Resistance in Humanized Mice. *J. Virol.* **2015**, *89*, 6761–6772. [CrossRef] [PubMed]
41. Rosenzweig, M.; Marks, D.F.; Hempel, D.; Heusch, M.; Kraus, G.; Wong-Staal, F.; Johnson, R.P. Intracellular immunization of rhesus CD34+ hematopoietic progenitor cells with a hairpin ribozyme protects T cells and macrophages from simian immunodeficiency virus infection. *Blood* **1997**, *90*, 4822–4831. [PubMed]
42. An, D.S.; Donahue, R.E.; Kamata, M.; Poon, B.; Metzger, M.; Mao, S.H.; Bonifacino, A.; Krouse, A.E.; Darlix, J.L.; Baltimore, D.; et al. Stable reduction of CCR5 by RNAi through hematopoietic stem cell transplant in non-human primates. *Proc. Natl. Acad. Sci. USA* **2007**, *104*, 13110–13115. [CrossRef] [PubMed]
43. DiGiusto, D.L.; Krishnan, A.; Li, L.; Li, H.; Li, S.; Rao, A.; Mi, S.; Yam, P.; Stinson, S.; Kalos, M.; et al. RNA-based gene therapy for HIV with lentiviral vector-modified CD34(+) cells in patients undergoing transplantation for AIDS-related lymphoma. *Sci. Transl. Med.* **2010**, *2*, 36ra43. [CrossRef] [PubMed]

44. Mitsuyasu, R.T.; Merigan, T.C.; Carr, A.; Zack, J.A.; Winters, M.A.; Workman, C.; Bloch, M.; Lalezari, J.; Becker, S.; Thornton, L.; et al. Phase 2 gene therapy trial of an anti-HIV ribozyme in autologous CD34+ cells. *Nat. Med.* **2009**, *15*, 285–292. [CrossRef] [PubMed]
45. Podsakoff, G.M.; Engel, B.C.; Carbonaro, D.A.; Choi, C.; Smogorzewska, E.M.; Bauer, G.; Selander, D.; Csik, S.; Wilson, K.; Betts, M.R.; et al. Selective survival of peripheral blood lymphocytes in children with HIV-1 following delivery of an anti-HIV gene to bone marrow CD34(+) cells. *Mol. Ther.* **2005**, *12*, 77–86. [CrossRef] [PubMed]
46. Aiuti, A.; Cattaneo, F.; Galimberti, S.; Benninghoff, U.; Cassani, B.; Callegaro, L.; Scaramuzza, S.; Andolfi, G.; Mirolo, M.; Brigida, I.; et al. Gene therapy for immunodeficiency due to adenosine deaminase deficiency. *N. Engl. J. Med.* **2009**, *360*, 447–458. [CrossRef] [PubMed]
47. Boztug, K.; Schmidt, M.; Schwarzer, A.; Banerjee, P.P.; Diez, I.A.; Dewey, R.A.; Bohm, M.; Nowrouzi, A.; Ball, C.R.; Glimm, H.; et al. Stem-cell gene therapy for the Wiskott-Aldrich syndrome. *N. Engl. J. Med.* **2010**, *363*, 1918–1927. [CrossRef] [PubMed]
48. Hacein-Bey-Abina, S.; Hauer, J.; Lim, A.; Picard, C.; Wang, G.P.; Berry, C.C.; Martinache, C.; Rieux-Laucat, F.; Latour, S.; Belohradsky, B.H.; et al. Efficacy of gene therapy for X-linked severe combined immunodeficiency. *N. Engl. J. Med.* **2010**, *363*, 355–364. [CrossRef] [PubMed]
49. De Ravin, S.S.; Wu, X.; Moir, S.; Anaya-O'Brien, S.; Kwatemaa, N.; Littel, P.; Theobald, N.; Choi, U.; Su, L.; Marquesen, M.; et al. Lentiviral hematopoietic stem cell gene therapy for X-linked severe combined immunodeficiency. *Sci. Transl. Med.* **2016**, *8*, 335ra357. [CrossRef] [PubMed]
50. Uchida, N.; Sutton, R.E.; Friera, A.M.; He, D.; Reitsma, M.J.; Chang, W.C.; Veres, G.; Scollay, R.; Weissman, I.L. HIV, but not murine leukemia virus, vectors mediate high efficiency gene transfer into freshly isolated G0/G1 human hematopoietic stem cells. *Proc. Natl. Acad. Sci. USA* **1998**, *95*, 11939–11944. [CrossRef] [PubMed]
51. Cattoglio, C.; Pellin, D.; Rizzi, E.; Maruggi, G.; Corti, G.; Miselli, F.; Sartori, D.; Guffanti, A.; Di Serio, C.; Ambrosi, A.; et al. High-definition mapping of retroviral integration sites identifies active regulatory elements in human multipotent hematopoietic progenitors. *Blood* **2010**, *116*, 5507–5517. [CrossRef] [PubMed]
52. Piras, F.; Riba, M.; Petrillo, C.; Lazarevic, D.; Cuccovillo, I.; Bartolaccini, S.; Stupka, E.; Gentner, B.; Cittaro, D.; Naldini, L.; et al. Lentiviral vectors escape innate sensing but trigger p53 in human hematopoietic stem and progenitor cells. *EMBO Mol. Med.* **2017**, *9*, 1198–1211. [CrossRef] [PubMed]
53. Yu, S.F.; von Ruden, T.; Kantoff, P.W.; Garber, C.; Seiberg, M.; Ruther, U.; Anderson, W.F.; Wagner, E.F.; Gilboa, E. Self-inactivating retroviral vectors designed for transfer of whole genes into mammalian cells. *Proc. Natl. Acad. Sci. USA* **1986**, *83*, 3194–3198. [CrossRef] [PubMed]
54. Scala, S.; Basso-Ricci, L.; Dionisio, F.; Pellin, D.; Giannelli, S.; Salerio, F.A.; Leonardelli, L.; Cicalese, M.P.; Ferrua, F.; Aiuti, A.; et al. Dynamics of genetically engineered hematopoietic stem and progenitor cells after autologous transplantation in humans. *Nat. Med.* **2018**, *24*, 1683–1690. [CrossRef] [PubMed]
55. Nasimuzzaman, M.; Lynn, D.; Ernst, R.; Beuerlein, M.; Smith, R.H.; Shrestha, A.; Cross, S.; Link, K.; Lutzko, C.; Nordling, D.; et al. Production and purification of high-titer foamy virus vector for the treatment of leukocyte adhesion deficiency. *Mol. Ther. Methods Clin. Dev.* **2016**, *3*, 16004. [CrossRef] [PubMed]
56. Vanhee, S.; Vandekerckhove, B. Pluripotent stem cell based gene therapy for hematological diseases. *Crit. Rev. Oncol. Hematol.* **2016**, *97*, 238–246. [CrossRef] [PubMed]
57. Li, Y.; Chan, L.; Nguyen, H.V.; Tsang, S.H. Personalized Medicine: Cell and Gene Therapy Based on Patient-Specific iPSC-Derived Retinal Pigment Epithelium Cells. *Adv. Exp. Med. Biol.* **2016**, *854*, 549–555. [CrossRef] [PubMed]
58. Ou, Z.; Niu, X.; He, W.; Chen, Y.; Song, B.; Xian, Y.; Fan, D.; Tang, D.; Sun, X. The Combination of CRISPR/Cas9 and iPSC Technologies in the Gene Therapy of Human beta-thalassemia in Mice. *Sci. Rep.* **2016**, *6*, 32463. [CrossRef] [PubMed]
59. Carpenter, L.; Malladi, R.; Yang, C.T.; French, A.; Pilkington, K.J.; Forsey, R.W.; Sloane-Stanley, J.; Silk, K.M.; Davies, T.J.; Fairchild, P.J.; et al. Human induced pluripotent stem cells are capable of B-cell lymphopoiesis. *Blood* **2011**, *117*, 4008–4011. [CrossRef] [PubMed]
60. Higaki, K.; Hirao, M.; Kawana-Tachikawa, A.; Iriguchi, S.; Kumagai, A.; Ueda, N.; Bo, W.; Kamibayashi, S.; Watanabe, A.; Nakauchi, H.; et al. Generation of HIV-Resistant Macrophages from IPSCs by Using Transcriptional Gene Silencing and Promoter-Targeted RNA. *Mol. Ther. Nucleic Acids* **2018**, *12*, 793–804. [CrossRef] [PubMed]

61. Nishimura, T.; Kaneko, S.; Kawana-Tachikawa, A.; Tajima, Y.; Goto, H.; Zhu, D.; Nakayama-Hosoya, K.; Iriguchi, S.; Uemura, Y.; Shimizu, T.; et al. Generation of rejuvenated antigen-specific T cells by reprogramming to pluripotency and redifferentiation. *Cell Stem Cell* **2013**, *12*, 114–126. [CrossRef] [PubMed]
62. Huyghe, J.; Magdalena, S.; Vandekerckhove, L. Fight fire with fire: Gene therapy strategies to cure HIV. *Expert Rev. Anti Infect. Ther.* **2017**, *15*, 747–758. [CrossRef] [PubMed]
63. Deeks, S.G. HIV: Shock and kill. *Nature* **2012**, *487*, 439–440. [CrossRef] [PubMed]
64. Darcis, G.; Kula, A.; Bouchat, S.; Fujinaga, K.; Corazza, F.; Ait-Ammar, A.; Delacourt, N.; Melard, A.; Kabeya, K.; Vanhulle, C.; et al. An In-Depth Comparison of Latency-Reversing Agent Combinations in Various In Vitro and Ex Vivo HIV-1 Latency Models Identified Bryostatin-1+JQ1 and Ingenol-B+JQ1 to Potently Reactivate Viral Gene Expression. *PLoS Pathog.* **2015**, *11*, e1005063. [CrossRef] [PubMed]
65. Archin, N.M.; Liberty, A.L.; Kashuba, A.D.; Choudhary, S.K.; Kuruc, J.D.; Crooks, A.M.; Parker, D.C.; Anderson, E.M.; Kearney, M.F.; Strain, M.C.; et al. Administration of vorinostat disrupts HIV-1 latency in patients on antiretroviral therapy. *Nature* **2012**, *487*, 482–485. [CrossRef] [PubMed]
66. Elliott, J.H.; Wightman, F.; Solomon, A.; Ghneim, K.; Ahlers, J.; Cameron, M.J.; Smith, M.Z.; Spelman, T.; McMahon, J.; Velayudham, P.; et al. Activation of HIV transcription with short-course vorinostat in HIV-infected patients on suppressive antiretroviral therapy. *PLoS Pathog.* **2014**, *10*, e1004473. [CrossRef] [PubMed]
67. Rasmussen, T.A.; Tolstrup, M.; Brinkmann, C.R.; Olesen, R.; Erikstrup, C.; Solomon, A.; Winckelmann, A.; Palmer, S.; Dinarello, C.; Buzon, M.; et al. Panobinostat, a histone deacetylase inhibitor, for latent-virus reactivation in HIV-infected patients on suppressive antiretroviral therapy: A phase 1/2, single group, clinical trial. *Lancet HIV* **2014**, *1*, e13–e21. [CrossRef]
68. Cary, D.C.; Peterlin, B.M. Targeting the latent reservoir to achieve functional HIV cure. *F1000Research* **2016**, *5*, F1000. [CrossRef] [PubMed]
69. Chen, T.; Hudnall, S.D. Anatomical mapping of human herpesvirus reservoirs of infection. *Mod. Pathol.* **2006**, *19*, 726–737. [CrossRef] [PubMed]
70. Sengupta, S.; Siliciano, R.F. Targeting the Latent Reservoir for HIV-1. *Immunity* **2018**, *48*, 872–895. [CrossRef] [PubMed]
71. Hutter, G.; Nowak, D.; Mossner, M.; Ganepola, S.; Mussig, A.; Allers, K.; Schneider, T.; Hofmann, J.; Kucherer, C.; Blau, O.; et al. Long-term control of HIV by CCR5 Delta32/Delta32 stem-cell transplantation. *N. Engl. J. Med.* **2009**, *360*, 692–698. [CrossRef] [PubMed]
72. Allers, K.; Hutter, G.; Hofmann, J.; Loddenkemper, C.; Rieger, K.; Thiel, E.; Schneider, T. Evidence for the cure of HIV infection by CCR5Delta32/Delta32 stem cell transplantation. *Blood* **2011**, *117*, 2791–2799. [CrossRef] [PubMed]
73. Archin, N.M.; Margolis, D.M. Emerging strategies to deplete the HIV reservoir. *Curr. Opin. Infect. Dis.* **2014**, *27*, 29–35. [CrossRef] [PubMed]
74. Berzofsky, J.A.; Ahlers, J.D.; Janik, J.; Morris, J.; Oh, S.; Terabe, M.; Belyakov, I.M. Progress on new vaccine strategies against chronic viral infections. *J. Clin. Invest.* **2004**, *114*, 450–462. [CrossRef] [PubMed]
75. Noto, A.; Trautmann, L. Developing Combined HIV Vaccine Strategies for a Functional Cure. *Vaccines* **2013**, *1*, 481–496. [CrossRef] [PubMed]
76. Reeves, R.K.; Gillis, J.; Wong, F.E.; Johnson, R.P. Vaccination with SIVmac239Deltanef activates CD4+ T cells in the absence of CD4 T-cell loss. *J. Med. Primatol.* **2009**, *38* (Suppl. 1), 8–16. [CrossRef]
77. Whitney, J.B.; Ruprecht, R.M. Live attenuated HIV vaccines: Pitfalls and prospects. *Curr. Opin. Infect. Dis.* **2004**, *17*, 17–26. [CrossRef] [PubMed]
78. Sutton, M.S.; Burns, C.M.; Weiler, A.M.; Balgeman, A.J.; Braasch, A.; Lehrer-Brey, G.; Friedrich, T.C.; O'Connor, S.L. Vaccination with Live Attenuated Simian Immunodeficiency Virus (SIV) Protects from Mucosal, but Not Necessarily Intravenous, Challenge with a Minimally Heterologous SIV. *J. Virol.* **2016**, *90*, 5541–5548. [CrossRef] [PubMed]
79. Abel, K.; Compton, L.; Rourke, T.; Montefiori, D.; Lu, D.; Rothaeusler, K.; Fritts, L.; Bost, K.; Miller, C.J. Simian-human immunodeficiency virus SHIV89.6-induced protection against intravaginal challenge with pathogenic SIVmac239 is independent of the route of immunization and is associated with a combination of cytotoxic T-lymphocyte and alpha interferon responses. *J. Virol.* **2003**, *77*, 3099–3118. [PubMed]

80. Nilsson, C.; Makitalo, B.; Thorstensson, R.; Norley, S.; Binninger-Schinzel, D.; Cranage, M.; Rud, E.; Biberfeld, G.; Putkonen, P. Live attenuated simian immunodeficiency virus (SIV)mac in macaques can induce protection against mucosal infection with SIVsm. *AIDS* **1998**, *12*, 2261–2270. [CrossRef] [PubMed]
81. Connor, R.I.; Montefiori, D.C.; Binley, J.M.; Moore, J.P.; Bonhoeffer, S.; Gettie, A.; Fenamore, E.A.; Sheridan, K.E.; Ho, D.D.; Dailey, P.J.; et al. Temporal analyses of virus replication, immune responses, and efficacy in rhesus macaques immunized with a live, attenuated simian immunodeficiency virus vaccine. *J. Virol.* **1998**, *72*, 7501–7509. [PubMed]
82. Sugimoto, C.; Watanabe, S.; Naruse, T.; Kajiwara, E.; Shiino, T.; Umano, N.; Ueda, K.; Sato, H.; Ohgimoto, S.; Hirsch, V.; et al. Protection of macaques with diverse MHC genotypes against a heterologous SIV by vaccination with a deglycosylated live-attenuated SIV. *PLoS ONE* **2010**, *5*, e11678. [CrossRef] [PubMed]
83. Tsukamoto, T.; Yamamoto, H.; Matano, T. CD8(+) Cytotoxic-T-Lymphocyte Breadth Could Facilitate Early Immune Detection of Immunodeficiency Virus-Derived Epitopes with Limited Expression Levels. *mSphere* **2019**, *4*. [CrossRef] [PubMed]
84. Duarte, R.F.; Salgado, M.; Sanchez-Ortega, I.; Arnan, M.; Canals, C.; Domingo-Domenech, E.; Fernandez-de-Sevilla, A.; Gonzalez-Barca, E.; Moron-Lopez, S.; Nogues, N.; et al. CCR5 Delta32 homozygous cord blood allogeneic transplantation in a patient with HIV: A case report. *Lancet HIV* **2015**, *2*, e236–e242. [CrossRef]
85. Xu, L.; Yang, H.; Gao, Y.; Chen, Z.; Xie, L.; Liu, Y.; Liu, Y.; Wang, X.; Li, H.; Lai, W.; et al. CRISPR/Cas9-Mediated CCR5 Ablation in Human Hematopoietic Stem/Progenitor Cells Confers HIV-1 Resistance In Vivo. *Mol. Ther.* **2017**, *25*, 1782–1789. [CrossRef] [PubMed]
86. Symonds, G.; Bartlett, J.S.; Kiem, H.P.; Tsie, M.; Breton, L. Cell-Delivered Entry Inhibitors for HIV-1: CCR5 Downregulation and Blocking Virus/Membrane Fusion in Defending the Host Cell Population. *AIDS Patient Care STDS* **2016**, *30*, 545–550. [CrossRef] [PubMed]
87. Shi, B.; Li, J.; Shi, X.; Jia, W.; Wen, Y.; Hu, X.; Zhuang, F.; Xi, J.; Zhang, L. TALEN-Mediated Knockout of CCR5 Confers Protection Against Infection of Human Immunodeficiency Virus. *J. Acquir. Immune Defic. Syndr.* **2017**, *74*, 229–241. [CrossRef] [PubMed]
88. Shimizu, S.; Yadav, S.S.; An, D.S. Stable Delivery of CCR5-Directed shRNA into Human Primary Peripheral Blood Mononuclear Cells and Hematopoietic Stem/Progenitor Cells via a Lentiviral Vector. *Methods Mol. Biol.* **2016**, *1364*, 235–248. [CrossRef] [PubMed]
89. Sather, B.D.; Romano Ibarra, G.S.; Sommer, K.; Curinga, G.; Hale, M.; Khan, I.F.; Singh, S.; Song, Y.; Gwiazda, K.; Sahni, J.; et al. Efficient modification of CCR5 in primary human hematopoietic cells using a megaTAL nuclease and AAV donor template. *Sci. Transl. Med.* **2015**, *7*, 307ra156. [CrossRef] [PubMed]
90. Saydaminova, K.; Ye, X.; Wang, H.; Richter, M.; Ho, M.; Chen, H.; Xu, N.; Kim, J.S.; Papapetrou, E.; Holmes, M.C.; et al. Efficient genome editing in hematopoietic stem cells with helper-dependent Ad5/35 vectors expressing site-specific endonucleases under microRNA regulation. *Mol. Ther. Methods Clin. Dev.* **2015**, *1*, 14057. [CrossRef] [PubMed]
91. Manotham, K.; Chattong, S.; Setpakdee, A. Generation of CCR5-defective CD34 cells from ZFN-driven stop codon-integrated mesenchymal stem cell clones. *J. Biomed. Sci.* **2015**, *22*, 25. [CrossRef] [PubMed]
92. Burke, B.P.; Levin, B.R.; Zhang, J.; Sahakyan, A.; Boyer, J.; Carroll, M.V.; Colon, J.C.; Keech, N.; Rezek, V.; Bristol, G.; et al. Engineering Cellular Resistance to HIV-1 Infection In Vivo Using a Dual Therapeutic Lentiviral Vector. *Mol. Ther. Nucleic Acids* **2015**, *4*, e236. [CrossRef] [PubMed]
93. Wolstein, O.; Boyd, M.; Millington, M.; Impey, H.; Boyer, J.; Howe, A.; Delebecque, F.; Cornetta, K.; Rothe, M.; Baum, C.; et al. Preclinical safety and efficacy of an anti-HIV-1 lentiviral vector containing a short hairpin RNA to CCR5 and the C46 fusion inhibitor. *Mol. Ther. Methods Clin. Dev.* **2014**, *1*, 11. [CrossRef] [PubMed]
94. Holt, N.; Wang, J.; Kim, K.; Friedman, G.; Wang, X.; Taupin, V.; Crooks, G.M.; Kohn, D.B.; Gregory, P.D.; Holmes, M.C.; et al. Human hematopoietic stem/progenitor cells modified by zinc-finger nucleases targeted to CCR5 control HIV-1 in vivo. *Nat. Biotechnol.* **2010**, *28*, 839–847. [CrossRef] [PubMed]
95. Liang, M.; Kamata, M.; Chen, K.N.; Pariente, N.; An, D.S.; Chen, I.S. Inhibition of HIV-1 infection by a unique short hairpin RNA to chemokine receptor 5 delivered into macrophages through hematopoietic progenitor cell transduction. *J. Gene Med.* **2010**, *12*, 255–265. [CrossRef] [PubMed]
96. Shimizu, S.; Hong, P.; Arumugam, B.; Pokomo, L.; Boyer, J.; Koizumi, N.; Kittipongdaja, P.; Chen, A.; Bristol, G.; Galic, Z.; et al. A highly efficient short hairpin RNA potently down-regulates CCR5 expression in systemic lymphoid organs in the hu-BLT mouse model. *Blood* **2010**, *115*, 1534–1544. [CrossRef] [PubMed]

97. Anderson, J.S.; Javien, J.; Nolta, J.A.; Bauer, G. Preintegration HIV-1 inhibition by a combination lentiviral vector containing a chimeric TRIM5 alpha protein, a CCR5 shRNA, and a TAR decoy. *Mol. Ther.* **2009**, *17*, 2103–2114. [CrossRef] [PubMed]
98. Anderson, J.; Akkina, R. Complete knockdown of CCR5 by lentiviral vector-expressed siRNAs and protection of transgenic macrophages against HIV-1 infection. *Gene Ther.* **2007**, *14*, 1287–1297. [CrossRef] [PubMed]
99. Anderson, J.; Li, M.J.; Palmer, B.; Remling, L.; Li, S.; Yam, P.; Yee, J.K.; Rossi, J.; Zaia, J.; Akkina, R. Safety and Efficacy of a Lentiviral Vector Containing Three Anti-HIV Genes-CCR5 Ribozyme, Tat-rev siRNA, and TAR Decoy-in SCID-hu Mouse-Derived T Cells. *Mol. Ther.* **2007**, *15*, 1182–1188. [CrossRef] [PubMed]
100. Li, M.J.; Kim, J.; Li, S.; Zaia, J.; Yee, J.K.; Anderson, J.; Akkina, R.; Rossi, J.J. Long-term inhibition of HIV-1 infection in primary hematopoietic cells by lentiviral vector delivery of a triple combination of anti-HIV shRNA, anti-CCR5 ribozyme, and a nucleolar-localizing TAR decoy. *Mol. Ther.* **2005**, *12*, 900–909. [CrossRef] [PubMed]
101. Bai, J.; Rossi, J.; Akkina, R. Multivalent anti-CCR ribozymes for stem cell-based HIV type 1 gene therapy. *AIDS Res. Hum. Retroviruses* **2001**, *17*, 385–399. [CrossRef] [PubMed]
102. Bai, J.; Gorantla, S.; Banda, N.; Cagnon, L.; Rossi, J.; Akkina, R. Characterization of anti-CCR5 ribozyme-transduced CD34+ hematopoietic progenitor cells in vitro and in a SCID-hu mouse model in vivo. *Mol. Ther.* **2000**, *1*, 244–254. [CrossRef] [PubMed]
103. Jamieson, A.C.; Miller, J.C.; Pabo, C.O. Drug discovery with engineered zinc-finger proteins. *Nat. Rev. Drug Discov.* **2003**, *2*, 361–368. [CrossRef] [PubMed]
104. Tebas, P.; Stein, D.; Tang, W.W.; Frank, I.; Wang, S.Q.; Lee, G.; Spratt, S.K.; Surosky, R.T.; Giedlin, M.A.; Nichol, G.; et al. Gene editing of CCR5 in autologous CD4 T cells of persons infected with HIV. *N. Engl. J. Med.* **2014**, *370*, 901–910. [CrossRef] [PubMed]
105. Scarborough, R.J.; Gatignol, A. HIV and Ribozymes. *Adv. Exp. Med. Biol.* **2015**, *848*, 97–116. [CrossRef] [PubMed]
106. Cornu, T.I.; Mussolino, C.; Cathomen, T. Refining strategies to translate genome editing to the clinic. *Nat. Med.* **2017**, *23*, 415–423. [CrossRef] [PubMed]
107. Swamy, M.N.; Wu, H.; Shankar, P. Recent advances in RNAi-based strategies for therapy and prevention of HIV-1/AIDS. *Adv. Drug Deliv. Rev.* **2016**, *103*, 174–186. [CrossRef] [PubMed]
108. Symonds, G.P.; Johnstone, H.A.; Millington, M.L.; Boyd, M.P.; Burke, B.P.; Breton, L.R. The use of cell-delivered gene therapy for the treatment of HIV/AIDS. *Immunol. Res.* **2010**, *48*, 84–98. [CrossRef] [PubMed]
109. Ledger, S.; Howe, A.; Turville, S.; Aggarwal, A.; Savkovic, B.; Ong, A.; Wolstein, O.; Boyd, M.; Millington, M.; Gorry, P.R.; et al. Analysis and dissociation of anti-HIV effects of shRNA to CCR5 and the fusion inhibitor C46. *J. Gene Med.* **2018**, *20*, e3006. [CrossRef] [PubMed]
110. Vangelista, L.; Vento, S. The Expanding Therapeutic Perspective of CCR5 Blockade. *Front. Immunol.* **2017**, *8*, 1981. [CrossRef] [PubMed]
111. Lim, J.K.; Louie, C.Y.; Glaser, C.; Jean, C.; Johnson, B.; Johnson, H.; McDermott, D.H.; Murphy, P.M. Genetic deficiency of chemokine receptor CCR5 is a strong risk factor for symptomatic West Nile virus infection: A meta-analysis of 4 cohorts in the US epidemic. *J. Infect. Dis.* **2008**, *197*, 262–265. [CrossRef] [PubMed]
112. Glass, W.G.; McDermott, D.H.; Lim, J.K.; Lekhong, S.; Yu, S.F.; Frank, W.A.; Pape, J.; Cheshier, R.C.; Murphy, P.M. CCR5 deficiency increases risk of symptomatic West Nile virus infection. *J. Exp. Med.* **2006**, *203*, 35–40. [CrossRef] [PubMed]
113. Liu, Y.P.; Westerink, J.T.; ter Brake, O.; Berkhout, B. RNAi-inducing lentiviral vectors for anti-HIV-1 gene therapy. *Methods Mol. Biol.* **2011**, *721*, 293–311. [CrossRef] [PubMed]
114. Kumar, P.; Ban, H.S.; Kim, S.S.; Wu, H.; Pearson, T.; Greiner, D.L.; Laouar, A.; Yao, J.; Haridas, V.; Habiro, K.; et al. T cell-specific siRNA delivery suppresses HIV-1 infection in humanized mice. *Cell* **2008**, *134*, 577–586. [CrossRef] [PubMed]
115. ter Brake, O.; Legrand, N.; von Eije, K.J.; Centlivre, M.; Spits, H.; Weijer, K.; Blom, B.; Berkhout, B. Evaluation of safety and efficacy of RNAi against HIV-1 in the human immune system (Rag-2(-/-)gammac(-/-)) mouse model. *Gene Ther.* **2009**, *16*, 148–153. [CrossRef] [PubMed]
116. Suzuki, K.; Shijuuku, T.; Fukamachi, T.; Zaunders, J.; Guillemin, G.; Cooper, D.; Kelleher, A. Prolonged transcriptional silencing and CpG methylation induced by siRNAs targeted to the HIV-1 promoter region. *J. RNAi Gene Silenc.* **2005**, *1*, 66–78. [PubMed]

117. Santat, L.; Paz, H.; Wong, C.; Li, L.; Macer, J.; Forman, S.; Wong, K.K.; Chatterjee, S. Recombinant AAV2 transduction of primitive human hematopoietic stem cells capable of serial engraftment in immune-deficient mice. *Proc. Natl. Acad. Sci. USA* **2005**, *102*, 11053–11058. [CrossRef] [PubMed]
118. Li, M.J.; Bauer, G.; Michienzi, A.; Yee, J.K.; Lee, N.S.; Kim, J.; Li, S.; Castanotto, D.; Zaia, J.; Rossi, J.J. Inhibition of HIV-1 infection by lentiviral vectors expressing Pol III-promoted anti-HIV RNAs. *Mol. Ther.* **2003**, *8*, 196–206. [CrossRef]
119. Akkina, R.; Banerjea, A.; Bai, J.; Anderson, J.; Li, M.J.; Rossi, J. siRNAs, ribozymes and RNA decoys in modeling stem cell-based gene therapy for HIV/AIDS. *Anticancer Res.* **2003**, *23*, 1997–2005. [PubMed]
120. Banerjea, A.; Li, M.J.; Bauer, G.; Remling, L.; Lee, N.S.; Rossi, J.; Akkina, R. Inhibition of HIV-1 by lentiviral vector-transduced siRNAs in T lymphocytes differentiated in SCID-hu mice and CD34+ progenitor cell-derived macrophages. *Mol. Ther.* **2003**, *8*, 62–71. [CrossRef]
121. Bauer, G.; Valdez, P.; Kearns, K.; Bahner, I.; Wen, S.F.; Zaia, J.A.; Kohn, D.B. Inhibition of human immunodeficiency virus-1 (HIV-1) replication after transduction of granulocyte colony-stimulating factor-mobilized CD34+ cells from HIV-1-infected donors using retroviral vectors containing anti-HIV-1 genes. *Blood* **1997**, *89*, 2259–2267. [PubMed]
122. Rosenzweig, M.; Marks, D.F.; Hempel, D.; Lisziewicz, J.; Johnson, R.P. Transduction of CD34+ hematopoietic progenitor cells with an antitat gene protects T-cell and macrophage progeny from AIDS virus infection. *J. Virol.* **1997**, *71*, 2740–2746. [PubMed]
123. Yu, M.; Leavitt, M.C.; Maruyama, M.; Yamada, O.; Young, D.; Ho, A.D.; Wong-Staal, F. Intracellular immunization of human fetal cord blood stem/progenitor cells with a ribozyme against human immunodeficiency virus type 1. *Proc. Natl. Acad. Sci. USA* **1995**, *92*, 699–703. [CrossRef] [PubMed]
124. Wang, G.; Zhao, N.; Berkhout, B.; Das, A.T. CRISPR-Cas based antiviral strategies against HIV-1. *Virus Res.* **2018**, *244*, 321–332. [CrossRef] [PubMed]
125. Herrera-Carrillo, E.; Berkhout, B. Attacking HIV-1 RNA versus DNA by sequence-specific approaches: RNAi versus CRISPR-Cas. *Biochem. Soc. Trans.* **2016**, *44*, 1355–1365. [CrossRef] [PubMed]
126. Hu, W.; Kaminski, R.; Yang, F.; Zhang, Y.; Cosentino, L.; Li, F.; Luo, B.; Alvarez-Carbonell, D.; Garcia-Mesa, Y.; Karn, J.; et al. RNA-directed gene editing specifically eradicates latent and prevents new HIV-1 infection. *Proc. Natl. Acad. Sci. USA* **2014**, *111*, 11461–11466. [CrossRef] [PubMed]
127. Wang, Z.; Pan, Q.; Gendron, P.; Zhu, W.; Guo, F.; Cen, S.; Wainberg, M.A.; Liang, C. CRISPR/Cas9-Derived Mutations Both Inhibit HIV-1 Replication and Accelerate Viral Escape. *Cell Rep.* **2016**, *15*, 481–489. [CrossRef] [PubMed]
128. Man, D.; Sansbury, B.; Bialk, P.; Bloh, K.; Kolb, E.A.; Kmiec, E.B. Target Site Mutagenesis during Crispr/ Cas 9/Single-Stranded- Oligonucleotide Directed Gene Editing for Sickle Cell Anemia. *Blood* **2016**, *128*, 4706.
129. White, M.K.; Kaminski, R.; Young, W.B.; Roehm, P.C.; Khalili, K. CRISPR Editing Technology in Biological and Biomedical Investigation. *J. Cell. Biochem.* **2017**, *118*, 3586–3594. [CrossRef] [PubMed]
130. Yin, C.; Zhang, T.; Qu, X.; Zhang, Y.; Putatunda, R.; Xiao, X.; Li, F.; Xiao, W.; Zhao, H.; Dai, S.; et al. In Vivo Excision of HIV-1 Provirus by saCas9 and Multiplex Single-Guide RNAs in Animal Models. *Mol. Ther.* **2017**, *25*, 1168–1186. [CrossRef] [PubMed]
131. Kunze, C.; Borner, K.; Kienle, E.; Orschmann, T.; Rusha, E.; Schneider, M.; Radivojkov-Blagojevic, M.; Drukker, M.; Desbordes, S.; Grimm, D.; et al. Synthetic AAV/CRISPR vectors for blocking HIV-1 expression in persistently infected astrocytes. *Glia* **2018**, *66*, 413–427. [CrossRef] [PubMed]
132. Ahlenstiel, C.L.; Suzuki, K.; Marks, K.; Symonds, G.P.; Kelleher, A.D. Controlling HIV-1: Non-Coding RNA Gene Therapy Approaches to a Functional Cure. *Front. Immunol.* **2015**, *6*, 474. [CrossRef] [PubMed]
133. Suzuki, K.; Juelich, T.; Lim, H.; Ishida, T.; Watanebe, T.; Cooper, D.A.; Rao, S.; Kelleher, A.D. Closed chromatin architecture is induced by an RNA duplex targeting the HIV-1 promoter region. *J. Biol. Chem.* **2008**, *283*, 23353–23363. [CrossRef] [PubMed]
134. Ahlenstiel, C.; Mendez, C.; Lim, S.T.; Marks, K.; Turville, S.; Cooper, D.A.; Kelleher, A.D.; Suzuki, K. Novel RNA Duplex Locks HIV-1 in a Latent State via Chromatin-mediated Transcriptional Silencing. *Mol. Ther. Nucleic Acids* **2015**, *4*, e261. [CrossRef] [PubMed]
135. Suzuki, K.; Hattori, S.; Marks, K.; Ahlenstiel, C.; Maeda, Y.; Ishida, T.; Millington, M.; Boyd, M.; Symonds, G.; Cooper, D.A.; et al. Promoter Targeting shRNA Suppresses HIV-1 Infection In vivo Through Transcriptional Gene Silencing. *Mol. Ther. Nucleic Acids* **2013**, *2*, e137. [CrossRef] [PubMed]

136. Zhang, M.; Poh, T.Y.; Louache, F.; Sundell, I.B.; Yuan, J.; Evans, S.; Koka, P.S. Rescue of multi-lineage hematopoiesis during HIV-1 infection by human c-mpl gene transfer and reconstitution of CD34+ progenitor cells in vivo. *J. Stem Cells* **2009**, *4*, 161–177. [PubMed]
137. Anderson, J.; Akkina, R. Human immunodeficiency virus type 1 restriction by human-rhesus chimeric tripartite motif 5alpha (TRIM 5alpha) in CD34(+) cell-derived macrophages in vitro and in T cells in vivo in severe combined immunodeficient (SCID-hu) mice transplanted with human fetal tissue. *Hum. Gene Ther.* **2008**, *19*, 217–228. [CrossRef] [PubMed]
138. Dimitrova, D.I.; Yang, X.; Reichenbach, N.L.; Karakasidis, S.; Sutton, R.E.; Henderson, E.E.; Rogers, T.J.; Suhadolnik, R.J. Lentivirus-mediated transduction of PKR into CD34(+) hematopoietic stem cells inhibits HIV-1 replication in differentiated T cell progeny. *J. Interferon Cytokine Res.* **2005**, *25*, 345–360. [CrossRef] [PubMed]
139. Leibman, R.S.; Richardson, M.W.; Ellebrecht, C.T.; Maldini, C.R.; Glover, J.A.; Secreto, A.J.; Kulikovskaya, I.; Lacey, S.F.; Akkina, S.R.; Yi, Y.; et al. Supraphysiologic control over HIV-1 replication mediated by CD8 T cells expressing a re-engineered CD4-based chimeric antigen receptor. *PLoS Pathog.* **2017**, *13*, e1006613. [CrossRef] [PubMed]
140. Liu, L.; Patel, B.; Ghanem, M.H.; Bundoc, V.; Zheng, Z.; Morgan, R.A.; Rosenberg, S.A.; Dey, B.; Berger, E.A. Novel CD4-Based Bispecific Chimeric Antigen Receptor Designed for Enhanced Anti-HIV Potency and Absence of HIV Entry Receptor Activity. *J. Virol.* **2015**, *89*, 6685–6694. [CrossRef] [PubMed]
141. Hammer, O. CD19 as an attractive target for antibody-based therapy. *MAbs* **2012**, *4*, 571–577. [CrossRef] [PubMed]
142. Pulsipher, M.A. Are CAR T cells better than antibody or HCT therapy in B-ALL? *Hematol. Am. Soc. Hematol. Educ. Program* **2018**, *2018*, 16–24. [CrossRef] [PubMed]
143. Miliotou, A.N.; Papadopoulou, L.C. CAR T-cell Therapy: A New Era in Cancer Immunotherapy. *Curr. Pharm. Biotechnol.* **2018**, *19*, 5–18. [CrossRef] [PubMed]
144. Davila, M.L.; Brentjens, R.J. CD19-Targeted CAR T cells as novel cancer immunotherapy for relapsed or refractory B-cell acute lymphoblastic leukemia. *Clin. Adv. Hematol. Oncol.* **2016**, *14*, 802–808. [PubMed]
145. Quintas-Cardama, A. CD19 directed CAR T cell therapy in diffuse large B-cell lymphoma. *Oncotarget* **2018**, *9*, 29843–29844. [CrossRef] [PubMed]
146. Ali, A.; Kitchen, S.G.; Chen, I.S.Y.; Ng, H.L.; Zack, J.A.; Yang, O.O. HIV-1-Specific Chimeric Antigen Receptors Based on Broadly Neutralizing Antibodies. *J. Virol.* **2016**, *90*, 6999–7006. [CrossRef] [PubMed]
147. Liu, B.; Zou, F.; Lu, L.; Chen, C.; He, D.; Zhang, X.; Tang, X.; Liu, C.; Li, L.; Zhang, H. Chimeric Antigen Receptor T Cells Guided by the Single-Chain Fv of a Broadly Neutralizing Antibody Specifically and Effectively Eradicate Virus Reactivated from Latency in CD4+ T Lymphocytes Isolated from HIV-1-Infected Individuals Receiving Suppressive Combined Antiretroviral Therapy. *J. Virol.* **2016**, *90*, 9712–9724. [CrossRef] [PubMed]
148. Hale, M.; Mesojednik, T.; Romano Ibarra, G.S.; Sahni, J.; Bernard, A.; Sommer, K.; Scharenberg, A.M.; Rawlings, D.J.; Wagner, T.A. Engineering HIV-Resistant, Anti-HIV Chimeric Antigen Receptor T Cells. *Mol. Ther.* **2017**, *25*, 570–579. [CrossRef] [PubMed]
149. Zhen, A.; Kamata, M.; Rezek, V.; Rick, J.; Levin, B.; Kasparian, S.; Chen, I.S.; Yang, O.O.; Zack, J.A.; Kitchen, S.G. HIV-specific Immunity Derived From Chimeric Antigen Receptor-engineered Stem Cells. *Mol. Ther.* **2015**, *23*, 1358–1367. [CrossRef] [PubMed]
150. Zhen, A.; Peterson, C.W.; Carrillo, M.A.; Reddy, S.S.; Youn, C.S.; Lam, B.B.; Chang, N.Y.; Martin, H.A.; Rick, J.W.; Kim, J.; et al. Long-term persistence and function of hematopoietic stem cell-derived chimeric antigen receptor T cells in a nonhuman primate model of HIV/AIDS. *PLoS Pathog.* **2017**, *13*, e1006753. [CrossRef] [PubMed]
151. Luo, X.M.; Maarschalk, E.; O'Connell, R.M.; Wang, P.; Yang, L.; Baltimore, D. Engineering human hematopoietic stem/progenitor cells to produce a broadly neutralizing anti-HIV antibody after in vitro maturation to human B lymphocytes. *Blood* **2009**, *113*, 1422–1431. [CrossRef] [PubMed]
152. Poznansky, M.C.; La Vecchio, J.; Silva-Arietta, S.; Porter-Brooks, J.; Brody, K.; Olszak, I.T.; Adams, G.B.; Ramstedt, U.; Marasco, W.A.; Scadden, D.T. Inhibition of human immunodeficiency virus replication and growth advantage of CD4+ T cells and monocytes derived from CD34+ cells transduced with an intracellular antibody directed against human immunodeficiency virus type 1 Tat. *Hum. Gene Ther.* **1999**, *10*, 2505–2514. [CrossRef] [PubMed]

153. Prince, A.M.; Reesink, H.; Pascual, D.; Horowitz, B.; Hewlett, I.; Murthy, K.K.; Cobb, K.E.; Eichberg, J.W. Prevention of HIV infection by passive immunization with HIV immunoglobulin. *AIDS Res. Hum. Retroviruses* **1991**, *7*, 971–973. [CrossRef] [PubMed]
154. Morris, L.; Mkhize, N.N. Prospects for passive immunity to prevent HIV infection. *PLoS Med.* **2017**, *14*, e1002436. [CrossRef] [PubMed]
155. Balazs, A.B.; West, A.P., Jr. Antibody gene transfer for HIV immunoprophylaxis. *Nat. Immunol.* **2013**, *14*, 1–5. [CrossRef] [PubMed]
156. Balazs, A.B.; Chen, J.; Hong, C.M.; Rao, D.S.; Yang, L.; Baltimore, D. Antibody-based protection against HIV infection by vectored immunoprophylaxis. *Nature* **2011**, *481*, 81–84. [CrossRef] [PubMed]
157. Badamchi-Zadeh, A.; Tartaglia, L.J.; Abbink, P.; Bricault, C.A.; Liu, P.T.; Boyd, M.; Kirilova, M.; Mercado, N.B.; Nanayakkara, O.S.; Vrbanac, V.D.; et al. Therapeutic Efficacy of Vectored PGT121 Gene Delivery in HIV-1-Infected Humanized Mice. *J. Virol.* **2018**, *92*. [CrossRef] [PubMed]
158. Cyranoski, D.; Ledford, H. Genome-edited baby claim provokes international outcry. *Nature* **2018**, *563*, 607–608. [CrossRef] [PubMed]
159. Frankel, M.S.; Chapman, A.R. Genetic technologies. Facing inheritable genetic modifications. *Science* **2001**, *292*, 1303. [CrossRef] [PubMed]
160. Zhang, X.H.; Tee, L.Y.; Wang, X.G.; Huang, Q.S.; Yang, S.H. Off-target Effects in CRISPR/Cas9-mediated Genome Engineering. *Mol. Ther. Nucleic Acids* **2015**, *4*, e264. [CrossRef] [PubMed]
161. Keep off-target effects in focus. *Nat. Med.* **2018**, *24*, 1081. [CrossRef] [PubMed]
162. Aryal, N.K.; Wasylishen, A.R.; Lozano, G. CRISPR/Cas9 can mediate high-efficiency off-target mutations in mice in vivo. *Cell Death Dis.* **2018**, *9*, 1099. [CrossRef] [PubMed]
163. Ormond, K.E.; Mortlock, D.P.; Scholes, D.T.; Bombard, Y.; Brody, L.C.; Faucett, W.A.; Garrison, N.A.; Hercher, L.; Isasi, R.; Middleton, A.; et al. Human Germline Genome Editing. *Am. J. Hum. Genet.* **2017**, *101*, 167–176. [CrossRef] [PubMed]

pharmaceutics

MDPI

Review

Eradication of Human Immunodeficiency Virus Type-1 (HIV-1)-Infected Cells

Nejat Düzgüneş * and Krystyna Konopka

Department of Biomedical Sciences, Arthur A. Dugoni School of Dentistry, University of the Pacific, 155 Fifth Street, Room 412, San Francisco, CA 94103, USA; kkonopka@pacific.edu

* Correspondence: nduzgunes@pacific.edu

Received: 12 March 2019; Accepted: 24 May 2019; Published: 1 June 2019

Abstract: Predictions made soon after the introduction of human immunodeficiency virus type-1 (HIV-1) protease inhibitors about potentially eradicating the cellular reservoirs of HIV-1 in infected individuals were too optimistic. The ability of the HIV-1 genome to remain in the chromosomes of resting CD4+ T cells and macrophages without being expressed (HIV-1 latency) has prompted studies to activate the cells in the hopes that the immune system can recognize and clear these cells. The absence of natural clearance of latently infected cells has led to the recognition that additional interventions are necessary. Here, we review the potential of utilizing suicide gene therapy to kill infected cells, excising the chromosome-integrated HIV-1 DNA, and targeting cytotoxic liposomes to latency-reversed HIV-1-infected cells.

Keywords: suicide gene therapy; CRISPR/Cas9; broadly neutralizing antibodies; cytotoxic liposomes; lentivirus

1. Antiretroviral Therapy

Following cell entry, human immunodeficiency virus type 1 (HIV-1) integrates its genome into the host cell chromosome. While some infected cells produce new virions, others are infected latently and do not express any viral envelope glycoproteins (Env; gp120/gp41) on their surface [1,2]. Treatment with antiviral drugs can inhibit the binding of the viral glycoproteins to the co-receptors on host cells (e.g., maraviroc), the fusion of the viral membrane with host cells (e.g., enfuvirtide), reverse transcription of the RNA genome of the virus (e.g., abacavir, lamivudine), integration of the reverse transcribed, double-stranded DNA into the host cell chromosome (e.g., raltegravir), and the viral protease involved in viral polyprotein cleavage and virus maturation (e.g., indinavir, darunavir). These treatments, however, do not eliminate the source of the virus, which is the HIV provirus integrated into the cellular chromosome.

Various therapeutic approaches have been tried in an attempt to eradicate the source of the virus. Soon after the introduction of protease inhibitors for HIV therapy, a mathematical analysis of the decay of blood levels of the virus following the co-administration of nelfinavir, zidovudine, and lamivudine suggested that cell-free virions and productively infected CD4+ T cells would be eliminated in less than two months [3]. The analysis also suggested that lymphocytes latently infected with an infectious provirus could be "completely eliminated after 2.3–3.1 years of treatment with a 100%-inhibitory antiretroviral regimen." Nevertheless, the possibility was raised that longer treatment periods might be needed because of the "possible existence of undetected viral compartments or sanctuary sites", as well as the persistence of infected mononuclear cells that could not be activated to produce virions [3]. Therapeutic interventions have also included early antiretroviral treatment during seroconversion, structured treatment interruptions, and targeted toxins; however, these approaches have not been able to eradicate the virus [4,5].

2. HIV-1 Latency

HIV-1 remains latent in some infected cells. These cells do not express any viral proteins on their surface, and they do not present any peptides in association with Class I major histocompatibility (MHC) molecules [6–8]. These latently infected cells, widely believed to be memory CD4+ T cells, and possibly cells of the monocyte-macrophage lineage [9–11], are hidden from recognition by the cellular immune system and render HIV infection intrinsically incurable with current antiretroviral therapy alone [12]. It is thought that the low-level viremia that continues despite therapy and the short-term viral blips (RNA below 200 copies/mL) do not depend on the presence of new drug-resistance mutations due to active replication but rather arise from viral release from stable reservoirs [12,13].

HIV latency is a multifactorial process. In most cases, the HIV-1 proviral DNA, reverse transcribed from the viral RNA genome, integrates into the host cell genome in regions that are being transcribed actively. It is ironic that most latent proviral DNA in patients who are on combination antiretroviral therapy (cART) and have suppressed viral replication are found in actively transcribed segments of the cellular chromosomes [13]. Several mechanisms of transcriptional interference may be involved in suppressing the expression of the proviral DNA: (i) the cytoplasmic sequestration of transcription factors, including NF-κB and NFAT, may inhibit viral gene expression; (ii) if the host promoter is located upstream of the provirus, the RNA polymerase (Pol II) may "read through" the HIV-1 promoter (5′-LTR) and displace the transcription factors necessary for viral transcription; (iii) if the proviral DNA has integrated into the host chromosome in the opposite orientation of the host gene, the RNA Pol II complexes of the provirus and the host may collide, and RNA transcription stops; (iv) resting CD4+ T cells have low levels of Cyclin T1, which would otherwise form P-TEFb and also play a role in HIV-1 transcription and Tat-mediated transactivation—this would therefore result in a lack of transcription of proviral DNA; (v) methylation of DNA and compaction of chromatin may also contribute to transcriptional silencing and hence HIV-1 latency [13].

A groundbreaking study by Lehrman et al. [14] showed that inhibition of histone deacetylase (the chromatin remodeling enzyme that helps maintain the latency of integrated HIV-1) by valproic acid, together with cART supplemented with enfuvirtide (ENF), reduced the frequency of resting cell infection. This finding suggested that, with new approaches, it may be possible to reduce or eliminate the reservoir of HIV in infected individuals. A subsequent study employing raltegravir or enfuvirtide in addition to standard cART and valproic acid in a slightly larger patient pool showed, however, that there was no effect on the low-level viremia measured by single-copy plasma HIV RNA [15]. This approach relied on cell death upon activation of the latently infected cells, or by recognition and destruction by the immune system. Margolis et al. [16] indicated that "a major approach to HIV eradication envisions antiretroviral suppression, paired with targeted therapies to enforce the expression of viral antigen from quiescent HIV-1 genomes, *and immunotherapies to clear latent infection*." Sengupta and Siliciano [17] reiterated this point, stating "neither viral cytopathic effects nor CTL-mediated lysis may occur upon latency reversal *without additional interventions*." In this mini-review, we describe most of these "additional interventions", including some approaches from our laboratory.

3. Latency Reversal

Richman et al. [9] emphasized that latency is likely to be established and maintained by blocks at different steps in the HIV-1 replicative cycle, which may complicate efforts to eradicate these cells. One of these blocks was inhibited by the administration of the histone deacetylase inhibitor, vorinostat, to HIV-infected patients, as shown by cellular acetylation and an increase in HIV RNA expression in resting CD4+ cells [18]. Thus, the latency of HIV-1 in resting CD4+ cells in patients can be reversed by a tolerable dose of vorinostat. The low number of latently infected cells in patients has hindered studies on HIV-1 activation. Latently infected T cell lines are thus useful in experiments identifying the agents that can activate latent HIV-1. Another problem with the activation of latently infected cells is the percentage of cells that are actually induced to synthesize viral proteins. The site of chromosomal insertion of HIV-1 DNA affects the response to activating agents; for example, phytohemagglutinin and vorinostat reactivated proviruses at distinct genomic locations [19]. Since the activation of the cells will produce infectious virions, it will be necessary to concomitantly employ a cART regimen used in the studies of Lehrman et al. [20] (2005) and Archin et al. [15].

Latency reversing agents include various protein kinase C activators, such as diacylglycerol lactones [21], anti-tumor-promoting phorbol esters [4,22] histone deacetylase inhibitors [23], interleukin-7 [14], toll-like receptor-1 and-7 agonists [24], bryostatin analogs [25] and other compounds [26,27]. Spina et al. [26] compared the response to numerous activators of primary T cell models, J-Lat cell lines, and patient-derived infected cells, and found that the different cellular systems responded differently to the activators than the latently infected T cells obtained from patients. Nevertheless, protein kinase C agonists and phytohemagglutinin were able to activate latent HIV in all the cellular models developed in different laboratories. A library of marine natural products was used to identify four latency reversing agents, including aplysiatoxin, which induced the expression of HIV-1 provirus in cell lines and primary cell models at concentrations 900-fold lower than that of prostratin without significant effects on cell viability [28]. The expression of provirus-derived RNA in single cells upon the reversal of latency—even before the synthesis of viral proteins—may be useful in delineating the mechanisms of these agents [29].

Another approach to reversing latency is the use of the CRISPR/Cas9-derived systems [SunTag and synergistic activation mediator (SAM) systems] that recruit several transcriptional activation domains to an optimal target region within the HIV 5′ long terminal repeat (LTR) [30]. Transcriptional activation of proviral genomes was observed in various latently infected cell lines at levels comparable to or higher than treatment with established latency reversal agents and led to the production of infectious virions. A similar system that comprised an RNA-guided dCas9-VP64 activator targeting the junction between two NF-κB transcription factor-binding sites within the LTR enhancer region induced transcriptional activation of latent HIV-1 infection in all latency models tested [31].

In the "shock and kill" studies described above, it was expected that HIV-1-specific CD8+ cytotoxic T lymphocytes would recognize HIV-1 antigens on the activated CD4+ T cells and kill them. However, some latency reversal agents were found to also adversely affect the function of CD8+ cells [32,33]. Another consideration in this approach, which is also relevant to the other methods described below, is the potential of stimulating a deleterious, systemic inflammatory response [33]. Huang et al. [34] showed that following the ex vivo exposure of CD4+ T lymphocytes from ART-treated individuals to several latency-reversing agents and autologous CD8+ T lymphocytes reduced cellular HIV DNA but could not deplete the replication-competent virus. This observation may indicate that cells containing replication-competent HIV are resistant to cytotoxic T cells.

4. Suicide Gene Therapy

The insertion of potentially cytotoxic genes that can be activated by characteristic proteins of HIV-1 may be a useful approach to eliminating HIV-1-infected cells. Plasmids expressing the diphtheria toxin A-fragment (DT-A) under the control of HIV-1 LTR sequences (−167 to +80) could be *trans*-activated by the viral Tat protein, but the toxin gene was also expressed to some extent without Tat [35]. When *cis*-acting negative regulatory elements from the *env* region of the viral genome were incorporated in the 3′ untranslated region of these plasmids, basal expression from the LTR was minimized. The DT-A gene could then be *trans*-activated at a maximal level in the presence of both Tat and Rev proteins from the virus. CD4+ H9 cells that were transduced with a retrovirus to express this gene construct and then infected with HIV-1 could limit HIV-1 production [36]. The intracellularly "immunized" cells were protected against clinical HIV-1 strains up to 59 days [37]. Co-transfection of HeLa cells with the plasmid expressing DT-A and the proviral HIV-1 clone, HXBΔBgl, completely inhibited virus production by the cells [38].

Our laboratory is working on developing an HIV-1-specific promoter that minimizes basal expression (e.g., in the absence of the HIV-1 transactivator, Tat) and that drives the expression of suicide genes that would induce cell death specifically in HIV-1-infected cells. As an initial attempt at modifying the LTR, we generated several clones with deletions in parts of the LTR (Figure 1). For example, LTR2 had the modulatory region deleted, whereas in LTR3, the NF-κB binding site was deleted. In a model system employing Tat-expressing HeLa cells and luciferase-expressing, wild-type, and mutant LTR-driven plasmids, we showed that the mutant LTR2, from which the modulatory elements at the 5′ end of the LTR were deleted, was 100-fold more responsive to the presence of Tat (which would only be present in infected cells that are actively producing the virus) than in control cells not expressing any Tat (e.g., uninfected cells or quiescent, latently-infected cells) [39] (Figure 2). The LTR2 promoter may now be used to drive the expression of the herpes simplex virus thymidine kinase (HSV-*tk*) gene preferentially in infected cells, resulting in cytotoxicity upon administration of the anti-herpesvirus drug, ganciclovir (Figure 3). To enable this construct to be delivered to all HIV-1-infected cells, it would most likely have to be incorporated into an HIV-1-based lentiviral vector. The advantage of this system over the DT-A-based plasmid described above is that it can be turned on and off. Thus, "treatment" can be initiated by the administration of ganciclovir and interrupted by ceasing it. We previously studied cytotoxicity induced by the HSV-*tk* system in oral squamous cell carcinoma and cervical carcinoma cells [40–43], as well as in an animal model of oral cancer [44,45].

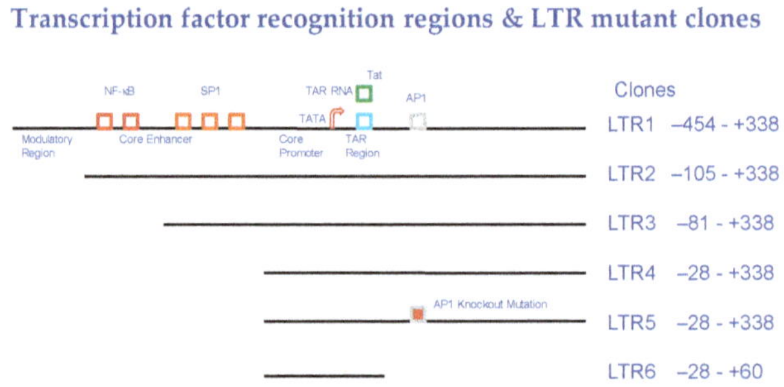

Figure 1. Transcription factor recognition regions of human immunodeficiency virus type-1 (HIV-1) long terminal repeat (LTR) and mutated LTR sequences used in HIV-1-specific gene expression studies shown in Figure 2.

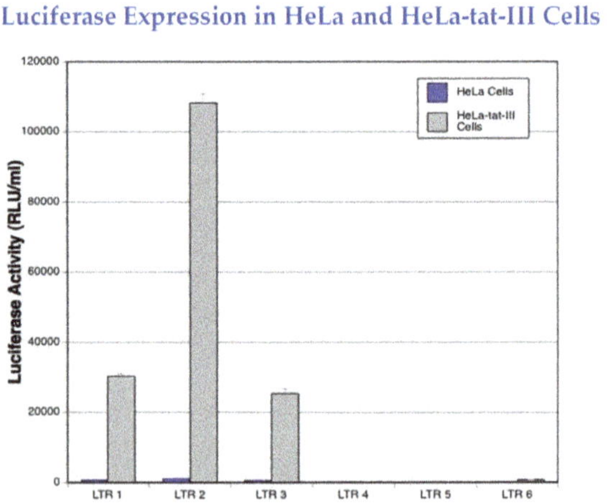

Figure 2. Comparison of luciferase gene expression from wild type LTR and LTR mutant clones (shown in Figure 1) in HeLa cells and HeLa-tat-III cells that constitutively express the HIV-1 Tat protein (data from [39]).

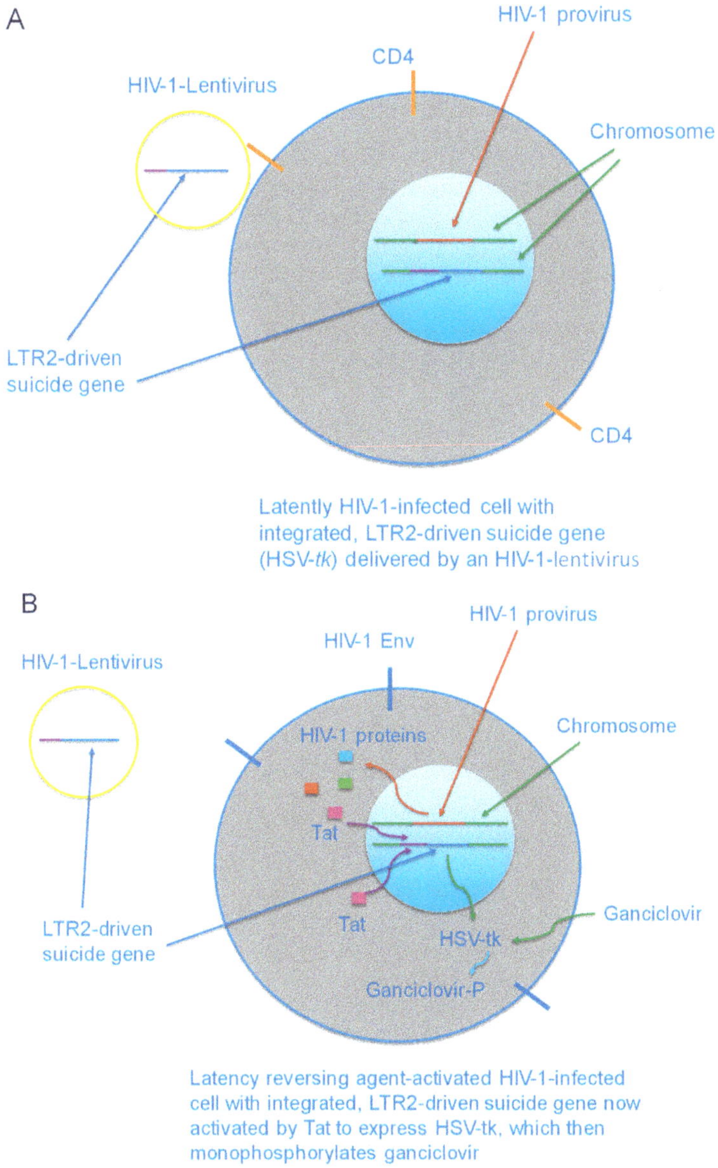

Figure 3. (**A**) HIV-1-lentivirus delivery of LTR2-driven suicide gene (HSV-*tk*) to a latently HIV-1-infected CD4+ cell. The suicide gene is then integrated into the chromosome of the cell. (**B**) Latency reversing agent-activated HIV-1-infected cell with integrated LTR2-driven suicide gene. The cell produces viral proteins, including Tat, which activates the LTR2 to express HSV-*tk*, which then monophosphorylates ganciclovir that has been delivered to the cell.

There is still a finite amount of luciferase expression from LTR1, LTR2, and LTR3 in the absence of Tat (Figure 2). Further genetic engineering of the HIV-1 promoter will be necessary to eliminate even this basal level of gene expression.

Garg and Joshi [46] cloned the *tk* gene into the vector, pNL-GFPRRESA, which includes the full LTR promoter that also expresses GFP in the presence of Tat. Following the selection of cells expressing GFP and treatment with ganciclovir, virus production and the number of virus-infected cells was reduced, demonstrating the feasibility of this system.

5. Excision of Chromosome-Integrated HIV-1 DNA

The modification and the development of the clustered regularly interspaced palindromic repeat (CRISPR)/Cas9 endonuclease system, originally identified in certain bacteria as an adaptive immune system, has led to a molecular tool that can target specific sequences in DNA [47,48]. This method could be particularly useful in cells that are not activated to produce HIV proteins and hence are not amenable to become targets of the immune system (as in the proposed "shock and kill" method), or in suicide gene activation by the Tat protein (*vide supra*). This system has been applied to alter the HIV-1 genome and block its expression. CRISPR/Cas9 components targeting various HIV1-derived sequences were transfected into T cells with integrated, LTR-driven GFP and TAR sequences. Upon stimulation of the cells, LTR-driven gene expression was inhibited significantly [49]. Sequence analysis confirmed that the targeted LTR and TAR sequences were cleaved.

Following the cleavage by the enzyme Cas9 of specific DNA sequences determined by intracellularly delivered guide RNA, the non-homologous end joining (NHEJ) machinery of the cells causes insertions and deletions (termed "indels") at the cleaved site, thereby causing the impairment of DNA function at the site. In the case of HIV-1-infected cells, these indels can inactivate the virus, but they can also produce replication-competent virions that have a slightly different proviral DNA sequence. These sequences may now be resistant to recognition by the same guide RNA [50], demonstrating that the CRISPR/Cas9 system can both inactivate HIV-1 and generate mutant virus [51].

Transcription activator-like effector nucleases (TALENs) [52] were employed to target the LTR site used with the CRISPR/Cas9 system above. The intracellular introduction of mRNA encoding the specific TALEN caused about 80% of the target DNA to be removed [53].

The Cas9/guide RNA (gRNA) system was utilized to target the HIV-1 LTR U3 region and excised a 9709 bp fragment of integrated proviral DNA from its 5′ to 3′ LTRs [54]. This resulted in inactivation of viral gene expression and replication in various cell types, including a microglial cell line and a promonocytic cell line, without causing genotoxicity or off-target effects in the host cells. The same approach was applied to latently infected human CD4+ T-cells to cleave the chromosome-integrated proviral DNA, and whole-genome sequencing of the treated cells indicated that there was no effect on cell viability, cell cycle, and apoptosis. The co-expression of Cas9 and the targeting gRNAs in cells from which HIV-1 had been eradicated protected the cells against *de novo* HIV-1 infection [55]. In a study with transgenic rodents with the HIV-1 genome, a short version of the Cas9 endonuclease was used in conjunction with a multitude of gRNAs that targeted the viral 5′-LTR and the *gag* gene and delivered in an adeno-associated virus [56]. This treatment resulted in the generation of a 978 bp HIV-1 DNA fragment in various organs and in circulating lymphocytes, indicating the cleavage of part of the proviral DNA.

6. Cytotoxic Liposomes Targeted to HIV-1-Infected Cells

In addition to the delivery of an HIV-1-activated suicide gene into latently-infected cells, our laboratory is focusing on the killing of such cells by targeting liposomes encapsulating cytotoxic drugs to infected cells whose latency has been reversed and that now express Env on their surface. For this purpose, we are coupling broadly neutralizing anti-Env antibodies (bNAb) [57–59], CD4-immunoadhesin [60], or CD4-derived peptides [61] as ligands to target the activated cells (Figure 4). Such liposomes are expected to be internalized, as shown for liposomes targeted to cancer cells [62,63] (*vide infra*) and kill the infected cells. To prevent an immune reaction to the antibodies, they can be engineered to be "humanized" for eventual clinical use, as in the case of a number of antibody-based drugs, including anti-HER2 [64].

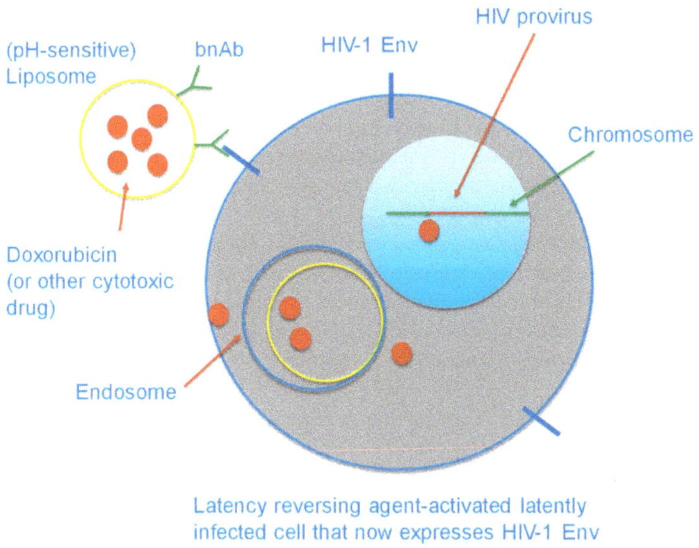

Figure 4. Cytotoxic liposome targeted to cell surface Env, which is expressed following treatment of a latently infected cell with a latency reversing agent. The targeting ligand is a broadly neutralizing anti-Env antibody. The liposome is endocytosed after binding to cell surface Env. The liposome may be engineered to be pH-sensitive so as to destabilize the endosome membrane at mildly acidic pH achieved in the endosome lumen and to enhance drug delivery to the cytoplasm.

Sterically stabilized liposomes containing poly(ethylene glycol) (PEG)-conjugated lipids and loaded with the cytotoxic DNA-intercalating anticancer drug, doxorubicin, are currently approved for the treatment of Kaposi's sarcoma, ovarian cancer, breast cancer, and multiple myeloma [65]. These liposomes have prolonged circulation in the bloodstream and can extravasate into tissues, including lymph nodes [66]. Liposomes administered subcutaneously are cleared via the local lymph nodes [66,67], thus localizing in tissues where HIV-1 is either replicating or hiding in latently infected cells [68–70]. Subcutaneous injection of liposomes carrying indinavir resulted in a 21–126 fold higher accumulation of the drug in all tissues compared to the free drug [71]. Indinavir delivered subcutaneously in liposomes to HIV-1-infected macaques localized in lymph nodes and caused a significant reduction in viral load [72]. Liposomes encapsulating the HIV-1 protease inhibitor, L-689,502, reduced the EC_{50} of the drug by 3-5-fold in infected macrophages [73]. The protease inhibitor PI1 encapsulated in liposomes targeted to gp120 expressed on infected cells via the antibody F105 had a 10-fold higher anti-HIV activity than the free drug at 100 nM [74]. However, it should be noted that, although the activity of antiviral agents can be enhanced by delivery in liposomes, and passive targeting to lymph nodes may reduce the need for daily administration of the drugs, this approach will not lead to the eradication of HIV-1-infected cells.

Nevertheless, liposomes containing cytotoxic drugs and targeted via anti-HER2 (ErbB2) monoclonal antibody fragments have been utilized in cancer chemotherapy. For example, they enhanced doxorubicin uptake in HER2-overexpressing cells in culture by up to 700-fold and resulted in tumor regression in five different tumor xenograft animal models [62], indicating their superior ability to kill tumor cells. Doxorubicin liposomes targeted to cell surface CD44 receptors on B16F10 melanoma cells had a 5-6-fold higher rate constant of cell killing than the free drug for a given amount of intracellular doxorubicin [63], showing the effectiveness of targeting. Sterically stabilized liposomes can also be rendered pH-sensitive to facilitate or enhance the intracellular delivery of cytotoxic drugs [75,76].

Since latently infected cells do not express the viral glycoproteins on their surface, they need to be activated by "latency reversing agents" to produce HIV-1 and express Env and hence will be recognizable by the targeted liposomes. These liposomes do not have to be administered for prolonged periods of time, since they will eliminate HIV reservoirs, unlike current treatment modalities with lifelong administration of antiviral agents. Thus, the inconvenience of subcutaneous, intravenous, intraperitoneal, or spinal delivery of liposomes is likely to be tolerable by patients who are likely to be cured of their HIV infection, perhaps after a series of injections.

The activated, previously latently infected cells are thus expected to express the HIV-1 Env protein on their surface, be recognized by the targeted cytotoxic liposomes, and be killed as a result of the intracellular delivery of the cytotoxic drug.

The fact that doxorubicin encapsulated in sterically stabilized liposomes is already approved for clinical use in the treatment of various cancers supports the feasibility of our approach [65,77]. An additional route of liposome administration for lymph node accumulation is intraperitoneal injection [78]. Anti-HLA-DR-bearing sterically stabilized liposomes accumulate in the lymph node cortex following subcutaneous injection [79]. Liposomes can also be injected into the spinal cord, as demonstrated in an animal model [80], enabling them to reach HIV-1-infected macrophages/microglia in the central nervous system [81]. Although this appears to be a difficult procedure, it is considerably more applicable than complete eradication of the immune system for bone marrow transplantation of HIV-1-resistant CCR5Δ32/Δ32 hematopoietic stem cells, a method that was applied in curing the "Berlin patient" and the "London patient" [82,83]. Therefore, after we demonstrate that targeted cytotoxic liposomes can specifically eliminate HIV-1-infected cells in culture, it will be possible to apply this method in vivo in HIV-1 infection models and eventually in patients.

7. Concluding Remarks

HIV-1 latency, the ability of the HIV-1 genome to remain in the chromosomes of resting CD4+ T cells and macrophages without being expressed, has been an important challenge in attaining HIV-1 remission and an eventual cure. It is astounding that a virus with just a few genes has evolved a way to reverse transcribe and integrate its genome into host cell chromosomes, but that virologists and molecular biologists throughout the world have not yet come up with an effective solution to excise or inactivate the viral genome or to specifically kill the infected cells. One possible reason for this is that the prevailing dogma in the scientific community soon after the identification of HIV-1 and HIV-2 as the etiologic agents of acquired immune deficiency syndrome (AIDS) was that it was impossible to cure HIV/AIDS and that our efforts should be focused on preventing viral replication. Nevertheless, in 1993, we proposed to use oligonucleotide-conjugated endonucleases [84,85] to cleave the chromosome-integrated HIV-1 provirus. In 2004, we also proposed to use triple-helix forming oligonucleotides [86,87] with an intervening spacer that would hybridize with the viral LTRs at the beginning and the end of the provirus, thereby connecting the two LTRs via the spacer and potentially inducing DNA repair mechanisms that would remove the looped DNA. We expect that the approaches we reviewed here (suicide gene therapy to kill infected cells, excision of chromosome-integrated HIV-1 DNA, and cytotoxic liposomes targeted to latency-reversed HIV-1-infected cells) will be developed further to be able to treat infected patients.

Author Contributions: Conceptualization, N.D. and K.K.; writing—original draft preparation, N.D.; writing—review and editing, N.D. and K.K.

Funding: This research received no external funding.

Acknowledgments: We thank Matt Milnes, Senait Gebremedhin and Amy Au for their work on the LTR mutants described here.

Conflicts of Interest: The authors declare no conflict of interest.

References

1. Sundquist, W.I.; Kräusslich, H.-G. HIV-1 assembly, budding, and maturation. *Cold Spring Harb. Perspect. Med.* **2012**, *2*, a006924. [CrossRef] [PubMed]

2. Baumgärtel, V.; Müller, B.; Lamb, D.C. Quantitative live-cell imaging of human immunodeficiency virus (HIV-1) assembly. *Viruses* **2012**, *4*, 777–799. [CrossRef]
3. Perelson, A.S.; Essunger, P.; Cao, Y.; Vesanen, M.; Hurley, A.; Saksela, K.; Markowitz, M.; Ho, D.D. Decay characteristics of HIV-1-infected compartments during combination therapy. *Nat. Cell Boil.* **1997**, *387*, 188–191. [CrossRef]
4. Kulkosky, J.; Sullivan, J.; Xu, Y.; Souder, E.; Hamer, D.H.; Pomerantz, R.J. Expression of latent HAART-persistent HIV Type 1 induced by novel cellular activating agents. *AIDS Hum. Retrovir.* **2004**, *20*, 497–505. [CrossRef] [PubMed]
5. Shehu-Xhilaga, M.; Tachedjian, G.; Crowe, S.; Kedzierska, K. Antiretroviral compounds: Mechanisms underlying failure of HAART to eradicate HIV-1. *Med. Chem.* **2005**, *12*, 1705–1719. [CrossRef]
6. Chun, T.W.; Engel, D.; Berrey, M.M.; Shea, T.; Corey, L.; Fauci, A.S. Early establishment of a pool of latently infected, resting CD4+ T cells during primary HIV-1 infection. *Proc. Natl. Acad. Sci. USA* **1998**, *95*, 8869–8873. [CrossRef]
7. Blankson, J.N.; Persaud, D.; Siliciano, R.F. The challenge of viral reservoirs in HIV-1 infection. *Annu. Med.* **2002**, *53*, 557–593. [CrossRef]
8. Siliciano, J.D.; Siliciano, R.F. A long-term latent reservoir for HIV-1: Discovery and clinical implications. *J. Antimicrob. Chemother.* **2004**, *54*, 6–9. [CrossRef] [PubMed]
9. Richman, D.D.; Margolis, D.M.; Delaney, M.; Greene, W.C.; Hazuda, D.; Pomerantz, R.J. The challenge of finding a cure for HIV infection. *Science* **2009**, *323*, 1304–1307. [CrossRef]
10. Murray, A.J.; Kwon, K.J.; Farber, D.L.; Siliciano, R.F. The latent reservoir for HIV-1: How immunologic memory and clonal expansion contribute to HIV-1 persistence. *J. Immunol.* **2016**, *197*, 407–417. [CrossRef]
11. Siliciano, R.F.; Greene, W.C. HIV latency. *Cold Spring Harb. Perspect. Med.* **2011**, *1*, a007096. [CrossRef]
12. Siliciano, R.F. Scientific rationale for antiretroviral therapy in 2005: Viral reservoirs and resistance evolution. *Top. HIV Med. Publ. Int. AIDS Soc. USA* **2005**, *13*, 96–100.
13. Ruelas, D.S.; Greene, W.C. An integrated overview of HIV-1 latency. *Cell* **2013**, *155*, 519–529. [CrossRef]
14. Lehrman, G.; Ylisastigui, L.; Bosch, R.J.; Margolis, D.M. Interleukin-7 induces HIV type 1 outgrowth from peripheral resting CD4+ T cells. *JAIDS J. Acquir. Immune Defic. Syndr.* **2004**, *36*, 1103–1104. [CrossRef] [PubMed]
15. Archin, N.M.; Cheema, M.; Parker, D.; Wiegand, A.; Bosch, R.J.; Coffin, J.M.; Eron, J.; Cohen, M.; Margolis, D.M. Antiretroviral intensification and valproic acid lack sustained effect on residual HIV-1 viremia or resting CD4+ cell infection. *PLoS ONE* **2010**, *5*, e9390. [CrossRef] [PubMed]
16. Margolis, D.M.; Garcia, J.V.; Hazuda, D.J.; Haynes, B.F. Latency reversal and viral clearance to cure HIV-1. *Science* **2016**, *353*, aaf6517. [CrossRef]
17. Sengupta, S.; Siliciano, R.F. Targeting the latent reservoir for HIV-1. *Immunity* **2018**, *48*, 872–895. [CrossRef] [PubMed]
18. Archin, N.M.; Bateson, R.; Tripathy, M.K.; Crooks, A.M.; Yang, K.H.; Dahl, N.P.; Kearney, M.F.; Anderson, E.M.; Coffin, J.M.; Strain, M.C.; et al. HIV-1 expression within resting CD4+ T cells after multiple doses of vorinostat. *J. Infect. Dis.* **2014**, *210*, 728–735. [CrossRef] [PubMed]
19. Chen, H.C.; Martinez, J.P.; Zorita, E.; Meyerhans, A.; Filion, G.J. Position effects influence HIV latency reversal. *Nat. Struct. Mol. Biol.* **2017**, *24*, 47–54. [CrossRef]
20. Lehrman, G.; Hogue, I.B.; Palmer, S.; Jennings, C.; A Spina, C.; Wiegand, A.; Landay, A.L.; Coombs, R.W.; Richman, D.D.; Mellors, J.W.; et al. Depletion of latent HIV-1 infection in vivo: A proof-of-concept study. *Lancet* **2005**, *366*, 549–555. [CrossRef]
21. Hamer, D.H.; Bocklandt, S.; McHugh, L.; Chun, T.-W.; Blumberg, P.M.; Sigano, D.M.; Marquez, V.E. Rational design of drugs that induce Human Immunodeficiency Virus replication. *J. Virol.* **2003**, *77*, 10227–10236. [CrossRef] [PubMed]
22. Bocklandt, S.; Blumberg, P.M.; Hamer, D.H. Activation of latent HIV-1 expression by the potent anti-tumor promoter 12-deoxyphorbol 13-phenylacetate. *Antivir. Res.* **2003**, *59*, 89–98. [CrossRef]
23. Demonte, D.; Quivy, V.; Colette, Y.; Van Lint, C. Administration of HDAC inhibitors to reactivate HIV-1 expression in latent cellular reservoirs: Implications for the development of therapeutic strategies. *Biochem. Pharmacol.* **2004**, *68*, 1231–1238. [CrossRef]

24. Macedo, A.B.; Novis, C.L.; De Assis, C.M.; Sorensen, E.S.; Moszczynski, P.; Huang, S.-H.; Ren, Y.; Spivak, A.M.; Jones, R.B.; Planelles, V.; et al. Dual TLR2 and TLR7 agonists as HIV latency-reversing agents. *JCI Insight* **2018**, *3*. [CrossRef] [PubMed]
25. Marsden, M.D.; Wu, X.; Navab, S.M.; Loy, B.A.; Schrier, A.J.; DeChristopher, B.A.; Shimizu, A.J.; Hardman, C.T.; Ho, S.; Ramirez, C.M.; et al. Characterization of designed, synthetically accessible bryostatin analog HIV latency reversing agents. *Virology* **2018**, *520*, 83–93. [CrossRef]
26. Spina, C.A.; Anderson, J.; Archin, N.M.; Bosque, A.; Chan, J.; Famiglietti, M.; Greene, W.C.; Kashuba, A.; Lewin, S.R.; Margolis, D.M.; et al. An in-depth comparison of latent HIV-1 reactivation in multiple cell model systems and resting CD4+ T cells from aviremic patients. *PLoS Pathog.* **2013**, *9*, e1003834. [CrossRef] [PubMed]
27. Beliakova-Bethell, N.; Hezareh, M.; Wong, J.K.; Strain, M.C.; Lewinski, M.K.; Richman, D.D.; Spina, C.A. Relative efficacy of T cell stimuli as inducers of productive HIV-1 replication in latently infected CD4 lymphocytes from patients on suppressive cART. *Virology* **2017**, *508*, 127–133. [CrossRef] [PubMed]
28. Richard, K.; Williams, D.E.; De Silva, E.D.; Brockman, M.A.; Brumme, Z.L.; Andersen, R.J.; Tietjen, I. Identification of novel HIV-1 latency-reversing agents from a Library of Marine Natural Products. *Viruses* **2018**, *10*, 348. [CrossRef] [PubMed]
29. Ukah, O.B.; Puray-Chavez, M.; Tedbury, P.R.; Herschhorn, A.; Sodroski, J.G.; Sarafianos, S.G. Visualization of HIV-1 RNA Transcription from Integrated HIV-1 DNA in Reactivated Latently Infected Cells. *Viruses* **2018**, *10*, 534. [CrossRef]
30. Bialek, J.K.; Dunay, G.A.; Voges, M.; Schäfer, C.; Spohn, M.; Stucka, R.; Hauber, J.; Lange, U.C. Targeted HIV-1 latency reversal using CRISPR/Cas9-derived transcriptional activator systems. *PLoS ONE* **2016**, *11*, e0158294. [CrossRef]
31. Saayman, S.M.; Lazar, D.C.; Scott, T.A.; Hart, J.R.; Takahashi, M.; Burnett, J.C.; Planelles, V.; Morris, K.V.; Weinberg, M.S. Potent and targeted activation of latent HIV-1 using the CRISPR/dCas9 activator complex. *Mol. Ther.* **2016**, *24*, 488–498. [CrossRef] [PubMed]
32. Walker-Sperling, V.E.; Pohlmeyer, C.W.; Tarwater, P.M.; Blankson, J.N. The effect of latency reversal agents on primary CD8+ T cells: Implications for shock and kill strategies for human immunodeficiency virus eradication. *EBioMedicine* **2016**, *8*, 217–229. [CrossRef] [PubMed]
33. Clutton, G.T.; Jones, R.B. Diverse impacts of HIV latency-reversing agents on CD8+ T-cell function: Implications for HIV cure. *Front. Immunol.* **2018**, *9*, 1452. [CrossRef]
34. Huang, S.H.; Ren, Y.; Thomas, A.S.; Chan, D.; Mueller, S.; Ward, A.R.; Patel, S.; Bollard, C.M.; Cruz, C.R.; Karandish, S.; et al. Latent HIV reservoirs exhibit inherent resistance to elimination by CD8+ T cells. *J. Clin. Investig.* **2018**, *128*, 876–889. [CrossRef] [PubMed]
35. Harrison, G.S.; Maxwell, F.; Long, C.J.; Rosen, C.A.; Glode, L.M.; Maxwell, I.H. Activation of a Diphtheria Toxin A gene by expression of human immunodeficiency virus-1 Tat and Rev proteins in transfected cells. *Hum. Gene Ther.* **1991**, *2*, 53–60. [CrossRef] [PubMed]
36. Harrison, G.S.; Long, C.J.; Curiel, T.J.; Maxwell, F.; Maxwell, I.H. Inhibition of Human Immunodeficiency Virus-1 production resulting from transduction with a retrovirus containing an HIV-regulated diphtheria toxin A chain gene. *Hum. Gene Ther.* **1992**, *3*, 461–469. [CrossRef] [PubMed]
37. Curiel, T.J.; Cook, D.R.; Wang, Y.; Hahn, B.H.; Ghosh, S.K.; Harrison, G.S. Long-Term inhibition of clinical and laboratory human immunodeficiency virus strains in human T-cell lines containing an HIV-regulated diphtheria toxin A chain gene. *Hum. Gene Ther.* **1993**, *4*, 741–747. [CrossRef] [PubMed]
38. Konopka, K.; Harrison, G.S.; Felgner, P.L.; Düzgüneş, N. Cationic liposome-mediated expression of HIV-regulated luciferase and diphtheria toxin genes in HeLa cells infected with or expressing HIV. *Biochim. Biophys. Acta (BBA) Mol. Cell Res.* **1997**, *1356*, 185–197. [CrossRef]
39. Gebremedhin, S.; Au, A.; Konopka, K.; Milnes, M.; Düzgüneş, N. A gene therapy approach to eliminate HIV-1-infected cells. *J. Calif. Dent. Assoc.* **2012**, *40*, 402–406.
40. Young, M.; Overlid, N.; Konopka, K.; Düzgüneş, N. Gene therapy for oral cancer: Efficient delivery of a 'suicide gene' to murine oral cancer cells in physiological milieu. *J. Calif. Dent. Assoc.* **2005**, *33*, 967–971.
41. Gebremedhin, S.; Singh, A.; Koons, S.; Bernt, W.; Konopka, K.; Düzgüneş, N. Gene delivery to carcinoma cells via novel non-viral vectors: Nanoparticle tracking analysis and suicide gene therapy. *Eur. J. Pharm. Sci.* **2014**, *60*, 72–79. [CrossRef]

42. Düzgüneş, N.; Cheung, J.; Konopka, K. Non-viral suicide gene therapy in cervical, oral and pharyngeal carcinoma cells with CMV- and EEV-plasmids. *J. Gene Med.* **2018**, *20*, e3054. [CrossRef] [PubMed]
43. Düzgüneş, N.; Cheung, J.; Konopka, K. Suicide gene therapy of oral squamous cell carcinoma and cervical carcinoma in vitro. *Methods Mol. Biol.* **2019**, *1895*, 177–184.
44. Neves, S.; Faneca, H.; Bertin, S.; Konopka, K.; Düzgüneş, N.; Pierrefite-Carle, V.; Simoes, S.; De Lima, M.P. Transferrin lipoplex-mediated suicide gene therapy of oral squamous cell carcinoma in an immunocompetent murine model and mechanisms involved in the antitumoral response. *Cancer Gene Ther.* **2009**, *16*, 91–101. [CrossRef]
45. Faneca, H.; Düzgüneş, N.; Pedroso de Lima, M.C. Suicide gene therapy for oral squamous cell carcinoma. *Suicide Gene Ther.* **2019**, *1895*, 43–55.
46. Garg, H.; Joshi, A. Conditional cytotoxic anti-HIV gene therapy for selectable cell modification. *Hum. Gene Ther.* **2016**, *27*, 400–415. [CrossRef] [PubMed]
47. Mali, P.; Esvelt, K.M.; Church, G.M. Cas9 as a versatile tool for engineering biology. *Nat. Methods* **2013**, *10*, 957–963. [CrossRef] [PubMed]
48. Doudna, J.A.; Charpentier, E. The new frontier of genome engineering with CRISPR-Cas9. *Science* **2014**, *346*, 1258096. [CrossRef] [PubMed]
49. Ebina, H.; Misawa, N.; Kanemura, Y.; Koyanagi, Y. Harnessing the CRISPR/Cas9 system to disrupt latent HIV-1 provirus. *Sci. Rep.* **2013**, *3*, 2510. [CrossRef] [PubMed]
50. Wang, Z.; Pan, Q.; Gendron, P.; Zhu, W.; Guo, F.; Cen, S.; Wainberg, M.A.; Liang, C. CRISPR/Cas9-Derived mutations both inhibit HIV-1 replication and accelerate viral escape. *Cell Rep.* **2016**, *15*, 481–489. [CrossRef]
51. Wang, G.; Zhao, N.; Berkhout, B.; Das, A.T. CRISPR-Cas based antiviral strategies against HIV-1. *Virus Res.* **2018**, *244*, 321–332. [CrossRef] [PubMed]
52. Christian, M.; Cermak, T.; Doyle, E.L.; Schmidt, C.; Zhang, F.; Hummel, A.; Bogdanove, A.J.; Voytas, D.F. Targeting DNA double-strand breaks with TAL effector nucleases. *Genetics* **2010**, *186*, 757–761. [CrossRef] [PubMed]
53. Ebina, H.; Kanemura, Y.; Misawa, N.; Sakuma, T.; Kobayashi, T.; Yamamoto, T.; Koyanagi, Y. A high excision potential of TALENs for integrated DNA of HIV-based lentiviral vector. *PLoS ONE* **2015**, *10*, e0120047. [CrossRef] [PubMed]
54. Hu, W.; Kaminski, R.; Yang, F.; Zhang, Y.; Cosentino, L.; Li, F.; Luo, B.; Alvarez-Carbonell, D.; Garcia-Mesa, Y.; Karn, J.; et al. RNA-directed gene editing specifically eradicates latent and prevents new HIV-1 infection. *Proc. Natl. Acad. Sci. USA* **2014**, *111*, 11461–11466. [CrossRef] [PubMed]
55. Kaminski, R.; Chen, Y.; Fischer, T.; Tedaldi, E.; Napoli, A.; Zhang, Y.; Karn, J.; Hu, W.; Khalili, K. Elimination of HIV-1 genomes from human T-lymphoid cells by CRISPR/Cas9 gene editing. *Sci. Rep.* **2016**, *6*, 22555. [CrossRef] [PubMed]
56. Kaminski, R.; Bella, R.; Yin, C.; Otte, J.; Ferrante, P.; Gendelman, H.E.; Li, H.; Booze, R.; Gordon, J.; Hu, W.; et al. Excision of HIV-1 DNA by gene editing: A proof-of-concept in vivo study. *Gene Ther.* **2016**, *23*, 690–695. [CrossRef] [PubMed]
57. Walker, L.M.; Phogat, S.K.; Chan-Hui, P.-Y.; Wagner, D.; Phung, P.; Goss, J.L.; Wrin, T.; Simek, M.D.; Fling, S.; Mitcham, J.L.; et al. Broad and potent neutralizing antibodies from an African donor reveal a new HIV-1 vaccine target. *Science* **2009**, *326*, 285–289. [CrossRef]
58. Yee, M.; Konopka, K.; Balzarini, J.; Düzgüneş, N. Inhibition of HIV-1 Env-mediated cell-cell fusion by lectins, peptide T-20, and neutralizing antibodies. *Open Virol. J.* **2011**, *5*, 44–51. [CrossRef]
59. Stephenson, K.E.; Barouch, D.H. Broadly neutralizing antibodies for HIV eradication. *HIV/AIDS Rep.* **2016**, *13*, 31–37. [CrossRef]
60. Flasher, D.; Konopka, K.; Chamow, S.M.; Dazin, P.; Ashkenazi, A.; Pretzer, E.; Düzgüneş, N. Liposome targeting to human immunodeficiency virus type 1-infected cells via recombinant soluble CD4 and CD4 immunoadhesin (CD4-IgG). *Biochim. et Biophys. Acta (BBA) Biomembr.* **1994**, *1194*, 185–196. [CrossRef]
61. Slepushkin, V.A.; Salem, I.I.; Andreev, S.M.; Dazin, P.; Düzgüneş, N. Targeting of liposomes to HIV-1-infected cells by peptides derived from the CD4 receptor. *Biochem. Biophys. Commun.* **1996**, *227*, 827–833. [CrossRef] [PubMed]
62. Park, J.; Kirpotin, D.; Hong, K.; Shalaby, R.; Shao, Y.; Nielsen, U.; Marks, J.; Papahadjopoulos, D.; Benz, C. Tumor targeting using anti-her2 immunoliposomes. *J. Control. Release* **2001**, *74*, 95–113. [CrossRef]

63. Eliaz, R.E.; Nir, S.; Marty, C.; Szoka, F.C., Jr. Determination and modeling of kinetics of cancer cell killing by doxorubicin and doxorubicin encapsulated in targeted liposomes. *Cancer Res.* **2004**, *64*, 711–718. [CrossRef] [PubMed]
64. Li, R.; Hu, S.; Chang, Y.; Zhang, Z.; Zha, Z.; Huang, H.; Shen, G.; Liu, J.; Song, L.; Wei, W.; et al. Development and characterization of a humanized anti-HER2 antibody HuA21 with potent anti-tumor properties in breast cancer cells. *Int. J. Mol. Sci.* **2016**, *17*, 563. [CrossRef] [PubMed]
65. Gabizon, A.A.; Patil, Y.; La-Beck, N.M. New insights and evolving role of pegylated liposomal doxorubicin in cancer therapy. *Drug Resist. Updat.* **2016**, *29*, 90–106. [CrossRef] [PubMed]
66. Oussoren, C.; Storm, G. Liposomes to target the lymphatics by subcutaneous administration. *Adv. Drug Deliv. Rev.* **2001**, *50*, 143–156. [CrossRef]
67. Allen, T.M.; Hansen, C.B.; Guo, L.S.S. Subcutaneous administration of liposomes: A comparison with the intravenous and intraperitoneal routes of injection. *Biochim. Biophys. Acta (BBA) Biomembr.* **1993**, *1150*, 9–16. [CrossRef]
68. Pantaleo, G.; Graziosi, C.; Butini, L.; Pizzo, P.A.; Schnittman, S.M.; Kotler, D.P.; Fauci, A.S. Lymphoid organs function as major reservoirs for human immunodeficiency virus. *Proc. Natl. Acad. Sci. USA* **1991**, *88*, 9838–9842. [CrossRef] [PubMed]
69. Pantaleo, G.; Graziosi, C.; Demarest, J.F.; Butini, L.; Montroni, M.; Fox, C.H.; Orenstein, J.M.; Kotler, D.P.; Fauci, A.S. HIV infection is active and progressive in lymphoid tissue during the clinically latent stage of disease. *Nat. Cell Boil.* **1993**, *362*, 355–358. [CrossRef]
70. Embretson, J.; Zupancic, M.; Ribas, J.L.; Burke, A.; Racz, P.; Tenner-Racz, K.; Haase, A.T. Massive covert infection of helper T lymphocytes and macrophages by HIV during the incubation period of AIDS. *Nat. Cell Boil.* **1993**, *362*, 359–362. [CrossRef]
71. Désormeaux, A.; Bergeron, M.G. Lymphoid tissue targeting of anti-HIV drugs using liposomes. *Methods Enzymol.* **2005**, *391*, 330–351. [PubMed]
72. Kinman, L.; Brodie, S.J.; Tsai, C.C.; Bui, T.; Larsen, K.; Schmidt, A.; Anderson, D.; Morton, W.R.; Hu, S.-L.; Ho, R.J.Y. Lipid–drug association enhanced HIV-1 protease inhibitor Indinavir localization in lymphoid tissues and viral load reduction: A proof of concept study in HIV-2287-infected macaques. *JAIDS J. Acquir. Immune Defic. Syndr.* **2003**, *34*, 387–397. [CrossRef]
73. Pretzer, E.; Flasher, D.; Düzgüneş, N. Inhibition of human immunodeficiency virus type-1 replication in macrophages and H9 cells by free or liposome-encapsulated L-689,502, an inhibitor of the viral protease. *Antivir. Res.* **1997**, *34*, 1–15. [CrossRef]
74. Clayton, R.; Öhagen, Å.; Nicol, F.; Del Vecchio, A.M.; Jonckers, T.H.; Goethals, O.; Van Loock, M.; Michiels, L.; Grigsby, J.; Xu, Z.; et al. Sustained and specific in vitro inhibition of HIV-1 replication by a protease inhibitor encapsulated in gp120-targeted liposomes. *Antivir. Res.* **2009**, *84*, 142–149. [CrossRef] [PubMed]
75. Slepushkin, V.A.; Simões, S.; Dazin, P.; Newman, M.S.; Guo, L.S.; de Lima, M.C.P.; Düzgüneş, N. Sterically stabilized pH-sensitive liposomes. Intracellular delivery of aqueous contents and prolonged circulation in vivo. *J. Biol. Chem.* **1997**, *272*, 2382–2388. [CrossRef] [PubMed]
76. Slepushkin, V.; Simões, S.; de Lima, M.C.; Düzgüneş, N. Sterically stabilized pH-sensitive liposomes. *Methods Enzymol.* **2004**, *387*, 134–147. [PubMed]
77. Gabizon, A.; Shmeeda, H.; Grenader, T. Pharmacological basis of pegylated liposomal doxorubicin: Impact on cancer therapy. *Eur. J. Pharm. Sci.* **2012**, *45*, 388–398. [CrossRef]
78. Phillips, W.T.; Medina, L.A.; Klipper, R.; Goins, B. A novel approach for the increased delivery of pharmaceutical agents to peritoneum and associated lymph nodes. *J. Pharmacol. Exp. Ther.* **2002**, *303*, 11–16. [CrossRef] [PubMed]
79. Bestman-Smith, J.; Gourde, P.; Désormeaux, A.; Tremblay, M.J.; Bergeron, M.G. Sterically stabilized liposomes bearing anti-HLA-DR antibodies for targeting the primary cellular reservoirs of HIV-1. *Biochim. Biophys. Acta (BBA) Biomembr.* **2000**, *1468*, 161–174. [CrossRef]
80. Kim, S.; Scheerer, S.; Geyer, M.A.; Howell, S.B. Direct cerebrospinal fluid delivery of an antiretroviral agent using multivesicular liposomes. *J. Infect. Dis.* **1990**, *162*, 750–752. [CrossRef]
81. Bissel, S.J.; Wiley, C.A. Human immunodeficiency virus infection of the brain: Pitfalls in evaluating infected/affected cell populations. *Brain Pathol.* **2004**, *14*, 97–108. [CrossRef] [PubMed]

82. Hütter, G.; Nowak, D.; Mossner, M.; Ganepola, S.; Müßig, A.; Allers, K.; Schneider, T.; Hofmann, J.; Kücherer, C.; Blau, O.; et al. Long-term control of HIV by CCR5Delta32/Delta32 stem-cell transplantation. *N. Engl. J. Med.* **2009**, *360*, 692–698. [CrossRef] [PubMed]
83. Gupta, R.K.; Abdul-Jawad, S.; McCoy, L.E.; Mok, H.P.; Peppa, D.; Salgado, M.; Martinez-Picado, J.; Nijhuis, M.; Wensing, A.M.J.; Lee, H.; et al. HIV-1 remission following CCR5Δ32/Δ32 haematopoietic stem-cell transplantation. *Nat. Cell Boil.* **2019**, *568*, 1. [CrossRef] [PubMed]
84. Corey, D.; Schultz, P. Generation of a hybrid sequence-specific single-stranded deoxyribonuclease. *Science* **1987**, *238*, 1401–1403. [CrossRef] [PubMed]
85. Pei, D.; Corey, D.R.; Schultz, P.G. Site-specific cleavage of duplex DNA by a semisynthetic nuclease via triple-helix formation. *Proc. Natl. Acad. Sci. USA* **1990**, *87*, 9858–9862. [CrossRef]
86. Guieysse, A.; Praseuth, D.; François, J.; Helene, C. Inhibition of replication initiation by triple helix-forming oligonucleotides. *Biochem. Biophys. Commun.* **1995**, *217*, 186–194. [CrossRef]
87. Faria, M.; Wood, C.; Perrouault, L.; Nelson, J.S.; Winter, A.; White, M.R.H.; Helene, C.; Giovannangeli, C. Targeted inhibition of transcription elongation in cells mediated by triplex-forming oligonucleotides. *Proc. Natl. Acad. Sci. USA* **2000**, *97*, 3862–3867. [CrossRef]

pharmaceutics

Article

Tenofovir Hot-Melt Granulation using Gelucire® to Develop Sustained-Release Vaginal Systems for Weekly Protection against Sexual Transmission of HIV

Fernando Notario-Pérez [1], Raúl Cazorla-Luna [1], Araceli Martín-Illana [1], Roberto Ruiz-Caro [1], Juan Peña [2] and María-Dolores Veiga [1,*]

1 Departamento de Farmacia Galénica y Tecnología Alimentaria, Facultad de Farmacia, Universidad Complutense de Madrid, 28040 Madrid, Spain; fnotar01@ucm.es (F.N.-P.); racazorl@ucm.es (R.C.-L.); aracelimartin@ucm.es (A.M.-I.); rruizcar@ucm.es (R.R.-C.)

2 Departamento de Química en Ciencias Farmacéuticas, Facultad de Farmacia, Universidad Complutense de Madrid, 28040 Madrid, Spain; juanpena@ima.ucm.es

* Correspondence: mdveiga@ucm.es; Tel.: +34-913-942-091; Fax: +34-913-941-736

Received: 15 February 2019; Accepted: 18 March 2019; Published: 20 March 2019

Abstract: Hot-melt granulation is a technique used to obtain granules by dispersing a drug in polymers at a high temperature. Tenofovir, an antiretroviral drug with proven activity as a vaginal microbicide, was dispersed in melted Gelucire® (or a mixture of different Gelucire®) to obtain drug-loaded granules. Studies performed on the granules proved that the drug is not altered in the hot-melt granulation process. The granules obtained were included in a matrix formed by the hydrophilic polymers hydroxypropylmethylcellulose and chitosan to obtain vaginal tablets that combine different mechanisms of controlled release: The Gelucire® needs to soften to allow the release of the Tenofovir, and the hydrophilic polymers must form a gel so the drug can diffuse through it. The studies performed with the tablets were swelling behavior, Tenofovir release, and ex vivo mucoadhesion. The tablets containing granules obtained with Tenofovir and Gelucire® 43/01 in a ratio of 1:2 in a matrix formed by hydroxypropylmethylcellulose and chitosan in a ratio of 1.9:1 were selected as the optimal formulation, since they release Tenofovir in a sustained manner over 216h and remain attached to the vaginal mucosa throughout. A weekly administration of these tablets would therefore offer women protection against the sexual transmission of HIV.

Keywords: controlled release; ex vivo bioadhesion; Gelucire®; Human Immunodeficiency Virus; mucoadhesive vaginal compacts; Tenofovir

---

1. Introduction

The hot-melt granulation technique consists of dissolving or dispersing an active principle in one or more melted polymers in order, after cooling, to obtain granules with a defined structure [1]. The melt entropy of the different components determines whether they will crystallize simultaneously, in turn conditioning the microstructure of the granules in the solid state [2].

The most common applications of this technique in pharmaceutical development are targeted drug release, masking the taste of active substances, and particularly improving the solubility of poorly soluble drugs [3]. In the latter case, solid dispersions of the drug are usually prepared in an inert hydrophilic matrix, causing the active principle to convert from its crystalline state to an amorphous form and improving its solubility [4]. However, this technique could also be applied in reverse to delay the dissolution of water-soluble drugs, which would aid the development of sustained-release systems capable of releasing the drug at a given rate, thereby achieving a constant concentration of the active principle at the site of action [3,4]. Some references to this possibility can be found in the literature: Mini-matrices have been developed where the drug is found in polymers as diverse as ethyl cellulose and xanthan gum [5], polyvinylpyrrolidone (PVP) [6], thermoplastic polyurethane [7] or ethylene vinyl acetate and polyethylene oxide [8]. One alternative yet to be explored would be to develop these systems using as carriers various Gelucire®, a group of amphiphilic excipients commonly used to develop sustained-release matrices [9,10]. Another advantage is that Gelucire® is commercialized with different melting points and hydrophilic-lipophilic balances (HLB), offering a wide range of options from which to select the most suitable for our purposes [11].

This last option—developing systems for the sustained release of drugs—raises the possibility of applying this technique to produce a vaginal microbicide to prevent the sexual transmission of the human immunodeficiency virus (HIV). A vaginal microbicide is any suitably formulated compound which, when applied in the vagina before intercourse, has the ability to prevent or reduce the transmission of sexually-transmitted diseases. The latest published data reflect the gender inequality in HIV protection, with an estimated 18.6 million women and girls living with the virus, and three young women infected every four minutes [12]. This explains the growing interest in recent decades in developing this type of formulations as a tool to protect women from the sexual acquisition of HIV.

Tenofovir (TFV) is an antiretroviral drug, specifically a nucleotide analogue reverse transcriptase inhibitor, approved by the Food and Drug Administration and recommended by the World Health Organization for its use as a first-line drug for the treatment of HIV [13]. It has also been extensively studied for the development of vaginal microbicides, mainly since the publication in 2010 of the results of the CAPRISA 004 trial, in which a 1% TFV gel showed a 39% effectiveness in protecting against HIV infection [14]. However, the fact that it required a daily application reduced the effectiveness of the formulation, so the current trend is to develop sustained-release formulations that require less frequent administration [15]. TFV is a hydrophilic drug that is soluble in aqueous media, such as vaginal fluid, requiring techniques to obtain sustained-release formulations that slow its dissolution rate in the vaginal environment.

The aim of this work is to prepare granules of TFV by hot-melt granulation using one or more Gelucire® containing mono-, di- and triglyceride esters of fatty acids (C8 to C18), –Gelucire® 39/01 (G39) and Gelucire® 43/01 (G43)–, which are characterized by their low HLB value (HLB = 1) and low melting point (39 °C and 43 °C respectively). However, rather than manufacturing matrices, the granules are incorporated in a hydrophilic matrix formed by a combination of hydroxypropylmethyl cellulose (HPMC) and chitosan (CH), two polymers that are capable of swelling in the presence of vaginal fluid and have proven to be an excellent choice for the development of sustained drug-delivery systems [16]. Vaginally administered tablets could then be developed in which the release of the drug would be conditioned by multiple factors. A Gelucire® with a predominantly lipophilic rather than hydrophilic character would prevent the dissolution of the granules in the vaginal medium, while their low melting point would cause Gelucire® to soften at body temperature and allow the vaginal fluid to diffuse through it and dissolve the drug; alternatively, the combination of HPMC and CH would swell with the vaginal fluid and form a gel which would initially delay the arrival of the vaginal fluid at the granules, and subsequently hinder the diffusion of the dissolved drug through the gel to reach the vaginal environment. The aim is to develop mucoadhesive systems that offer weekly protection against HIV infection and encourage greater adherence to the use of the microbicide, due to their lower frequency of administration, and consequently achieve greater protective efficacy.

2. Materials and Methods

2.1. Materials

Gelucire® 39/01 (G39, lot: 3E3701-2) and Gelurice® 43/01 (G43, lot: 1E5203-2) were a gift from Gattefossé (Saint-Priest, France). Tenofovir (TFV, lot: FT104801401) was supplied by Carbosynth Limited (Berkshire, UK). Chitosan, whose properties were characterized experimentally determining that it has a molecular weight of 32.1 kDa—being therefore a low-molecular-weight chitosan-, with an intrinsic viscosity of 24.75 dL/g and a degree of deacetylation of 54. 7% [17] (CH, lot: 0055790), was provided by Guinama (La Pobla de Vallbona, Spain). Hydroxypropylmethylcellulose—Methocel® K 100 M (HPMC; lot: SB13012N31) was kindly supplied by Colorcon Ltd. (Kent, UK). Anhydrous calcium hydrogen phosphate—Emprove®—(ACDP; lot: K93487944416) was supplied by Merck (Darmstadt, Germany). Polyvinylpyrrolidone—Kollidon® 30—(PVP; lot: 98-0820) was purchased from BASF Aktiengesellschaft (Ludwigshafen, Germany). Magnesium stearate PRS-CODEX (MgSt; lot: 85269 ALP) was acquired from Panreac (Barcelona, Spain).

All other reagents in this study were of analytical grade and used without further purification. Demineralized water was used in all cases.

2.2. *Preparation of the Granules*

The granules were prepared using G39, G43 and a mixture of both in equal proportions in the carrier/TFV ratios, shown in Table 1. The granules were obtained by first melting the Gelucire® in an oil bath (Selecta® Univeba-401, Barcelona, Spain), taking care to ensure the temperature never exceeded 50°C. Once completely melted, the TFV was incorporated and the mixture was stirred until the drug was homogeneously dispersed. The mixture was then spread over a smooth non-stick surface and left to stand at room temperature for a few minutes to solidify. Finally, the solidified mixture was forced through a 1 mm mesh to obtain the granules. Thus, this ensured that all granules obtained had a size of < 1mm in diameter.

Table 1. Granule composition (mg).

Granules	Tenofovir	Gelucire® 39	Gelucire® 43
TFV-1-G43-2	30		60
TFV-1-G43-1	30		30
TFV-2-G43-1	30		15
TFV-1-G41-2	30	30	30
TFV-1-G39-2	30	60	

2.3. *Characterization of the Granules*

2.3.1. Infrared Spectroscopy

Fourier Transform Infrared Attenuated Total Reflection Spectroscopy (FTIR-ATR) was used to characterize the pure materials and prepared granules with a Perkin-Elmer spectrophotometer equipped with a MIRacle™ accessory designed for FTIR-ATR measurements (Perkin-Elmer, Waltham, MA, USA).

2.3.2. Thermal Analysis

Hot-stage microscopy studies were performed between 30 °C and 350 °C. Approximately 1 mg of each sample was placed on a microscope slide with a cover on a Kofler stage and heated at a rate of 2 °C/min. A thermogalen microscope fitted with the Kofler stage (Leica, Wetzlar, Germany) was used for the microscopy examinations.

Thermogravimetric analysis (TGA) and differential scanning calorimetry (DSC) were done in an SDT-Q 600 TA Instruments TG/DTA analyzer (TA Instruments, New Castle, DE, USA). In this case, about 5–10 mg of samples were placed in a pinholed aluminium sample pan with a lid and heated in atmospheric air to between 30 °C and 500 °C at a rate of 10 °C/min.

2.3.3. X-ray Diffraction

An automated Philips X'Pert-MPD X-ray diffractometer with Bragg–Brentano geometry (Malvern Panalytical Ltd., Royston, UK) was used to obtain the X-ray diffraction patterns of pure materials and all granulated systems. Monochromatized Cu-Kα radiation (λ = 1.5406 Å) was irradiated over the samples at 45 kV, 40 mA and a time per step of 2 s, and analyzed between 2θ angles of 5° and 50°.

An X-ray thermodiffraction of the TFV was also performed using an X'Pert PRO MPD in theta-theta configuration with an Anton Paar HTK1200 high-temperature camera (Malvern Panalytical Ltd., Royston, UK). Monochromatized Cu-Kα radiation (λ = 1.5406 Å) was irradiated over the sample at 45 kV and 40 mA. The sample was heated to 5 °C/min and analyzed between 2θ angles of 5° and 50° at 25 °C, 150 °C, 175 °C and 225 °C. The sample was then cooled at the same rate and analyzed at 200 °C, 150 °C and 25 °C under the same conditions.

2.4. Preparation of the Tablets

Vaginal tablets were developed with a combination of hydrophilic polymers (HPMC and CH) containing the granules developed previously. Each granulated system was in turn included in three different combinations of HPMC and CH. HPMC, CH and ACDP were mixed, and the mixture was then wetted with a solution of PVP in ethanol to produce a mass, which was granulated using a 1 mm mesh. Three granulates were obtained (C1, with a ratio of 1:1 HPMC/CH; C2, with a ratio of 1.9:1 HPMC/CH; and C3, with a ratio of 1:1.9 HPMC/CH) and dried at room temperature for 24 h. Lastly, each granulate was mixed with the TFV granules for a final content of 30 mg of drug and 290 mg of hydrophilic polymer per tablet. All granulates were prepared by forcing them through the same mesh to ensure that all granules—polymeric and TFV-loaded granules—had the same size (<1 mm in diameter), and the blends were homogeneous. MgSt was added to the total dried granules before tableting. The final content of each batch is shown in Table 2.

Table 2. Tablet composition (mg/unit).

Batch	HPMC	CH	ACDP	PVP	TFV	G43	G39	MgSt
TFV-1-G43-2 C1	145	145	45	27	30	60		3
TFV-1-G43-2 C2	190	100	45	27	30	60		3
TFV-1-G43-2 C3	100	190	45	27	30	60		3
TFV-1-G43-1 C1	145	145	45	27	30	30		3
TFV-1-G43-1 C2	190	100	45	27	30	30		3
TFV-1-G43-1 C3	100	190	45	27	30	30		3
TFV-2-G43-1 C1	145	145	45	27	30	15		3
TFV-2-G43-1 C2	190	100	45	27	30	15		3
TFV-2-G43-1 C3	100	190	45	27	30	15		3
TFV-1-G41-2 C1	145	145	45	27	30	30	30	3
TFV-1-G41-2 C2	190	100	45	27	30	30	30	3
TFV-1-G41-2 C3	100	190	45	27	30	30	30	3
TFV-1-G39-2 C1	145	145	45	27	30		60	3
TFV-1-G39-2 C2	190	100	45	27	30		60	3
TFV-1-G39-2 C3	100	190	45	27	30		60	3

HPMC: Hydroxypropylmethylcellulose; CH: Chitosan; ACDP: Anhydrous calcium hydrogen phosphate.

The tablets were produced using a press similar to the one used for preparing solid samples for analysis by IR spectroscopy. A constant pressure of 5 tons was applied for four minutes using a punch. The resulting tablets were cylindrical in shape with a diameter of 13 mm and a height of 2.5–2.8 mm

2.5. Assessment of the Tablets

2.5.1. Swelling Behavior

The method described by Ruiz-Caro et al. [18] was used to analyze the swelling behavior of each batch in simulated vaginal fluid (SVF) [19]. The test was done in triplicate in a shaking water bath (Selecta® UNITRONIC320 OR, Barcelona, Spain) at 37 °C and 15 opm. The discs were removed from the medium at predetermined times, placed on filter paper to eliminate excess liquid and weighed.

The swelling ratio (SR) was calculated following Equation (1) where T_s and T_d correspond to the swollen and dry tablet weights respectively:

$$SR = \frac{T_s - T_d}{T_d} \cdot 100. \quad (1)$$

Since the initial tablet weight varies depending on the amount of Gelucire® it contains, the *SR* of different batches was compared by determining the adjusted swelling ratio (*ASR*), where the weight increment is related to the amount of swellable polymer, as shown in Equation (2) where *SR*, T_d and *SP* correspond respectively to the swelling ratio determined by Equation (1), the dry tablet weight, and the amount of swellable polymer in the tablet:

$$ASR = \frac{SR \cdot T_d}{SP}. \quad (2)$$

2.5.2. Release Study

The release of TFV from the batches was evaluated with the method described by Sánchez-Sánchez et al. [20]. The tablets were immersed in 80 mL of SVF in a borosilicate glass bottle and placed in a shaking water bath (Selecta® UNITRONIC320 OR, Barcelona, Spain) at 37 °C and 15 opm. The test was performed in triplicate. Release tests are performed in sink conditions because the solubility of TFV in SVF, measured at room temperature, is 4 mg/mL. 5 mL samples were taken and filtered at given times, and the medium was replaced with the same volume of SVF at the same temperature. The TFV released was quantified by UV spectroscopy at a wavelength of 260 nm in a Shimadzu® UV-1700 (Kyoto, Japan).

The similarity factor (f_2) (Equation (3)) was calculated in order to statistically demonstrate the differences between the batches.

$$f_2 = 50 \times \log\left\{\left[1 + \left(\frac{1}{n}\right)\sum_{j=1}^{n} W_j\left|R_j - T_j\right|^2\right]^{-0.5} \times 100\right\}. \quad (3)$$

This is a model-independent index described by Moore and Flanner [21], where n is the number of samples for each dissolution test, R_j and T_j are the drug release percentage at each time for the reference and test product respectively, and W_j is a weight factor, which in this study is equal to 1. A value of $f_2 > 65$ means similarity between the profiles of over 95%, while $f_2 < 65$ indicates that the profiles are significantly different [22].

Finally, the experimental drug-release data were fitted to different model-dependent methods (zero order, first order, Higuchi, Hixson-Crowell, Hopfenberg and Korsmeyer-Peppas models) to investigate the drug release kinetics [23].

2.5.3. Mucoadhesion Test

The method described by Notario-Pérez et al. [24] was used to evaluate ex vivo mucoadhesion time. A sample of freshly excised veal vaginal mucosa was obtained from a local slaughterhouse and attached to an 8.5 cm × 5 cm stainless steel plate with a cyanoacrylate adhesive. Each tablet was then placed over the mucosa and weight of 500 g was applied for 30 s. The system was placed at an angle of 60° in a beaker containing 150 mL of SVF and then in the shaking water bath (Selecta® UNITRONIC320 OR, Barcelona, Spain) at 37 °C and 15 opm. All batches were tested in duplicate.

3. Results and Discussion

3.1. Characterization of the Granules

3.1.1. Infrared Spectroscopy

The analysis of the granules by FTIR-ATR, and the subsequent comparison with the spectrum obtained by analyzing the raw materials used in their preparation, reveal any possible interactions that may have occurred at the molecular level during the granulation process. These interactions—if they occur—require an in-depth analysis to ensure that the therapeutic efficacy of the TFV has not been modified.

The two types of Gelucire® analyzed (G39 and G43) showed a very similar FTIR-ATR spectrum, which is to be expected given their close structural similarity (Figure 1B). These spectra are also clearly similar to those obtained by other authors with other types of Gelucire® [25]. Characteristics worth noting include a band at around 1650 cm^{-1} corresponding to the C–O bonds and another double band around 2800 cm^{-1}, due to the C=O bond [26].

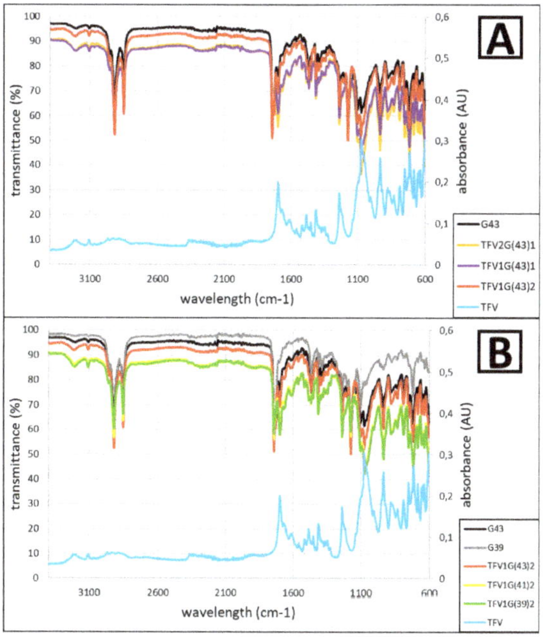

Figure 1. FTIR-ATR spectra of TFV, G43 and their combinations (**A**); and TFV, G43, G39 and their combinations in a 1:2 ratio of TFV/Gelucire® (**B**). The absorbance spectra (right axes) correspond to TFV.

TFV is notable for a band around 1700 cm^{-1}, corresponding to C=O stretching [27]. It also has bands at 3100 cm^{-1} and 3200 cm^{-1} corresponding to C–H and N–H bonds respectively [28] (Figure 1).

The results of the spectra obtained with the prepared granules are in all cases very similar, and always resemble the results that would be obtained by superimposing the spectra for the raw materials (Figure 1).

The FTIR-ATR analysis confirms that the raw materials—Gelucire® and TFV—do not interact with each other at a molecular level during the granulation process, thus maintaining the drug's properties.

3.1.2. Thermal Analysis

Thermal analysis was performed in order to determine the characteristics of the granules in greater detail. The TGA analysis (Figure 2) shows the weight lost by the granules during heating. The raw materials were also analyzed using this technique. In the case of TFV, about 5% of the weight is lost at around 90 °C (this is probably a loss of the water captured by the drug during storage), followed by a strong degradation that possibly corresponds to the drug decomposition, beginning at around 320 °C. At 500 °C, the sample maintains about 65% of its initial weight. In contrast, the Gelucire® samples show no water loss in the TGA analysis, due to the Gelucire®'s limited uptake of water at ambient humidity conditions [29], and there is no change in the TGA curves until around 240 °C. From this point until 370 °C the samples undergo a marked decomposition and lose over 80% of their weight. With further heating, Gelucire® samples continue losing weight more slowly and when the test ends at 500 °C, less than 5% of the initial weight remains. Both G43 and G39 show similar results, but should be noted that the final weight loss is lower in the G39 sample. These minor differences between the different Gelucire® products can be attributed to the slight variations in their composition [11]. Granules combining TFV and Gelucire® in all cases display an intermediate behavior between the raw materials used in their manufacture. The comparison of granules with different TFV/Gelucire® ratios show that the greater the amount of Gelucire® included in the sample, the higher the final weight loss (Figure 2A). The variations are almost negligible in granules prepared with different Gelucire®, although the inclusion of more G39 than G41 can still be seen to cause the granules to degrade more, as in the case of the raw materials. Finally, it should also be noted that no weight loss is observed in the TGA analysis of the granules at around 90 °C, as occurs in the TFV sample, which confirms that the weight loss is due to water, as the raw materials are heated and the water has already been removed during the preparation of the granules.

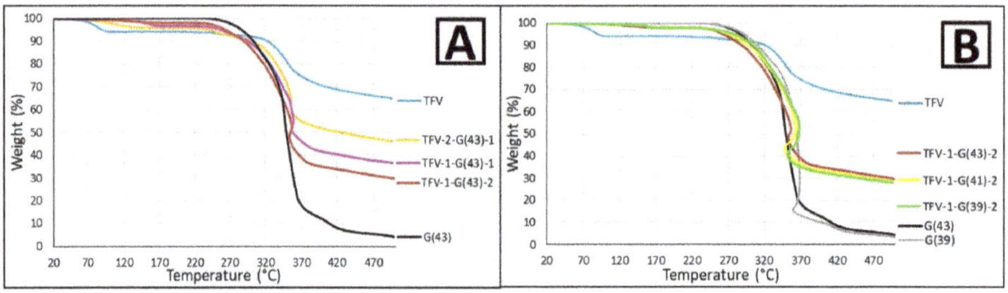

Figure 2. Thermogravimetric analysis (TGA) curves of TFV, G43 and their combinations (**A**); and TFV, G43, G39 and their combinations in a ratio of 1:2 of TFV/Gelucire® (**B**).

DSC analyses were performed to check if during the process of preparation of the solid dispersion it occurs the dissolution of the active principle in the molten vehicle, which would later be reversible by crystallization of the TFV during the cooling of the binary system, and the results are shown in Figure 3. The TFV DSC curve is characteristic, due to an endothermic peak at around 160 °C and an exothermic peak at around 220 °C, which may correspond to changes in the crystallinity of the drug. A stronger endothermic peak beginning at 295 °C corresponds to the TFV melting point [30]. Finally, drug decomposition is observed at around 315 °C.

An endothermic peak can be seen in both G43 and G39 at approximately their respective melting points. These melting points begin shortly after ambient temperature, as has already been observed by other researchers [31], and can be explained by the fact that Gelucire® are multicomponent mixtures with a semi-solid nature. Exothermic readings beginning at 230 °C are observed in both Gelucire®, and the main difference is in their decomposition, which in both cases begins after 400 °C but is much more vigorous for G43 than for G39.

When analyzing the DSC curves from the granules combining one or both Gelucire® with TFV, the main observation is an intermediate curve between the curves obtained from the raw materials. The three characteristic peaks of TFV appear in all the prepared granules, so mixing with Gelucire® apparently has no effect on the drug. Nevertheless, it should be noted that the inclusion of TFV appears to lower the melting point of Gelucire® (Figure 3).

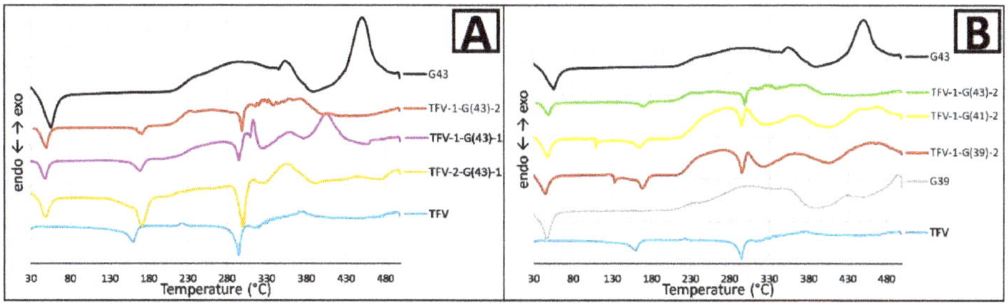

Figure 3. Differential scanning calorimetry (DSC) curves of TFV, G43 and their combinations (**A**); and TFV, G43, G39 and their combinations in a ratio of 1:2 of TFV/Gelucire® (**B**).

Finally, hot stage microscopy (HSM) was performed to complete the thermal analysis of the samples. TFV can be seen as a crystalline substance in the visualization. When heating the sample, some changes in the shape of particles are observed at 167 °C. The temperature continues rising and a similar behavior can be seen on reaching 220°C. These two changes may correspond to crystalline transitions of TFV, as was assumed to occur when analyzing the DSC curve. Finally, the sample begins to melt at around 265 °C.

Both Gelucire® appear as irregularly-shaped particles of a waxy substance with a rough surface. They are crystalline, since polarized light can pass through the samples. The start of the melting process is identified by HSM at a somewhat lower temperature than indicated by the supplier (40 °C for G43 and 36 °C for G39), as shown in the DSC analysis (Figure 3). G39 turns brown at around 310°C, as though the sample had charred, and it appears to start boiling.

Very similar results are obtained when analyzing the granules by HSM. The main changes are summarized in Figure 4. They all appear in the microscope as irregular particles that allow the partial passage of polarized light. In all the samples, the first change observed is the melting of the Gelucire®. The more G39 there is in the granules, the lower the melting point (Figure 4). When comparing samples containing only G41 and TFV, the granules with twice as much TFV as G41 have a higher melting point. However, the melting of G41 is more difficult to visual identify, due to its low proportion, so it probably occurs before it can be visually detected (when a halo of melted Gelucire® is seen around the TFV particles). Once the Gelucire® is completely melted, there are more particles that allow the passage of polarized light, corresponding to TFV crystals surrounded by melted Gelucire®. The first crystalline transition in the TFV raw material is not detected in any of the granules. However, an interesting phenomenon occurs at the temperature of the second crystalline transition of TFV. From 212 °C to 226 °C, depending on the sample, the TFV crystals that allow the passage of polarized light to disappear, probably because after the change in its crystallinity TFV is able to dissolve into the melted Gelucire®. This ability of melted Gelucire® to dissolve drugs has also been noted by other researchers [31]. However, some TFV crystals are still in evidence, so the amount of molten Gelucire® is insufficient to dissolve all the TFV present in the sample. A little later, at between 224 °C and 251°C, the sample carbonizes and turns brown, probably because the dissolved TFV has burned. Between 262–270 °C, the remaining particles that still allow the passage of polarized light disappear completely in some samples, corresponding to the melting of the undissolved TFV, which occurs at the temperature at which the TFV raw material melts. Finally, at around 290 °C, the Gelucire® can be seen to be boiling. After analyzing all the information from the thermal analysis of the granules, it can be confirmed that no interaction between Gelucire® and TFV occurs at the manufacturing temperatures; however, a more in-depth study was made using X-ray diffraction techniques to confirm the TFV transitions and some of the interactions that are assumed to occur at higher temperatures.

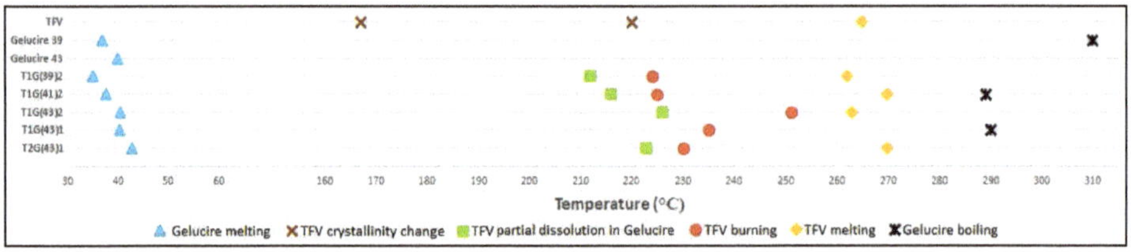

Figure 4. Diagram of the changes visualized during the observation of TFV, Gelucire® and the granules by hot stage microscopy.

3.1.3. X-ray Diffraction

X-ray spectrometry studies were performed to evaluate if these is any change in the crystalline state of the drug during the granulation process. These studies could also corroborate the hypotheses formulated after the thermal analysis of the granules. First the raw materials (TFV, G43 and G39) and the manufactured granules were analyzed to compare them and confirm that there is no interaction between the drug and the carriers during granulation.

The X-ray diffraction pattern for TFV reveals the four characteristic peaks of the drug at 7.3, 14.7, 18.3 and 23.4° 2θ [32]. This clearly confirms that TFV is a crystalline substance, as observed in HSM. There are hardly any differences between G43 and G39, and both are crystalline (as observed in HSM) with two intense peaks at 20.9 and 23.1° 2θ. This agrees with the results obtained for other Gelucire®, which also present peaks at between 20–25° 2θ [33]. A third peak of lower intensity can also be distinguished at around 6.8° 2θ, which is more clearly visible in G39 (Figure 5). The spectra observed in the granules is a mixture of the spectra obtained with their component raw materials, and the lower the proportion of each material in the composition of the granules, the lower the intensity of the characteristic peaks (Figure 5).

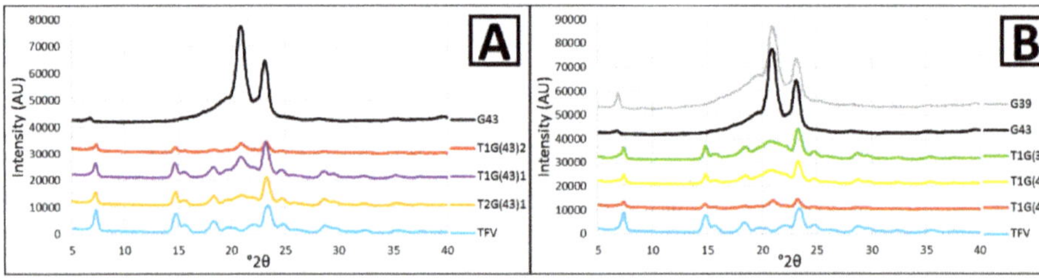

Figure 5. X-ray diffraction patterns of pure TFV, G43 and their combinations (**A**); and TFV, G43, G39 and their combinations in a ratio of 1:2 TFV/Gelucire® (**B**).

In conclusion, this analysis allows us to confirm that no drug-Gelucire® interaction occurs in the granulation process.

However, various changes were observed in TFV in DSC and HSM that have been attributed to the crystalline transition of the drug, although a further study was done to confirm this.

The X-Ray thermodiffraction technique revealed the changes in the drug's crystallinity caused by temperature. A study was done to confirm that the changes observed in DSC and HSM at 167 °C and 220 °C correspond to the crystalline transition of TFV. This consisted of measuring a sample of pure TFV by X-ray diffraction at room temperature (25 °C) and then after heating. The temperature was maintained for five minutes before and after the temperatures at which changes were observed, then a measurement was taken by X-ray diffraction. After reaching 225 °C, higher than the temperatures at which the previous change had occurred, the sample was cooled and new measurements were taken using the same procedure to determine whether the temperature-related changes are reversible.

The spectra observed when the sample was analyzed at 25 °C is exactly the same as in the previous study at room temperature (Figures 5 and 6). Although other authors have found that the crystallinity of TFV does not change with temperature, their studies reached only 80 °C [32,34]. In our study, the sample was heated to 150 °C before taking the second measurement, prior to the first change in the structure of TFV. Surprisingly, changes can be seen in TFV at this temperature. This is probably because the first transition has already begun, as the characteristic peaks have been displaced to 7.3° and 23.4° 2θ, the peak that was present at 14.7° 2θ has disappeared, and a series of peaks have appeared between 16–19° 2θ. At 175 °C, the spectrum is completely different to the one obtained at 25 °C, with a double peak between 8–9° 2θ, another two double peaks around 17–18° 2θ, and a band with multiple peaks between 19–20° 2θ that were not observed at lower temperatures. This clearly confirms that the first crystalline transition of TFV is complete (Figure 6). At 225 °C the spectrum is slightly different: The peak at 9° 2θ disappears and the intensity of the peak at 17° 2θ decreases. The second crystalline change is therefore complete at 225 °C, allowing the dissolution of TFV in its new state in the melted Gelucire®, as observed in the HSM studies. Finally, the sample was cooled and new measurements were taken, and no change was observed in X-ray diffraction, thus confirming that the crystalline changes in TFV are irreversible.

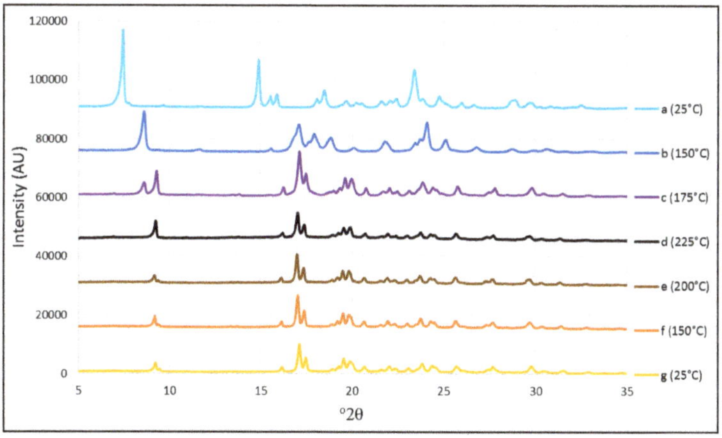

Figure 6. X-ray thermodiffraction patterns of pure TFV while heating to 25 °C (a), 150 °C (b), 175 °C (c) and 225 °C (d); and then while cooling to 200 °C (e), 150 °C (f) and 25 °C (g).

After the IR, thermal and X-ray analysis of the prepared granules, we can confirm that TFV is not altered in the hot-melt granulation process but conserves its therapeutic properties, thus enabling the use of the granules for prevention of HIV.

3.2. *Assessment of the Tablets*

3.2.1. Swelling Behavior

The evaluation of the swelling behavior of the formulations has a twofold importance: The capture of water is the main mechanism controlling the release of the drug, since it is unable to diffuse until the hydrophilic polymer swells and the water surrounds the granules [35]; and the swelling of the formulations can be seen as a way of improving adherence to prophylactic treatment, as the lower water capture would make the tablets more comfortable for women.

The inclusion of the drug granulated with G43 in the polymer matrix clearly improves the formulation in terms of swelling. As can be seen in Figure 7, the maximum swelling ratio is notably lower than in the formulations containing the drug without Gelucire® [16], and especially in the formulations in which HPMC is greater than or equal to CH, due to the predominantly hydrophobic character of Gelucire®. The formulation achieves complete erosion sooner with the addition of the granules: After 216 h instead of the 264 h for the original formulation. This behavior would imply greater adherence to prophylaxis, since the formulation would remain less time in the vaginal environment after the drug release, making it more comfortable for women. The more HPMC there is in the formulation, the higher the maximum swelling ratio. This was expected, as the swelling capacity of HPMC is higher than CH [24]. Finally, no clear differences ca be seen between the formulations containing granules with a different G43/TFV ratio (Figure 7).

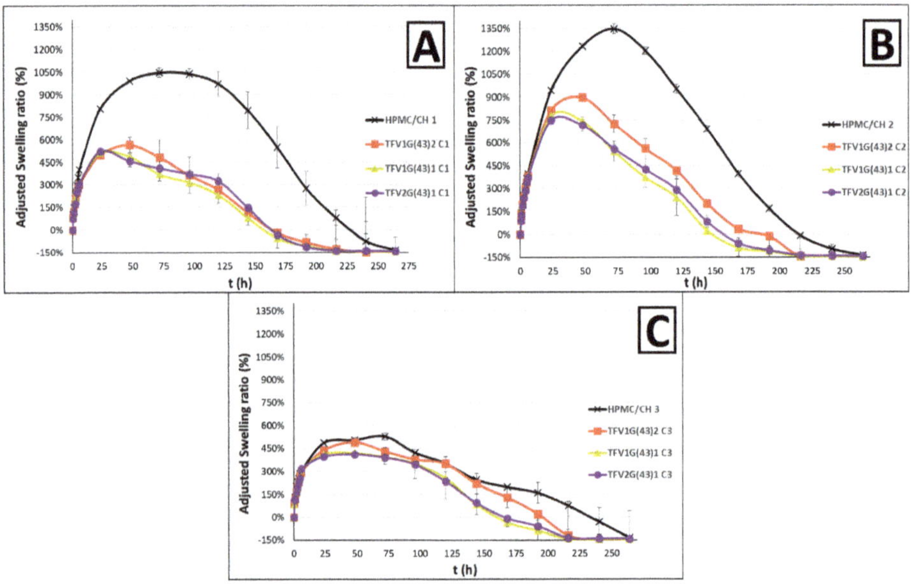

Figure 7. Adjusted swelling ratio profiles obtained from batches containing TFV/G43 drug-loaded granules in HPMC/CH matrices with equal ratios of both polymers (**A**); more HPMC than CH (**B**); and more CH than HPMC (**C**). They are all compared with standard formulations of HPMC/CH [16].

The use of different G39 or a mixture of G39 and G43 does not appear to change the swelling behavior of the formulations. This is to be expected since HPMC and CH are polymers with a capacity to capture water, and both Gelucire® act similarly as a barrier to water diffusion. Only the tablets with more HPMC than CH show a slight difference in the formulations containing G39 (or a G39/G43 mixture), which have a lower water capture than tablets with only G43. Nevertheless, an improvement is again observed compared to the formulations without Gelucire® (Figure 8).

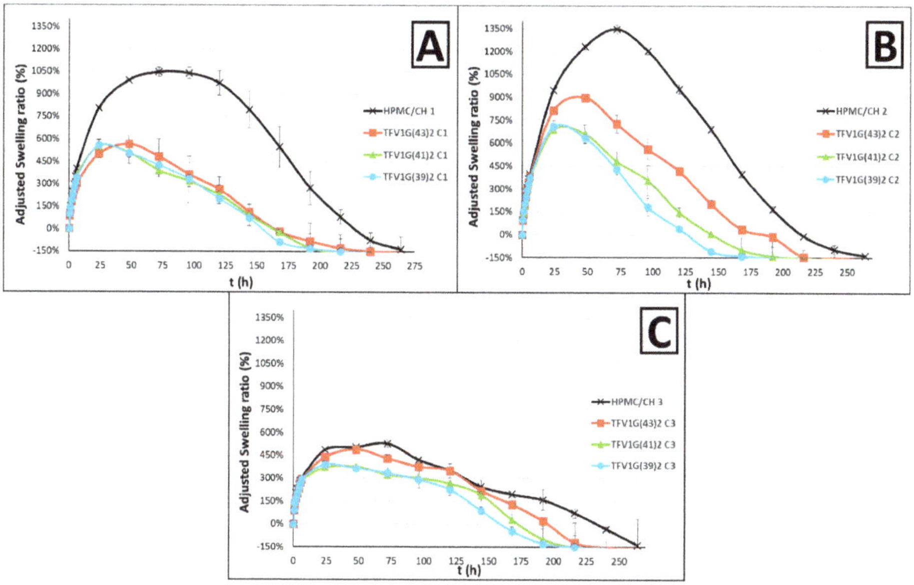

Figure 8. Adjusted swelling ratio profiles obtained from batches containing TFV/Gelucire® (in a 1:2 ratio) granules in HPMC/CH matrices with equal ratios of both polymers (**A**); more HPMC than CH (**B**); and more CH than HPMC (**C**). They are all compared with standard formulations of HPMC/CH [16].

The inclusion of the granules creates barriers within the tablet that hinder the diffusion of water and leads to lower water capture. This not only makes the tablets more comfortable, but would also probably prevent the loss of the drug, since it contains less water through which to diffuse. We can therefore formulate the hypothesis that this formulation would be more effective for the controlled release of TFV, although this must be confirmed in subsequent drug-release studies.

3.2.2. Release Study

As can be seen in Figure 9, the inclusion of mixed TFV/G43 granules in the formulations enhances the controlled release of the drug. These differences can be observed from the beginning of the drug release test, as the Gelucire® must soften as it forms granules with the drug to allow the release of the TFV. This softening is enabled by the presence of water and body temperature. Once the Gelucire® is softened and the drug diffuses through it, it is incorporated in a matrix of HPMC and CH polymers, which form a gel in the vaginal medium through which the drug must diffuse. As has been observed in the swelling test, the presence of the granules hinders the swelling of the polymers, which also causes the drug to be released from the tablet in a more sustained way.

In the samples containing granules with equal or more TFV than G43 (TFV2G(43)1 and TFV1G(43)1), the TFV drug release profiles obtained are quite similar. In these samples the main improvement is the slowdown of the release of TFV, since almost all the drug is released at 120 h (as occurred in the samples that do not use Gelucire®), although there are significant differences in the drug release behavior. In batches with TFV1G(43)2 granules, which include twice the amount of G43 as TFV, the improvement is much more marked (as an example, at 24 h approximately half the amount of drug has been released as in standard batches). This is because of the higher amount of G43 impedes softening and it takes longer for the drug to be released from the granules. Drug release is prolonged to 216–240 h, thus achieving 9−10 days of a controlled release of TFV. This is double the time taken without Gelucire® granules, and could represent a milestone in HIV transmission, since a degradable vaginal formulation capable of reaching these controlled release times has never previously been achieved.

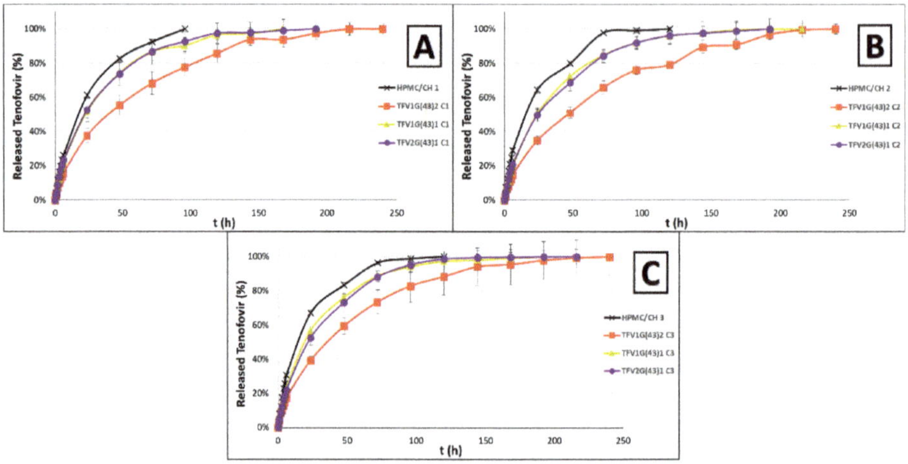

Figure 9. TFV release profiles obtained from batches including TFV/G43 drug-loaded granules in HPMC/CH matrices with equal ratios of both polymers (**A**), more HPMC than CH (**B**) and more CH than HPMC (**C**). All of them are compared with standard formulations of HPMC/CH [16].

It was also necessary to evaluate whether a Gelucire® with a melting point that was closer to body temperature would alter the release of the drug. Figure 10 shows that when G39 or a mixture of G39 and G41 is used to manufacture the granules, TFV is released faster than in tablets with TFV/G43 granules. These batches confirm that the softening of the Gelucire® is a crucial factor for controlling TFV release; this agrees with other authors, who also highlight that in the case of Gelucire® that melt above body temperature, the release of the drug depends on the composition and HLB value [36].

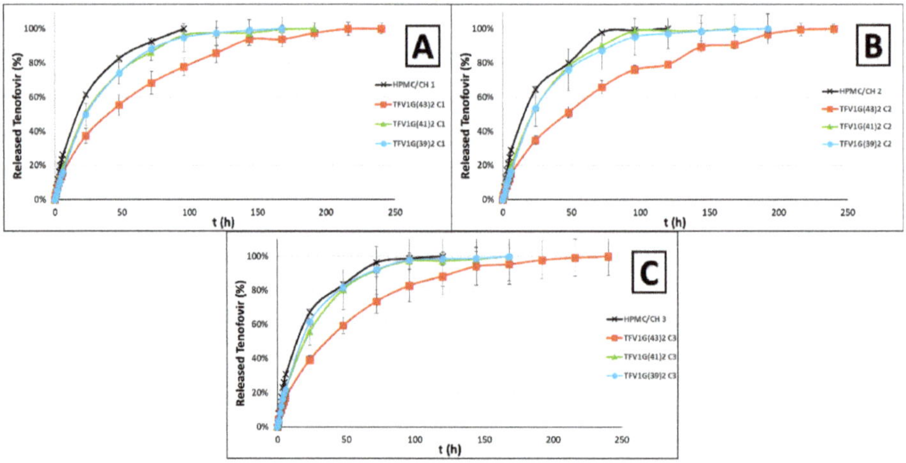

Figure 10. TFV release profiles obtained from batches containing TFV/Gelucire® (in a ratio of 1:2) granules in HPMC/CH matrices with equal ratios of both polymers (**A**); more HPMC than CH (**B**); and more CH than HPMC (**C**). All are compared with standard formulations of HPMC/CH [16].

The TFV1G(43)2 granules are therefore undoubtedly the best formulation. Finally, a comparison must be made of the formulations containing different proportions of HPMC/CH. Although the three proportions have similar TFV release profiles, the formulation with the most HPMC (TFV1G(43)2 C2) shows the greatest controlled release (Figure 10B).

Although the figures visually identify the improvements achieved with the formulations, the f_2 similarity factor was also calculated as a statistical analysis that could demonstrate these differences [37]. Table 3 shows the f_2 values for all the formulations.

Table 3. Similarity factor (f_2) values for the release profiles obtained from reference and problem formulations. Comparisons with significant difference ($f_2 < 65$) are in bold.

REFERENCE	PROBLEM	f_2	REFERENCE	PROBLEM	f_2	REFERENCE	PROBLEM	f_2
C1	T1G(43)2 C1	**44.6**	C1	C2	84.3	T1G(43)2 C1	T1G(43)2 C2	83.4
C1	T1G(43)1 C1	**64.7**	C1	C3	76.9	T1G(43)2 C1	T1G(43)2 C3	78.4
C1	T2G(43)1 C1	66.9	C2	C3	85.6	T1G(43)2 C2	T1G(43)2 C3	68.1
C1	T1G(41)2 C1	**60.4**	T1G(43)2 C1	T1G(43)1 C1	**54.5**	T1G(43)1 C1	T1G(43)1 C2	83.5
C1	T1G(39)2 C1	**60.5**	T1G(43)2 C1	T2G(43)1 C1	**53.7**	T1G(43)1 C1	T1G(43)1 C3	84.9
C2	T1G(43)2 C2	**40.9**	T1G(43)2 C1	T1G(41)2 C1	**56.3**	T1G(43)1 C2	T1G(43)1 C3	75.1
C2	T1G(43)1 C2	**55.8**	T1G(43)2 C1	T1G(39)2 C1	**57.2**	T2G(43)1 C1	T2G(43)1 C2	81.3
C2	T2G(43)1 C2	**56.7**	T1G(43)2 C2	T1G(43)1 C2	**50.0**	T2G(43)1 C1	T2G(43)1 C3	94.5
C2	T1G(41)2 C2	**60.5**	T1G(43)2 C2	T2G(43)1 C2	**51.0**	T2G(43)1 C2	T2G(43)1 C3	84.8
C2	T1G(39)2 C2	**54.7**	T1G(43)2 C2	T1G(41)2 C2	**48.4**	T1G(41)2 C1	T1G(41)2 C2	86.2
C3	T1G(43)2 C3	**43.3**	T1G(43)2 C2	T1G(39)2 C2	**50.0**	T1G(41)2 C1	T1G(41)2 C3	73.7
C3	T1G(43)1 C3	**60.5**	T1G(43)2 C3	T1G(43)1 C3	**54.6**	T1G(41)2 C2	T1G(41)2 C3	83.7
C3	T2G(43)1 C3	**56.5**	T1G(43)2 C3	T2G(43)1 C3	**59.1**	T1G(39)2 C1	T1G(39)2 C2	88.1
C3	T1G(41)2 C3	**61.9**	T1G(43)2 C3	T1G(41)2 C3	**52.4**	T1G(39)2 C1	T1G(39)2 C3	**62.7**
C3	T1G(39)2 C3	**64.1**	T1G(43)2 C3	T1G(39)2 C3	**49.3**	T1G(39)2 C2	T1G(39)2 C3	67.4

The first point to emphasize is that all the formulations developed in the present work improve those with the same proportion of HPMC/CH but without Gelucire® (except for formulation TFV2G(43)1 C1, which has a slight similarity with the formulation without Gelucire®). This highlights the role of Gelucire® in controlling the release of TFV from the tablets. However, when comparing formulations with the same granules but with a different proportion of the HPMC/CH mixture, significant differences were observed in only one case (batches TFV1G(39)2 C1 and TFV1G(39)2 C3). This reaffirms the fact that the combination of HPMC and CH forms a very robust mixed gel capable of delaying the release of TFV regardless of the proportion of polymer [16]. Although the granules' ability to control the release has been demonstrated, these results must be compared with formulations with the same proportion of HPMC/CH but different granules, in order to determine which granules are the most suitable. The comparison of these formulations confirms that the granules identified in Figure 9 as best controlling TFV release (TFV1G(43)2) always present significant differences with the other granules, regardless of the proportion of HPMC/CH in the tablets that include the granules.

The statistical analysis therefore shows that the TFV1G(43)2 granules are the best option for preparing our formulations, although it is impossible to specify the most appropriate proportion of hydrophilic polymers to include in these granules since they are all very similar.

Finally, the experimental data obtained in the drug release tests were adjusted to different mathematical models (zero order, first order, Higuchi, Hixson-Crowell, Hopfenberg and Korsmeyer-Peppas). The correlation coefficients (r^2) obtained after the adjustment are shown in Table 4.

Table 4. Correlation coefficients obtained when experimental data are fitted to different mathematical models.

Batch	Correlation Coefficients (r^2)					
	Zero Order	First Order	Higuchi	Hixson-Crowell	Hopfenberg	Korsmeyer-Peppas
TFV1G(43)2 C1	0.9557	0.7162	0.9975	0.9888	0.9824	0.9915
TFV1G(43)2 C2	0.9658	0.7139	0.9961	0.9925	0.9877	0.9876
TFV1G(43)2 C3	0.9567	0.7188	0.9971	0.9922	0.9858	0.9911
TFV1G(43)1 C1	0.9140	0.6436	0.9885	0.9749	0.9624	0.9795
TFV1G(43)1 C2	0.9298	0.6610	0.9911	0.9884	0.9771	0.9887
TFV1G(43)1 C3	0.9110	0.6514	0.9878	0.9835	0.9690	0.9896
TFV2G(43)1 C1	0.9190	0.6474	0.9920	0.9852	0.9723	0.9801
TFV2G(43)1 C2	0.9351	0.6671	0.9954	0.9911	0.9813	0.9864
TFV2G(43)1 C3	0.9306	0.6849	0.9939	0.9938	0.9834	0.9933
TFV1G(41)2 C1	0.9377	0.663	0.9930	0.9958	0.9877	0.9850
TFV1G(41)2 C2	0.9307	0.6604	0.9912	0.9963	0.9897	0.9889
TFV1G(41)2 C3	0.9148	0.6139	0.9886	0.9918	0.9775	0.9662
TFV1G(39)2 C1	0.9384	0.6522	0.9886	0.9948	0.9850	0.9862
TFV1G(39)2 C2	0.9270	0.6708	0.9854	0.9916	0.9796	0.9941
TFV1G(39)2 C3	0.9005	0.6437	0.9819	0.9879	0.9705	0.9983

As can be seen, all the formulations have a similar drug release profile and a good fit to the kinetics of Higuchi, Hixson-Crowell, Hopfenberg and Korsmeyer-Peppas (Table 4). However, there are slight differences between the batches that merit discussion. One observation is that all batches containing G43 tend to have a better fit to Higuchi kinetics. This model is often used to describe the behavior of matrix tablets, where the drug is released through a diffusion mechanism [38]. The most likely reason that these batches fit this kinetic more closely is that the G43 softens slowly, and the gelation of the outer layer of the hydrophilic polymer has already taken place when it occurs. This explains why the drug release is predominantly controlled by the diffusion mechanism (through both the softened Gelucire® and the gel formed by the HPMC/CH mixture).

The tablets containing the mixture of G43 and G39 have a better fit to the Hixson-Crowell kinetics, a mathematical model applied when the drug is released from the parallel planes of the dosage form, which maintains its shape while its size decreases [39]. In these cases the Gelucire® mixture softens a little faster than in the previous ones; one possible explanation for this drug-release mechanism is that the hydrophilic polymers gel from the outer to the inner layers, and the Gelucire® softens as the vaginal fluid reaches the granules through the gel. This explains why the drug is released in layers, first diffusing the TFV in the outer layers of the tablet, and then the drug in the innermost layers.

Finally, batches with G39 are better fitted to the Korsmeyer-Peppas kinetics (except batch TFV1G(39)2 C1, which has a better fit to the Hixson-Crowell kinetics, although it also fits the Korsmeyer-Peppas kinetics quite well) (Table 4). The Korsmeyer-Peppas model gives us the most information about the drug-release mechanism, since the drug is released in different ways depending on the value of n obtained in the adjustment [40]. Thus for cylindrical tablets, TFV release follows a pure diffusion process ($n \leq 0.45$) or an anomalous transport with simultaneous structural modification and diffusion ($0.45 < n < 0.89$). If there is a structural modification of the polymer matrix, the release is classified as transport case II ($n = 0.89$) or transport Supercase II ($n > 0.89$).

In our case, almost all the formulations can be seen to follow a simultaneous mechanism of structural modification and diffusion (Table 5), which is to be expected since it is first necessary for the hydrophilic polymer gel and the Gelucire® to soften (structural modification) and the TFV to subsequently diffuse before the drug can be released. However, there are three exceptions which present a Supercase II release and are precisely the three that include G39 in the granules (Table 5). In these cases, the melting temperature of the vehicle is so low that the G39 softens as soon as it is at body temperature, even before the vaginal fluid reaches the granules, implying that the structure of the tablet changes even before the HPMC/CH mixture begins to gel.

Table 5. TFV release kinetics from the different batches showing the kinetic constants for the models with the best fit.

Batch	Higuchi	Hixson-Crowell	Hopfenberg	Korsmeyer-Peppas	
	K_H	K_{HC}	K_{HF}	K_{KP}	n
TFV1G(43)2 C1	0.083	0.0040	0.0054	0.042	0.68
TFV1G(43)2 C2	0.080	0.0039	0.0052	0.036	0.71
TFV1G(43)2 C3	0.088	0.0045	0.0060	0.046	0.67
TFV1G(43)1 C1	0.102	0.0059	0.0075	0.045	0.82
TFV1G(43)1 C2	0.103	0.0060	0.0076	0.037	0.86
TFV1G(43)1 C3	0.106	0.0065	0.0081	0.048	0.81
TFV2G(43)1 C1	0.103	0.0061	0.0077	0.056	0.73
TFV2G(43)1 C2	0.108	0.0058	0.0074	0.047	0.78
TFV2G(43)1 C3	0.105	0.0065	0.0081	0.052	0.75
TFV1G(41)2 C1	0.107	0.0067	0.0083	0.038	0.85
TFV1G(41)2 C2	0.111	0.0076	0.0091	0.040	0.86
TFV1G(41)2 C3	0.110	0.0073	0.0089	0.043	0.88
TFV1G(39)2 C1	0.108	0.0066	0.0083	0.026	0.97
TFV1G(39)2 C2	0.109	0.0067	0.0084	0.029	0.93
TFV1G(39)2 C3	0.112	0.0075	0.0091	0.042	0.93

In addition to providing an insight into the mechanism of TFV release, this adjustment allows us to identify which formulation best controls the drug release based on the values of the release constants (K_H, K_{HC}, K_{HF} and K_{KP}). A comparison of these constants conclusively highlights the batches with the TFV1(G43)2 granules previously identified as the most suitable for controlled-release formulations. Taking the Higuchi kinetic as an example, which has the best fit for these formulations (Table 4), the batches that include these granules have a K_H value of 0.08, while the K_H value of the other batches is always greater than 0.1 (Table 5). This points to the greater control over the release of the TFV from these batches, but once again makes it impossible to select the ideal proportion of HPMC/CH, given the close similarity of the constants obtained with batches C1, C2 and C3.

3.2.3. Mucoadhesion Test

Finally, another crucial parameter to evaluate was the mucoadhesion residence time of the formulations. Mucoadhesion is important because the tablets must remain adhered to the vaginal mucosa during the time TFV is being released [41]. The detachment of the formulation would imply an incomplete prophylactic treatment and discomfort for women, leading to a decrease in the use of the formulations.

A comparison of formulations with different ratios of TFV/G43 reveals that the more G43 the formulation contains, the longer it remains adhered to the vaginal mucosa (Figure 11). As this excipient has no adhesive properties this bioadhesion cannot be directly attributed to G43, so the only possible explanation is that it acts as a structural agent that helps form a more robust gel—as observed in the swelling test (Figure 7)—and hence remains attached for longer. In terms of the influence of the type of Gelucire®, it can be seen that the lower the melting point of the Gelucire®, the shorter the mucoadhesion time. This is probably due to the easier and faster softening of G39, which again causes the tablets to erode faster. Finally, the C2 batches appear to be the most mucoadhesive, undoubtedly due to the higher mucoadhesion of HPMC than CH [24].

TFV1G(43)2 granules therefore confer the strongest adhesive properties on the tablets. Among these batches, it is worth highlighting TFV1G(43)2 C2, which remains adhered to the vaginal mucosa for almost 225 h.

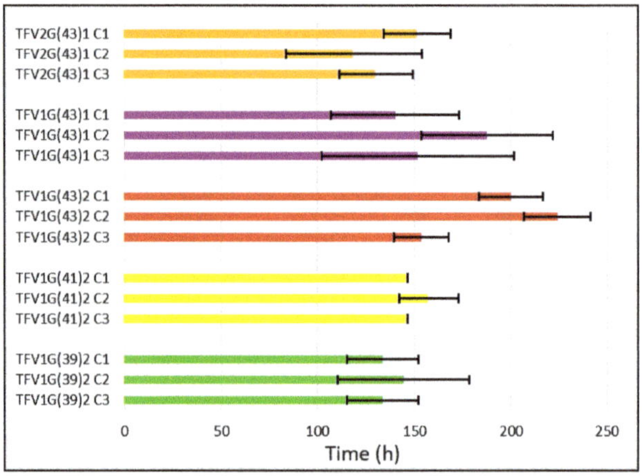

Figure 11. Mucoadhesion residence time in the simulated vaginal fluid in the batches.

The results of these studies, which are summarized in Figure 12, can be used to select the optimal formulation for preventing the sexual transmission of HIV.

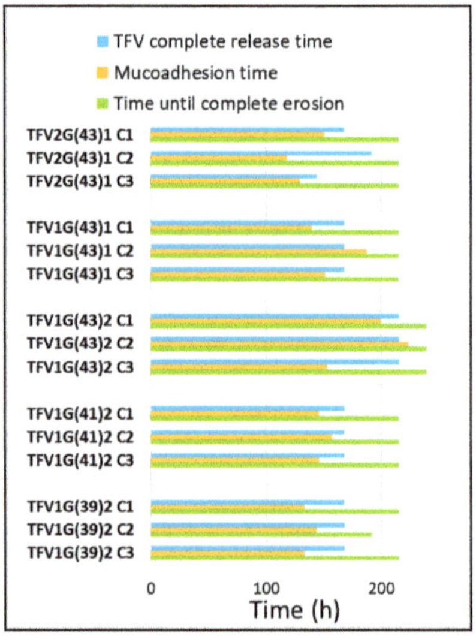

Figure 12. Summary of the data obtained from the drug release, mucoadhesion and swelling test. TFV complete release time (blue), mucoadhesion time (orange) and time until complete erosion (green) are shown for each batch.

The main parameter is undoubtedly the time taken to release the TFV from the formulation, since the aim of this work was to develop weekly sustained-release tablets whose administration could be markedly spaced in order to improve women's adherence to prophylactic treatment, which is currently the main problem with vaginal microbicides. Batches containing TFV1G(43)2 granules achieve the longest controlled release of TFV (216h), above the established target. However, the objective was not only to achieve a longer release time, but also to ensure the mucoadhesion of the formulation for as long as the drug is being released; there are only two formulations whose mucoadhesion time exceeds their complete TFV release time (batches TFV1G(43)1 C2 and TFV1G(43)2 C2). Finally, the formulation must not remain for long in the vaginal environment after the complete TFV release.

Batch TFV1G(43)2 C2 is conclusively selected as the optimal formulation for HIV prevention, since it is able to release TFV in a sustained manner for 216 h and remains attached to the vaginal mucosa throughout. Almost all the formulation has been eroded by the end of the process (with complete disintegration at 240 h), making it comfortable for women and enabling the next dose of the prophylactic treatment to be easily administered.

4. Conclusions

The combination of hydrophobic granules (prepared with Gelucire®) and hydrophilic matrices (HPMC and CH) in the development of vaginal mucoadhesive tablets allow the sustained-release of TFV.

The ratio TFV-Gelucire, as well as the proportion of HPMC and CH is critical for the development of the optimal formulation. Thus, the one with the best results contains TFV1G(43)2 granules in a mixture of HPMC and CH in a ratio of 1.9:1, which allows the sustained release of TFV for 216 h. The formulation remains adhered to the vaginal mucosa throughout this time, so a weekly administration of the tablets could protect women again the sexual transmission of HIV.

Author Contributions: Investigation, F.N.-P., R.C.-L., A.M.-I. and J.P.; Writing—original draft, F.N.-P.; Writing—review and editing, M.-D.V. and R.R.-C.

Funding: Funding: This work was supported by the Spanish Ministry of Economy, Industry and Competitiveness [grant number MAT2012-34552]; and by the Spanish Research Agency and the European Regional Development Fund (AEI/FEDER, UE) [grant number MAT2016-76416-R].

Acknowledgments: Fernando Notario-Pérez is beneficiary of a predoctoral researcher training fellowship from the Spanish Ministry of Economy and Business. Raúl Cazorla-Luna and Araceli Martín-Illana are beneficiaries of a university professor training fellowship from the Spanish Ministry of Education, Culture and Sport. We are grateful to the Carnes Barbero slaughterhouse (El Barraco, Avila, Spain) for supplying the biological samples. Acknowledgments to María Hernando, veterinarian of the Junta de Castilla y León, for verifying the suitability of the biological samples. TGA and DSC were characterized by the Escuela Politécnica de Cuenca (UCLM).

Conflicts of Interest: The authors declare no conflict of interest.

References

1. Chaudhary, R.S.; Amankwaa, E.; Kumar, S.; Hu, T.; Chan, M.; Sanghvi, P. Application of a hot-melt granulation process to enhance fenofibrate solid dose manufacturing. *Drug Develop. Ind. Pharm.* **2015**, *42*, 1137–1148. [CrossRef] [PubMed]
2. Jakson, K.A.; Hunt, J.D. Binary eutectic solidification. *Trans Met Soc AIME* **1966**, *236*, 843–852.
3. Patil, H.; Tiwari, R.V.; Repka, M.A. Hot-Melt Extrusion: From Theory to Application in Pharmaceutical Formulation. *AAPS PharmSciTech* **2015**, *17*, 20–42. [CrossRef] [PubMed]
4. Chen, Y.-C.; Ho, H.-O.; Chiou, J.-D.; Sheu, M.-T. Physical and dissolution characterization of cilostazol solid dispersions prepared by hot melt granulation (HMG) and thermal adhesion granulation (TAG) methods. *Int. J. Pharm.* **2014**, *473*, 458–468. [CrossRef] [PubMed]
5. Verhoeven, E.; De Beer, T.; Mooter, G.V.D.; Remon, J.; Vervaet, C. Influence of formulation and process parameters on the release characteristics of ethylcellulose sustained-release mini-matrices produced by hot-melt extrusion. *Eur. J. Pharm. Biopharm.* **2008**, *69*, 312–319. [CrossRef] [PubMed]
6. Özgüney, I.; Shuwisitkul, D.; Bodmeier, R. Development and characterization of extended release Kollidon® SR mini-matrices prepared by hot-melt extrusion. *Eur. J. Pharm. Biopharm.* **2009**, *73*, 140–145. [CrossRef]
7. Verstraete, G.; Mertens, P.; Grymonpré, W.; Van Bockstal, P.-J.; De Beer, T.; Boone, M.N.; Van Hoorebeke, L.; Remon, J.; Vervaet, C. A comparative study between melt granulation/compression and hot melt extrusion/injection molding for the manufacturing of oral sustained release thermoplastic polyurethane matrices. *Int. J. Pharm.* **2016**, *513*, 602–611. [CrossRef]

8. Almeida, A.; Saerens, L.; De Beer, T.; Remon, J.; Vervaet, C. Upscaling and in-line process monitoring via spectroscopic techniques of ethylene vinyl acetate hot-melt extruded formulations. *Int. J. Pharm.* **2012**, *439*, 223–229. [CrossRef]
9. El Hadri, M.; Achahbar, A.; El Khamkhami, J.; Khelifa, B.; Faivre, V.; Abbas, O.; Bresson, S. Lyotropic behavior of Gelucire 50/13 by XRD, Raman and IR spectroscopies according to hydration. *Chem. Phys. Lipids* **2016**, *200*, 11–23. [CrossRef]
10. Chambin, O.; Jannin, V. Interest of Multifunctional Lipid Excipients: Case of Gelucire® 44/14. *Drug Develop. Ind. Pharm.* **2005**, *31*, 527–534. [CrossRef]
11. Upadhyay, P.; Pandit, J.K.; Wahi, A.K. Gelucire: An alternative formulation technological tool for both sustained and fast release of drugs in treating diabetes mellitus type II disease. *J. Sci. Ind. Res.* **2013**, *72*, 776–780.
12. UNAIDS. When Women Lead Change Happens. Available online: http://www.unaids.org/sites/default/files/media_asset/when-women-lead-change-happens_en.pdf (accessed on 30 December 2017).
13. Aurpibul, L.; Puthanakit, T. Review of Tenofovir Use in HIV-infected Children. *Pediatric Infect. Dis. J.* **2015**, *34*, 383–391. [CrossRef]
14. Mertenskoetter, T.; Kaptur, P. Update on microbicide research and development-seeking new HIV prevention tools for women. *Eur. J. Med Res.* **2011**, *16*, 1–6. [CrossRef]
15. Hankins, C.A.; Dybul, M.R. The promise of pre-exposure prophylaxis with antiretroviral drugs to prevent HIV transmission: A review. *Curr. Opin. HIV AIDS* **2013**, *8*, 50–58. [CrossRef]
16. Notario-Pérez, F.; Cazorla-Luna, R.; Martín-Illana, A.; Ruiz-Caro, R.; Tamayo, A.; Rubio, J.; Veiga, M.-D. Optimization of tenofovir release from mucoadhesive vaginal tablets by polymer combination to prevent sexual transmission of HIV. *Carbohydr. Polym.* **2018**, *179*, 305–316. [CrossRef]
17. Cazorla-Luna, R.; Notario-Pérez, F.; Martín-Illana, A.; Ruiz-Caro, R.; Tamayo, A.; Rubio, J.; Veiga, M.D. Chitosan-Based Mucoadhesive Vaginal Tablets for Controlled Release of the Anti-HIV Drug Tenofovir. *Pharmaceutics* **2019**, *11*, 20. [CrossRef]
18. Ruiz-Caro, R.; Veiga-Ochoa, M.D. Characterization and Dissolution Study of Chitosan Freeze-Dried Systems for Drug Controlled Release. *Molecules* **2009**, *14*, 4370–4386. [CrossRef]
19. Owen, D.H.; Katz, D.F. A vaginal fluid simulant. *Contraception* **1999**, *59*, 91–95. [CrossRef]
20. Sánchez-Sánchez, M.-P.; Martín-Illana, A.; Ruiz-Caro, R.; Bermejo, P.; Abad, M.-J.; Carro, R.; Bedoya, L.-M.; Tamayo, A.; Rubio, J.; Fernández-Ferreiro, A.; et al. Chitosan and Kappa-Carrageenan Vaginal Acyclovir Formulations for Prevention of Genital Herpes. In Vitro and Ex Vivo Evaluation. *Mar. Drugs* **2015**, *13*, 5976–5992. [CrossRef]
21. Moore, J.W.; Flanner, H.H. Mathematical comparison of dissolution profiles. *Pharma Tec.* **1996**, *20*, 64–74.
22. Mamani, P.L.; Ruiz-Caro, R.; Veiga, M.D. Matrix Tablets: The Effect of Hydroxypropyl Methylcellulose/Anhydrous Dibasic Calcium Phosphate Ratio on the Release Rate of a Water-Soluble Drug Through the Gastrointestinal Tract I. In Vitro Tests. *AAPS PharmSciTech* **2012**, *13*, 1073–1083. [CrossRef]
23. Dash, S.; Murthy, P.N.; Nath, L.; Chowdhury, P. Kinetic modeling on drug release from controlled drug delivery systems. *Acta Pol. Pharm.* **2010**, *67*, 217–223.
24. Notario-Pérez, F.; Martín-Illana, A.; Cazorla-Luna, R.; Ruiz-Caro, R.; Bedoya, L.-M.; Tamayo, A.; Rubio, J.; Veiga, M.-D.; Harding, D.; Sashiwa, H. Influence of Chitosan Swelling Behaviour on Controlled Release of Tenofovir from Mucoadhesive Vaginal Systems for Prevention of Sexual Transmission of HIV. *Mar. Drugs* **2017**, *15*, 50. [CrossRef]
25. Ochiuz, L.; Grigoras, C.; Popa, M.; Stoleriu, I.; Munteanu, C.; Timofte, D.; Profire, L.; Grigoras, A.G. Alendronate-Loaded Modified Drug Delivery Lipid Particles Intended for Improved Oral and Topical Administration. *Molecules* **2016**, *21*, 858. [CrossRef]
26. Eloy, J.D.O.; Saraiva, J.; De Albuquerque, S.; Marchetti, J.M.; Albuquerque, S. Solid Dispersion of Ursolic Acid in Gelucire 50/13: A Strategy to Enhance Drug Release and Trypanocidal Activity. *AAPS PharmSciTech* **2012**, *13*, 1436–1445. [CrossRef]
27. Patil, S.; Kadam, C.; Pokharkar, V. QbD based approach for optimization of Tenofovir disoproxil fumarate loaded liquid crystal precursor with improved permeability. *J. Adv. Res.* **2017**, *8*, 607–616. [CrossRef]
28. Ramkumaar, G.R.; Srinivasan, S.; Bhoopathy, T.J.; Gunasekaran, S. Vibrational Spectroscopic Studies of Tenofovir Using Density Functional Theory Method. *J. Chem.* **2013**, *2013*, 1–12. [CrossRef]

29. Svensson, A.; Neves, C.; Cabane, B. Hydration of an amphiphilic excipient, Gelucire® 44/14. *Int. J. Pharm.* **2004**, *281*, 107–118. [CrossRef]
30. National Center for Biotechnology Information. PubChem Compound Database, CID=464205. Available online: https://pubchem.ncbi.nlm.nih.gov/compound/464205 (accessed on 1 July 2017).
31. Veiga, M.; Bernad, M.; Escobar, C. Thermal behaviour of drugs from binary and ternary systems. *Int. J. Pharm.* **1993**, *89*, 119–124. [CrossRef]
32. Agrahari, V.; Putty, S.; Mathes, C.; Murowchick, J.B.; Youan, B.-B.C. Evaluation of degradation kinetics and physicochemical stability of tenofovir. *Drug Test. Analysis* **2014**, *7*, 207–213. [CrossRef]
33. Albarahmieh, E.; Qi, S.; Craig, D.Q. Hot melt extruded transdermal films based on amorphous solid dispersions in Eudragit RS PO: The inclusion of hydrophilic additives to develop moisture-activated release systems. *Int. J. Pharm.* **2016**, *514*, 270–281. [CrossRef] [PubMed]
34. Zhang, T.; Zhang, C.; Agrahari, V.; Murowchick, J.B.; Oyler, N.A.; Youan, B.B.C. Spray drying tenofovir loaded mucoadhesive and pH-sensitive microspheres intended for HIV prevention. *Antivir. Res.* **2013**, *97*, 334. [CrossRef] [PubMed]
35. Yin, X.; Li, H.; Guo, Z.; Wu, L.; Chen, F.; De Matas, M.; Shao, Q.; Xiao, T.; York, P.; He, Y.; et al. Quantification of Swelling and Erosion in the Controlled Release of a Poorly Water-Soluble Drug Using Synchrotron X-ray Computed Microtomography. *AAPS J* **2013**, *15*, 1025–1034. [CrossRef] [PubMed]
36. Geeta, M.P.; Madhabhai, M.P. Design and in vitro evaluation of a novel vaginal drug delivery system based on gelucire. *Curr Drug Deliv* **2009**, *6*, 159–165. [PubMed]
37. Stevens, R.E.; Gray, V.; Dorantes, A.; Gold, L.; Pham, L. Scientific and Regulatory Standards for Assessing Product Performance Using the Similarity Factor, f2. *AAPS J* **2015**, *17*, 301–306. [CrossRef] [PubMed]
38. Paul, D. Elaborations on the Higuchi model for drug delivery. *Int. J. Pharm.* **2011**, *418*, 13–17. [CrossRef] [PubMed]
39. Hixson, A.W.; Crowell, J.H. Dependence of reaction velocity upon surface and agitation: Theoretical considerations. *Ind Eng Chem.* **1931**, *23*, 923–931. [CrossRef]
40. Korsmeyer, R.W.; Gurny, R.; Doelker, E.; Buri, P.; Peppas, N.A. Mechanisms of solute release from porous hydrophilic polymers. *Int. J. Pharm.* **1983**, *15*, 25–35. [CrossRef]
41. Caramella, C.M.; Rossi, S.; Ferrari, F.; Bonferoni, M.C.; Sandri, G. Mucoadhesive and thermogelling systems for vaginal drug delivery. *Adv. Drug Deliv. Rev.* **2015**, *92*, 39–52. [CrossRef]

pharmaceutics

MDPI

Review

Pharmaceutical Vehicles for Vaginal and Rectal Administration of Anti-HIV Microbicide Nanosystems

Letícia Mesquita [1,2,†], Joana Galante [1,2,3,†], Rute Nunes [1,2], Bruno Sarmento [1,2,4] and José das Neves [1,2,4,*]

1 i3S—Instituto de Investigação e Inovação em Saúde, Universidade do Porto, 4200-135 Porto, Portugal; leticia.s.mesquita@gmail.com (L.M.); joana.galante@i3s.up.pt (J.G.); rute.nunes@ineb.up.pt (R.N.); bruno.sarmento@ineb.up.pt (B.S.)
2 INEB—Instituto de Engenharia Biomédica, Universidade do Porto, 4200-135 Porto, Portugal
3 ICBAS—Instituto de Ciências Biomédicas Abel Salazar, Universidade do Porto, 4050-313 Porto, Portugal
4 CESPU, Instituto de Investigação e Formação Avançada em Ciências e Tecnologias da Saúde, 4585-116 Gandra, Portugal
* Correspondence: j.dasneves@ineb.up.pt; Tel.: +351-220408800
† These authors contributed equally to this work.

Received: 1 March 2019; Accepted: 22 March 2019; Published: 26 March 2019

Abstract: Prevention strategies play a key role in the fight against HIV/AIDS. Vaginal and rectal microbicides hold great promise in tackling sexual transmission of HIV-1, but effective and safe products are yet to be approved and made available to those in need. While most efforts have been placed in finding and testing suitable active drug candidates to be used in microbicide development, the last decade also saw considerable advances in the design of adequate carrier systems and formulations that could lead to products presenting enhanced performance in protecting from infection. One strategy demonstrating great potential encompasses the use of nanosystems, either with intrinsic antiviral activity or acting as carriers for promising microbicide drug candidates. Polymeric nanoparticles, in particular, have been shown to be able to enhance mucosal distribution and retention of promising antiretroviral compounds. One important aspect in the development of nanotechnology-based microbicides relates to the design of pharmaceutical vehicles that allow not only convenient vaginal and/or rectal administration, but also preserve or even enhance the performance of nanosystems. In this manuscript, we revise relevant work concerning the selection of vaginal/rectal dosage forms and vehicle formulation development for the administration of microbicide nanosystems. We also pinpoint major gaps in the field and provide pertinent hints for future work.

Keywords: antiretroviral drugs; dendrimers; dosage forms; HIV prevention; mucosal drug delivery; nanocarriers; nanomedicine; nanoparticles

---

1. Introduction

HIV/AIDS remains a huge healthcare burden. Despite continuous effort to circumvent the pandemic, the amount of new infections has been declining over recent years at a slower pace than required in order to meet the milestone set by the UNAIDS for 2020, i.e., less than 500,000 annual cases. During 2017 alone, 1.8 million people were estimated to be infected by HIV, mostly due to sexual exposure [1]. This alarming figure highlights that much more has to be done, particularly in the field of prevention. Alongside other approaches, topical pre-exposure prophylaxis (topical PrEP) holds considerable potential for tackling new HIV-1 infections [2]. This strategy comprises the use of microbicides, which can be defined as medical products intended to be administered in the vagina and/or rectum in order to avoid early steps of viral transmission upon sexual intercourse. The principle behind topical PrEP relies on the inhibition of HIV-1 at the mucosal level by one or several compounds, presenting more or less specific antiviral activity. Receptive partners are the ones potentially benefiting from microbicides, although reciprocal protection would also be desirable [3]. The field has come a long way but only mild success has been achieved. The most advanced microbicide in the development pipeline is the dapivirine (DPV) vaginal ring, which was shown to reduce HIV-1 acquisition in women by roughly 30% under placebo-controlled phase 3 clinical trials [4,5]. The product is now under regulatory evaluation and a final decision from the European Medicines Agency is expected soon [6]. Still, partial protection observed for more "conventional" products leaves a wide margin for advances in the development of microbicides.

Trailing on the exciting findings and achievements of nanomedicine, different nanotechnology-based microbicide systems were shown to present potentially advantageous features that could contribute to improved protection from HIV transmission. Several microbicide nanosystems, either with inherent antiviral activity or used as drug carriers, have been proposed over the years, often with promising pre-clinical results [7]. Still, one important feature has frequently been overlooked, namely the selection of dosage forms and the development of pharmaceutical vehicles (or platforms) that can incorporate microbicide nanosystems as active ingredients. The main objective is the development of final products that allow suitable vaginal and/or rectal self-administration by users. However, other questions can be further raised regarding functionality. For example, what are the interactions occurring upon incorporation (e.g., chemical or physical bonding between nanosystems and excipients from vehicles, structural changes to the matrix of polymeric vehicles, coating of nanosystems with components of vehicles) and how will these affect the behavior of nanosystems and vehicles, both during storage and in vivo? Do we want/need platforms that have additional roles beyond simple vaginal or rectal administration? Indeed, to which extent can a vehicle influence the original performance of nanosystems, especially in vivo? The intention of the present manuscript is to provide a brief and critical appraisal of the different vehicles that have been considered for vaginal and/or rectal administration of microbicide nanosystems, as well as hints for future work.

2. Microbicides for Preventing Sexual HIV Transmission

The idea behind the onset of modern microbicides is usually traced back to the early 1990s and, in particular, to the seminal commentary paper by Zena A. Stein [8]. In this, she called out to the need for the development of new methods that women could use in order to protect themselves against viral transmission. The following years saw the emergence of various drug candidates and vaginal microbicide products. This first generation of microbicides was characterized by the use of compounds featuring non-specific antiviral activity and simple formulations, mostly based on gels, although many other dosage forms have also been considered (e.g., creams, tablets, capsules, sponges, or films) [9]. The concept of using products already available in the market that could also present anti-HIV activity was particularly attractive. Nonoxynol-9, a non-ionic surfactant used in the composition of different vaginal spermicides and shown to present anti-HIV activity, was earlier suggested as a preferential target for development [10]. Results from clinical trials conducted during the late 1990s/early 2000s were, however, disappointing. Products based on surfactants and anionic/acidic polymers featuring non-specific antiviral activity not only lacked efficacy in protecting women against HIV acquisition but also, in some cases and paradoxically, increased transmission [11–16]. Additional studies confirmed that candidate microbicide compounds, such as nonoxynol-9 [17] and cellulose sulfate [18], can actually induce changes or even damage to the vaginal mucosa, thus facilitating HIV-1 transmission. Since then, a dramatic shift towards products based on potent antiretroviral drugs occurred in the field and particular emphasis was placed on safety evaluation. Additionally, formulation development of microbicide products benefited from increasing attention and was established as having a fundamental role in the performance of antiretroviral drug candidates [19].

Proof of a first partially successful microbicide was announced in 2010. Results from the CAPRISA 004 phase 2b clinical trial showed a 39% reduction in HIV-1 transmission in women using a 1% tenofovir (TFV) gel, as compared to placebo [20]. Partial protection was mainly attributed to poor adherence, which could have accounted for low or fluctuating levels of TFV throughout the timeframe of potential viral exposure and, thus, insufficient protection [21]. Recent evidence of differential drug metabolism by vaginal microbiota has been reported as another possibility [22]. Despite all the excitement around this first success, further studies testing similar TFV gels administered in alternative regimens failed to confirm the potential for protection [23,24]. Adherence issues were again held accountable [24,25].

Vaginal rings have the potential to abbreviate adherence issues observed with coitally-dependent products based on dosage forms, such as gels. This possibility motivated the development of the DPV ring, which allows maintaining sustained drug levels in the vagina for several weeks [26]. Somewhat surprisingly, data from two phase 3 clinical trials demonstrated only mild reduction (27–31%) in HIV acquisition by women [4,5]. Again, poor adherence was claimed as one of the main contributor to these results [27]. Intermittent use of the ring leads to a sharp decline in vaginal levels of DPV, which creates opportunity windows for viral transmission. Notably, adherence to product use is a key issue in microbicide protection. Post-hoc analysis of data from phase 3 clinical trials of the DPV ring indicates that higher protection rates were observed for women that used the product consistently [28]. Despite multiple factors being involved, acceptability and preferences regarding dosage forms and usage schemes are recognized as important in determining adherence [29]. Current thought in the field supports that a single product will not be enough to fulfil the requirements of a majority of women. Instead, a relatively vast panel of products from which to choose from, including multipurpose ones, will likely be required [30].

Microbicides for rectal use have followed on the footpath of vaginal products. Several antiretroviral drugs and dosage forms (e.g., suppositories or enemas) have been considered [31], although products based on TFV and presented as gels have been frequently selected as preferential. Some of the latter have reached as far as phase 2 clinical trials [32,33]. Product design has been regarded with equal or even greater importance than vaginal microbicides. The particularities of the colorectal compartment make it more susceptible to HIV transmission than the cervix/vagina [34] and a challenge to formulation. For example, the large surface area available for viral entrance that requires to be covered by a single product, the high susceptibility of the thin single columnar epithelium to toxic insult, or the possibility of extensive systemic absorption of the administered drug(s) are just some of the possible problems that need to be considered when developing a rectal microbicide [35].

3. Potential of Nanotechnology in Developing Microbicides

More than a trendy gimmick, the application of nanotechnology to the development of anti-HIV microbicides relies on solid research work, as previously revised by our group [7,36] and others [37,38]. General advantageous features are schematically presented in Figure 1. Although potentially applicable to both mucosal sites, work has been mostly conducted with the intention of vaginal use. Nanosystems intended for microbicide development may be classified in two major classes, as follows: (i) Nanosystems presenting intrinsic antiviral activity and/or competing with HIV for host targets and (ii) nanosystems acting as carriers for microbicide drugs [39]. In the first case, surface chemical functionalization of nanosystems is directly responsible for the inactivation of HIV or blockage of cell membrane receptors involved in viral infection. Dendrimers (often considered as polymers, not actual nanosystems) are the best representatives of this class, particularly SPL7013. This fourth-generation dendrimer is the active ingredient of VivaGel® (Starpharma, Australia), mainly owing its anti-HIV activity to the direct interaction of terminal naphthalene disulfonate groups with gp120, a viral surface glycoprotein essential for cell infection [40]. Despite promising early results, the development of VivaGel® as a microbicide has been discontinued following safety issues in clinical trials [41,42]. Meanwhile, other candidates have been under active development [43]. In particular, the G2-S16 dendrimer resulting from the systematic screening of a series of carbosilane dendrimers, proposed by Muñoz-Fernández and collaborators, was shown to be promising for further clinical testing [44,45]. Apart from dendrimers, other types of nanosystems with intrinsic activity have also been studied, namely oligomannose-coated gold nanoparticles (NPs) [46]. This type of nanosystems can block the C-type lectin receptor DC-SIGN present at macrophages and dendritic cells, which is involved in mediating HIV infection and translocation.

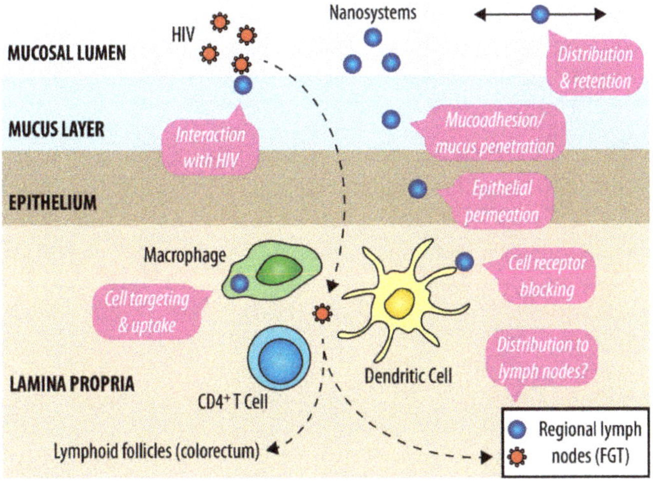

Figure 1. The potential of nanotechnology-based systems for microbicide development. Features of anti-HIV microbicide nanosystems at cervicovaginal or colorectal mucosal sites in the context of sexual HIV transmission are indicated in pink call-outs (see text for more details). FGT—female genital tract.

Interest in nanocarriers for the vaginal and/or rectal delivery of potent microbicide drugs has been growing over recent years. Fine tuning of the physicochemical properties of nanosystems, namely regarding size and surface properties, can provide active moieties with considerable advantage over "naked" counterparts (Figure 1). Apart from more general features, such as controlled drug release, protection of labile active molecules, or promotion of the solubility of hydrophobic compounds, nanocarriers may further (i) present enhanced distribution and retention along the mucosal site [47]; (ii) feature a variable ability to interact with mucus [48]; (iii) promote epithelial penetration and accumulation at mucosal tissues [49,50]; and/or (iv) target immune cells involved in viral transmission and increase intracellular drug accumulation [51]. One or more of these features should contribute to the improvement of local pharmacokinetics (PK) [52,53], i.e., prolong and/or increase drug levels at the organ/tissue/cell level and ultimately enhance protection from HIV transmission [54]. Antiretroviral drugs previously used in microbicide development, namely the viral transcriptase inhibitors TFV and DPV, have been selected as preferential compounds for assessing the potential of using nanocarriers [55,56]. Other more complex molecules have also been considered (e.g., fusion inhibitor peptides [57], RANTES analogues [58,59], or siRNA [60]). Polymeric NPs have been amongst the most popular nanocarriers used [61], although solid lipid NPs [62], liposomes [63], or nanolipogels [64] have been considered as alternatives. Furthermore, nanofibers attracted great interest as vaginal microbicides due to their ability to modulate drug release, provide fast distribution of incorporated compounds, and allow formulation of multiple active molecules in the same system [65].

Discussion on whether nanosystems should interact with mucus or not has generated substantial interest over recent years [66]. Contrary to the common idea that mucoadhesion is advantageous to improve mucosal residence, nanosystems presenting reduced size (typically under 500 nm or less) and an inert surface (usually conferred by dense coverage with 5–10 kDa polyethylene glycol (PEG)) can better traverse mucus and distribute throughout the vaginal and colorectal epithelial surface [67]. Furthermore, the ability of mucus-penetrating nanosystems to permeate mucus and reach areas closer to the epithelia, where mucus turnover is slower, can actually promote retention as compared to mucoadhesive counterparts [68]. Still, a case-by-case analysis as to which behavior may better sustain the residence and distribution of NPs following administration seems advisable, as shown by a recent report on the comparison between mucoadhesive and mucus-penetrating poly(lactic acid)-hyperbranched polyglycerol-based NPs [69]. Another issue needing further investigation is the ability of certain nanosystems to reach sites of interest beyond mucosae. For example, a few studies suggest that NPs can partially undergo cell-mediated or even free transport from the vagina to regional lymph nodes associated with the female mouse genital tract [70,71]. This may be relevant since HIV also endures a similar path following initial proliferation at the cervicovaginal mucosa [72]. Scanty evidence further indicates that the transport of vaginally administered NPs to the upper parts of the mouse genital tract may occur, albeit to a limited extent [73].

4. Vehicles for Microbicide Nanosystems

Pharmaceutical vehicles play an essential role in the delivery of active ingredients, encompassing both the general features of a defined dosage form and the individual properties of the specific formulation and manufacturing process. Broad definitions of "dosage form" adopted by regulatory bodies typically embrace the physical form of a product that contains one or more active ingredients and is intended to be administered to individuals in need [74,75]. For the purpose of this manuscript and from a technological perspective, microbicide nanosystems as a whole are considered as active ingredients. Various dosage forms have been traditionally considered for developing vehicles for vaginal and rectal drug administration and these have served as the usual starting point for the incorporation of nanosystems. Interested readers are further referred to compendia detailing the general principles of vaginal and rectal drug delivery and dosage forms [76,77]. Importantly, the design and production of vaginal or rectal vehicles should envision complying with the particularities of the mucosal site in which administration is intended. Many differences exist between both environments and some of the most relevant are summarized in Table 1. Safety is particularly relevant in the case of microbicides. Lessons learned the hard way in the early days of the field are now being translated into valuable guidance that aids in product development, particularly at preclinical stages [78]. Other features, such as manufacturing feasibility, cost, and acceptability by end-users must also be taken into consideration. Here we discuss the most relevant aspects of different vaginal and/or rectal vehicles, organized by dosage form, that were used for incorporation and delivery of microbicide nanosystems and detail on their relevance to the performance of such active ingredients.

Table 1. Typical characteristics of vaginal and rectal mucosae (revised in [35,76]).

Characteristics	Vaginal Mucosa [a]	Rectal Mucosa
Extension [b]	9–12 cm	15–20 cm [c]
Surface area	65–165 cm^2	200–400 cm^2
Epithelium	Stratified squamous	Simple columnar
pH of mucus	4–5	7–8
pH buffering capacity of mucus	Low	Low
Typical volume of mucus	0.5–1 mL [d]	1–3 mL
Mucin concentration in mucus	1–2%	<5%
Osmolality of mucus	Nearly isoosmolal [e]	Nearly isoosmolal [e]
Enzymatic activity	Low	Medium
Microbiota composition	Lactobacilli dominant	Variable
Involuntary motility	Low	Medium to high

[a] Considering healthy women of reproductive age; [b] at the longest axis; [c] total extension of the colorectum is around 150 cm; [d] largely increased upon sexual stimulation; [e] as compared to the osmolality of blood plasma (≈290 mOsm/Kg).

4.1. Suspensions

Aqueous liquids are, by far, the most common vehicles in which microbicide nanosystems have been suspended and tested, particularly for in vivo studies. Fluidity of these systems can be seen as an advantage since it promotes wide and fast distribution along the mucosal site upon administration and is particularly important in the case of rectal microbicides [79]. However, the chief reason for their use probably relates to the fact that nanosystems are usually produced/synthetized in aqueous media and, indeed, originate nanosuspensions. Formal formulation efforts are usually minimal since convenience, rather than the achievement of a putative pharmaceutical product, usually drives selection of this type of vehicles. Thus, for the purpose of this work, all nanosuspensions are considered as de facto dosage forms. Saline-based buffers (approximately isoosmolal and with the pH around neutral) have been the most commonly used vehicles [52,53,80,81]. Indeed, particular attention should be given to pH and osmolality, as these factors are recognized to potentially affect safety [82]. While vehicles featuring pH around neutrality appear to be suitable for rectal vehicles, the choice of such low hydrogen-ion concentrations is related to the physiology of animal models typically used to test microbicides [83]. Deviations from the normal vaginal pH of women of reproductive age (4–5) could, however, potentially affect performance.

Vehicles matching physiological osmolality are usually regarded as suitable. Still, Ensign et al. [84] showed that hypoosmolal vehicles can actually enhance the distribution and retention of 100 nm mucus-penetrating NPs in the vagina, without apparent mucosal damage. These findings were explained by the promotion of faster transport of NPs across mucus and towards the epithelial surface, as driven by osmotic convection. The effect was pronounced even when a slight variation from isoosmolality was observed (Figure 2) [84]. The same group also studied the effect of rectally administered 60 nm mucus-penetrating NPs and found that osmolality affected distribution and retention, although the effect was more complex and dependent on ion composition [85]. For instance, potassium-based, but not sodium-based, had significant impact on osmotic convection when testing for osmolality at or below normal levels (20–300 mOsm/kg). This effect was attributed to the role of ion transport on water absorption at the colon. While sodium promotes water uptake by tissue, the opposite is true for potassium [85]. Overall, these studies suggest that the performance of microbicide nanosystems can be enhanced or impaired by proper selection of liquid vehicles and at least their qualitative and quantitative composition should be reported.

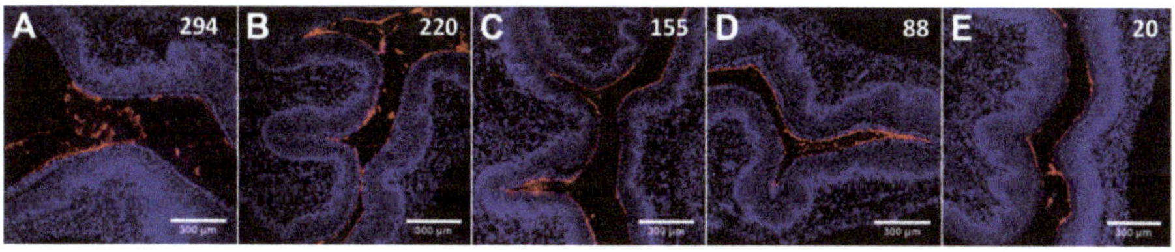

Figure 2. Vaginal distribution of fluorescent 100 nm mucus-penetrating particles administered in suspensions of varying osmolality. (**A–E**) Distribution of mucus-penetrating particles in transverse vaginal cryosections. Values of osmolality are presented in the individual images and have mOsm/kg as units. Images are representative of $n \geq 5$ mice. Adapted from [84], Copyright (2013), with permission from Elsevier.

A final aspect of liquid vehicles that should not be underestimated concerns their general low vaginal and rectal retention following application. Leakage is typical and, in the case of rectal administration, extensive if voluntary anal sphincter closure by users is not observed. The presence of material in the rectal ampulla induces an urge to defecate that needs to be firmly opposed upon administration. This problem is particularly noteworthy with increasing volumes being administered. Fractionated administration in a supine position may help mitigate leakage.

4.2. Gels

Gels are among the most popular dosage forms for vaginal and rectal drug delivery, largely due to their technological versatility, acceptability by users, and low production cost. These have been widely explored in microbicide formulation [86]. The semi-solid nature of gels as well as the mucoadhesive properties of various gelling agents may help with improving the vaginal and colorectal residence of nanosystems. Not surprisingly, this was the chosen dosage form selected for the SPL7013 dendrimer in order to obtain VivaGel®. The gel is based on carbomer 941 and is reported to contain 1% (*w*/*w*) of glycerin and propylene glycol, although fully detailed composition and basic properties (pH, osmolality, viscosity, dendrimer release, etc.) are not available [87]. A role for the gel on safety issues detected during clinical testing was not evident [41,42].

Selection of gel bases for the incorporation of microbicide nanosystems has been typically made in an empirical fashion. Most commonly used gelling agents include different carbomers [88] and hydroxyethylcellulose [44]. The common use of this last cellulose derivative in microbicide development has its origins in the established safety of the hydroxyethylcellulose-based formulation "Universal Placebo" [89]. However, a more systematic approach to the development of gels as platforms for microbicide nanosystems is recommended. Interactions between both are particularly relevant and may potentially impact the final protection outcomes. For example, Wang et al. [90] developed efforts in order to formulate different gels containing liposomes loaded with the microbicide candidate octylglycerol. Among other effects, the addition of gelling agents, such as carbomers, led to a mild reduction of the in vitro efficacy of octylglycerol-loaded liposomes against HIV-1. This effect was attributed to a reduction in the amount of octylglycerol that could be released, due to the increased viscosity. Still, the setup used for testing efficacy may have overlooked another relevant aspect of gels, namely the ability to provide a barrier to effective HIV diffusion. In a study by Lara et al. [91], the incorporation of microbicide polyvinylpyrrolidone (PVP)-coated silver nanoparticles (30–50 nm) into a carbomer-based commercial gel (Replens™, Lil' Drug Store Products, USA) slightly improved the inhibition of HIV-1 infection when the product provided a layer on top of cervical explants. The viscous layer presumably provided an additional barrier for virions to cross until reaching target cells present in tissue. Although limited animal data reported previously by other researchers appear to support this possibility [92], it is debatable for how long could such a barrier actually impact viral transmission. Additional research on this topic is required.

4.3. Thermosensitive Systems

Thermosensitive systems envisioned for vaginal or rectal drug administration are typically presented as liquid dosage forms that present the ability to undergo sol-gel transition at just below body temperature, originating semi-solid gels [93,94]. Rationale for use relies on their ease of administration and a wide initial distribution of a fluid along the mucosal cavity, while a semi-solid (often incorporating mucoadhesive polymers) provides improved local retention. Thermosensitive systems considered for vaginal and rectal drug delivery have been almost exclusively based on poloxamers due to their well-known temperature-dependent sol-gel transition, regulatory status (currently used in different pharmaceutical products), and excellent safety profile [95]. The utility of thermosensitive dosage forms is established for mucosal drug delivery, but their ability to influence the performance of nanosystems is not well understood.

Thermosensitive systems have been preferred for the incorporation and administration of microbicide drug-loaded NPs [96]. From a simple technological perspective, incorporation of nanosystems into a fluid system at room temperature is easier than in an already pre-formed gel. However, the possibilities and pitfalls for microbicide NPs-in-thermosensitive systems may extend beyond that. For example, drug release and gelation temperature may be affected. Timur et al. [97] found that the incorporation of TFV-loaded chitosan NPs ($\approx$545 nm) into a poloxamer-based thermosensitive vehicle allowed a mild delay in the release of the drug, as compared to NPs in suspension or the same thermosensitive platform containing free TFV. The gelation temperature was also affected by the incorporation of NPs [97].

One concern of using thermosensitive systems relates to the possibility that NPs are not able to escape the structured matrix formed upon sol-gel transition, thus affecting performance. Although impossible to generalize, Date et al. [98,99] showed that both poly(lactic-*co*-glycolic acid) (PLGA)- and cellulose acetate phthalate (CAP)-based NPs (around 80–100 nm) were able to be released from a poloxamer-based thermosensitive vaginal gel at 37 °C and be taken up by epithelial cells, as shown using an in vitro cell-based model. In the particular case of efavirenz (EFV)-loaded CAP NPs, the antiretroviral activity was demonstrated to be maintained following incorporation into the thermosensitive vehicle [99]. Importantly, the same thermosensitive gel, containing rilpivirine-loaded PLGA NPs (66 nm), was fully or partially (50%) effective in preventing HIV-1 transmission in a humanized mouse model when administered at 90 min or 24 h, respectively, before intravaginal viral challenge [54]. Although drug-loaded NPs in suspension were not actually tested in vivo, authors claimed that the protection observed even at 24 h post-administration was due to the extended residence of nanosystems provided by the thermosensitive vehicle. Microscopy analysis of vaginal sections, collected from animals treated intravaginally with fluorescent NPs-in-thermosensitive systems, indeed provided evidence that, although residual, NPs could still be found at 24 h post-administration. A subsequent study using the same thermosensitive vehicle to deliver tenofovir disoproxil fumarate (TDF)-loaded PLGA NPs (≈150 nm) to humanized mice also provided evidence of complete protection up to at least 24 h post-administration [100]. Again, the lack of a control group using drug-loaded NPs in a liquid vehicle limited the assessment of the true role of the thermosensitive system in the protection outcomes.

The use of NPs-in-thermosensitive vehicles seems to be particularly interesting for rectal use. Still, no specific study is available for the purpose of microbicide development. Apart from increased residence, wide distribution of microbicide nanosystems in the colorectum may be challenging. As stated previously, the liquid state preceding gel formation may allow enhanced distribution immediately upon intrarectal administration. Preliminary data from our group support that PLGA NPs (≈180 nm) administered to mice were able to provide roughly the same extent coverage of the colon when incorporated into phosphate buffered saline or a poloxamer-based enema [101]. Residence was enhanced in this last case, presumably due to the formation of a gel at body temperature.

4.4. Vaginal Films

Pharmaceutical films comprise thin, soft, flexible, and often translucent solid strips of polymeric nature, in which active ingredients are dissolved or dispersed [102]. Solvent-casting is the preferred technique for manufacturing films, although hot-melt extrusion can also be used. Upon contact with vaginal fluids, films typically dissolve rapidly and originate a low volume and gel-like fluid. However, films can also be designed to disintegrate slowly. Their use for vaginal drug administration, particularly in microbicide development, is well documented and supported by features such as the ability to (co-)formulate multiple drugs, the established technology, and good stability upon storage, among others [103,104]. Acceptability by women is usually high since the films are portable, easy to administer without the need of an applicator (contrary to liquids and semi-solids), and known to originate little vaginal leakage [105]. Although not so widely used as other vaginal dosage forms, a few contraceptive vaginal films are currently commercially available in selected countries.

The concept of using films as platforms for buccal/oral administration of nanosystems, namely drug-loaded polymeric NPs, has been around for over a decade [106], but its application to the development of vaginal microbicides is recent. General advantages for using films, namely the ones indicated above, continue to be valid, while more specific others may further apply [107]. In particular, the solid nature of films largely restrains drug leakage from nanosystems throughout storage, as well as other potential changes and microbial contamination due to the low amounts of water present (usually less than 10%). The tight polymeric matrix surrounding nanosystems also limits phenomena such as particle aggregation and segregation that can originate colloidal instability. Such possibilities cannot be ruled out for liquid/semi-solid systems and, indeed, often lead to stability issues [108]. The incorporation of NPs is usually conducted by mixing with an aqueous dispersion of film ingredients before casting and drying of films, but other strategies may also be adopted. For example, our group recently proposed a "sandwich-like" configuration film comprising TDF/emtricitabine-loaded Eudragit® L 100 NPs (≈400 nm) entrapped between two individual sheets of poly(vinyl alcohol)/hydroxypropyl methylcellulose (PVA/HPMC) film, previously prepared by solvent-casting [109]. Assembly of the final film was achieved by compression and the obtained system allowed for decreasing in vitro burst-release of both drugs.

Gu et al. [110] were the first to describe a NPs-in-film system for the development of a microbicide product. They produced siRNA-loaded, anti-human leukocyte antigen-DR (HLA-DR) antibody-functionalized PEG-PLGA-based NPs, with around a 230 nm mean diameter, and associated it to PVA/λ-carrageenan-based films using solvent-casting. The siRNA used was intended to silence the translation of SNAP-23, a host protein associated with HIV exocytosis, while the incorporation of the antibody envisioned targeting HLA-DR+ dendritic cells. Although successful, silencing was reduced when NPs-in-films were compared to NPs in aqueous suspension, thus suggesting that the film matrix could reduce the transport of NPs across the epithelial cell monolayer and towards target dendritic cells in an in vitro co-culture cell-based model (Figure 3). Even if these data seem little encouraging, the actual value of microbicide NPs-in-films is more likely to be observed in vivo.

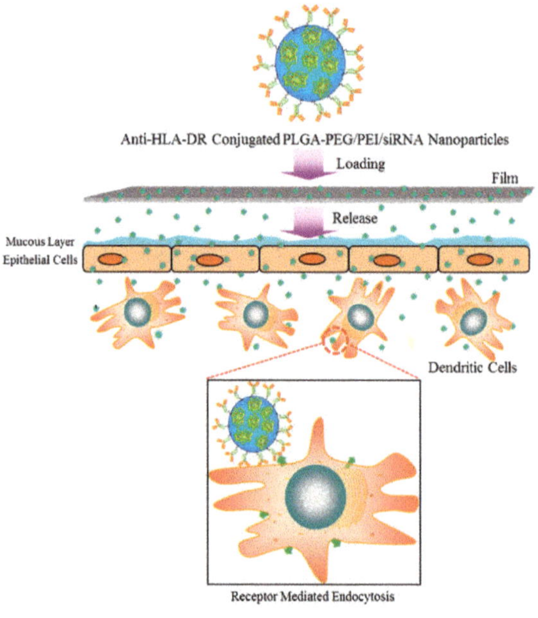

Figure 3. Schematic representation of targeted siRNA-loaded NPs formulated into a biodegradable film for targeting siRNA delivery into HLA-DR+ dendritic cells, using a co-culture cell model. Targeted siRNA-loaded NPs are homogeneously dispersed in a biodegradable film and, upon administration, the film is expected to disintegrate within the vaginal lumen, allowing siRNA-loaded NPs to penetrate across the vaginal mucosa and deliver siRNA in a targeted manner to HLA-DR+ mKG-1 dendritic cells. Reprinted with permission from [110]. Copyright (2015) American Chemical Society.

Our group has been particularly engaged in testing microbicide NPs-in-films in animal models. For example, we developed a hybrid film, based on PVA/HPMC, containing TFV (unassociated to NPs) and EFV-loaded PLGA NPs (≈275 nm) dissolved/dispersed in the film matrix [111]. Biodistribution experiments showed that, although substantial initial leakage occurred in all cases, the overall vaginal residence of NPs was significantly higher upon incorporation into films (Figure 4). The system containing TFV and EFV-loaded PLGA NPs allowed higher reaching and prolonged levels for both transcriptase inhibitor drugs in vaginal tissues and fluids following administration to mice, as compared to their administration in a liquid vehicle. Additionally, the tandem association of EFV to NPs and film allowed an additional improvement of local PK, as compared to the drug incorporated directly into the film matrix. These data appear to support that, indeed, the combination of NPs and films may provide an advantageous system for sustaining local levels of microbicide drugs. Furthermore, NPs-in-film tested were shown safe, following daily intravaginal administration to mice for 14 days [111,112]. Contrasting with the previous data, Srinivasan et al. [113] did not find the use of IQP-0528-loaded NPs-in-film to be advantageous, as compared to the free compound directly incorporated into the film. In fact, vaginal drug levels were shown to be slightly higher for this last formulation when administered intravaginally to pigtailed macaques. One possible reason may have been related with the intentionally slow and only partial release of IQP-0528 from NPs-in-film, as shown in vitro, which could have limited the ability of the drug to reach vaginal fluid and tissues before bulk vaginal leakage of the films [113]. Overall, these results seem to reinforce the need for fine tuning the technological properties of microbicide NPs-in-film systems.

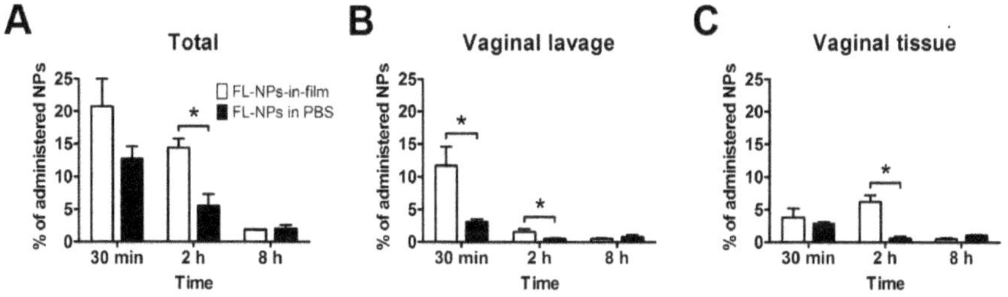

Figure 4. Distribution of fluorescent NPs (FL-NPs) at different times after intravaginal delivery in phosphate buffered saline (FL-NPs in PBS) or film (FL-NPs-in-film). Results are expressed as the (**A**) total amount of recovered particles or fractions retrieved from (**B**) vaginal lavage and (**C**) vaginal tissues. Columns and bars stand for mean and standard error of the mean values, respectively, and (*) indicates $p < 0.05$ (Student's t-test; $n = 3$). Adapted from [111], Copyright (2016), with permission from Elsevier.

4.5. Fiber Mats

Drug-loaded fiber mats or membranes captured substantial interest over recent years for developing vaginal microbicides [114]. Their rectal use has not yet been considered. Depending on the cross-section dimensions, fibers can be classified as nanofibers and constitute actual microbicide nanosystems. One of the advantages of fibers is the possibility of being used as such, namely in the form of differently shaped and sized mats (Figure 5). Similar to vaginal films in general appearance, but typically opaque, fiber mats are produced by electrospinning and present sufficient versatility to incorporate multiple drugs (namely for other applications, such as contraception) in various configurations. The fine-tuning of antiretroviral drug release from fibers and/or composite mats is also relatively simple to achieve [115]. Drug release is associated with fiber composition and structure and can be triggered by environmental changes pertinent to sexual intercourse, namely happening upon ejaculation (e.g., rising pH or the presence of components from semen) [116,117]. Depending on specific characteristics, namely concerning the ability to dissolve and the perception of which shape is easier to insert in the vagina, fiber mats appear to be well accepted by potential female users of microbicides [118].

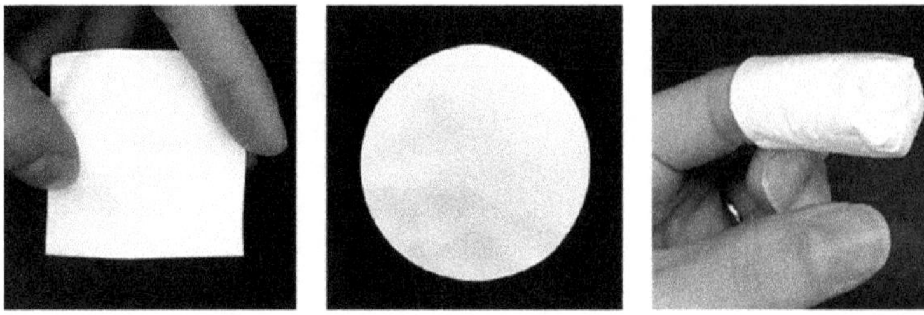

Figure 5. Examples of differently shaped, human sized PVA nanofiber mats (fiber cross-section diameter of ≈200–300 nm and mat thickness of 50–220 μm according to Krogstad et al. [119]). *Left*—square (25 cm^2); *center*—circle (diameter = 5 cm); *right*—capped tube (length = 4 cm, diameter = 2 cm). Adapted from [118] under the terms of the Creative Commons Attribution License 4.0 (Copyright 2018 Laborde et al.; doi:10.1371/journal.pone.0204821).

Fibers themselves can be used as vehicles for the administration of microbicide nanosystems. The concept is quite recent but available studies suggest great potential. Krogstad et al. [120] first proposed the use of two types of nanofibers, based either on PVA (≈250 nm cross-section diameter) or PVP (≈300 nm cross-section diameter), two mucoadhesive polymers, for the incorporation of PLGA-based NPs (≈170–180 nm). Composite nanofiber mats were rapidly wetted and disintegrated within a few seconds when exposed to an aqueous environment. Complete dissolution and release of over 85% of the content in NPs was observed to occur within 30 min under in vitro conditions. However, significant changes in size, polydispersity and zeta potential of the NPs were observed, particularly when PVA was considered, which suggests that coating with material from nanofibers occurs upon particle release. In an expanded series of distribution and retention studies using mice, composite nanofibers were not only able to provide extensive coverage of the vaginal mucosa with NPs, but also to significantly reduce leakage [120]. Mean recovery of NPs was around 45–50% from the total administered dose at 24 h, in the case of composite nanofibers, contrasting with under 5% for NPs administered in suspension. Strikingly, 40% of NPs were still recovered after three days in the case of PVA composite nanofibers. These results translated into enhanced PK when etravirine-loaded PLGA NPs incorporated into PVA nanofiber mats were tested. For example, fibers provided around 13-times higher drug levels in lavage up to three days post-administration, but roughly the same concentrations in vaginal tissue in the same time-frame, as compared to drug-loaded NPs in suspension and estimated by the area-under-the-curve values [120]. Relatively poor accumulation at tissues suggests that relatively rapid release of etravirine from NPs, low mobility of NPs towards the mucosa, or both, occurred. Still, overall data seems to support the utility of PVA nanofibers in enhancing the residence and PK of microbicide NPs.

An interesting approach to the use of fibers for the administration of microbicide NPs has been recently proposed by Kim et al. [121]. The researchers proposed electrospun fibers composed of a non-dissolving, pH-responsive polyurethane co-polymer, namely PEG-1,4-bis(2-hydroxyethyl)piperazine-4,4′-methylenebis(phenyl isocyanate)-propylene glycol, to be used for the controlled release of NPs. Fibers were not produced with the intention of constituting the actual dosage form, but rather to serve as a release limiting membrane that could be useful, as suggested by the authors, in the design of vaginal rings. The concept behind the responsive membrane was again based on the shift from acidic (4–5) to slightly alkaline (>7) values of pH observed in the vagina upon ejaculation. Membranes presenting in the dry state with a mean pore size and fiber diameter of 2.3 μm and 0.9 μm, respectively, were able to swell differently in a simulated vaginal fluid, depending on pH. This effect was shown to be mostly related to the differential protonation of the 1,4-bis(2-hydroxyethyl)piperazine (HEP) group, which could additionally lead to variable electrostatic interactions with charged NPs. In practice, and although pH-dependent behavior was also observed for a control polyurethane co-polymer without HEP, membranes allowed roughly 0% and 60% in vitro permeation after 24 h of model polystyrene NPs (≈200 nm) when tested at pH 4.5 and 7.0, respectively. Control of permeation for solid lipid NPs containing anti-CCR5 siRNA and presenting a mean diameter of around 270 nm was milder (around 30% and 60% at pH 4.5 and 7.0, respectively) [121]. Overall, developed fiber mats could be useful in the design of "smart" dosage forms able to release the majority of microbicide NPs only upon potential contact with the virus.

4.6. Other Potential Systems

Various additional dosage forms present the potential to be used for the vaginal or rectal administration of microbicide nanosystems. Strangely enough, some more "classical" vaginal and rectal dosage forms, such as tablets, soft capsules, inserts (suppositories and ovules), creams, and ointments, have not been described in the literature as vehicles for microbicide nanosystems. Technological feasibility to incorporate nanosystems in such platforms is recognized, but specific demonstration studies still need to be performed for nanotechnology-based microbicides. Foams may also be feasible and interesting platforms for the incorporation of microbicide nanosystems. In the only relevant study available, Vedha Hari et al. [122] reported on the basic technological features of EFV-loaded Eudragit® E 100 NPs (≈110–260 nm) incorporated into a foam formulation. The system was intended for vaginal use but the presence of sodium lauryl sulfate as a foaming agent may compromise the safety of the putative microbicide product.

Nanotechnology-based microbicides have been developed essentially as on-demand products, i.e., requiring administration within minutes to hours of coitus. Although challenging from a technological viewpoint, the development of coitally-independent delivery platforms that allow for prolonged intravaginal residence and sustained release of microbicide nanosystems could be an interesting new approach. For example, vaginal ring technology has come a long way over the last decade and appears to offer enough versatility for the appropriate incorporation and release of nanosystems [123]. Although the pH-sensitive polyurethane fiber membrane described in the previous sub-section was suggested for the design of rings, only the concept was presented and sustained release over at least several days was not demonstrated [121]. Alternatively, simpler rings based on designs previously explored for macromolecule delivery using, for example, pod-inserts [124] or exposed cores [125] could be seen as promising, but pioneering work is required.

5. Conclusions and Future Perspectives

The currently available body of knowledge supporting the development of nanotechnology-based microbicides is extensive and relies on solid grounds. Nanosystems with intrinsic activity against HIV-1 led the field early on (even reaching clinical testing), but have recently been overcome by the promising results obtained for antiretroviral drug-loaded nanocarriers. Still, much more has to be done until a robust candidate nanosystem can be targeted for advanced pre-clinical and clinical testing. Among different issues, suitable vehicles that can enable convenient administration of microbicide nanosystems require development. Several dosage forms have been proposed and tested. Proper design and characterization is paramount in order to guarantee that the original features of nanosystems are at least maintained or, ideally, enhanced. Understanding the interactions between microbicide nanosystems, administration platforms, and the mucosal environment is essential and not always broadly assessed in the work conducted so far. Thermosensitive systems, vaginal films, and fiber mats have been recently studied and hold the potential to provide improved distribution and retention of nanosystems following administration and, ultimately, improve PK at the mucosal level.

Still, many questions remain and various possibilities need to be explored. The stability of nanotechnology-based microbicides is usually overlooked, namely when incorporated into pharmaceutical vehicles. A particular concern has to deal with the release kinetics of drug payloads from nanocarriers and into the matrix of vehicles during manufacturing and storage. Product complexity and a lack of standards for evaluation pose important challenges that need to be addressed. These often translate into costly manufacturing and control, which are incompatible with achieving affordable microbicides. Another important issue in microbicide product development concerns users' preferences and acceptability. Products that meet women's and men's expectations are more likely to be consistently used and, thus, be able to provide proper protection. Furthermore, the idea of investing all efforts in developing a single microbicide product for universal use is illusive and outdated. The possibility to choose from a set of products which better suit individual lifestyles and situations seems to be paramount. The development of behaviorally congruent microbicides, i.e., products that can be incorporated in widespread practices associated with sex (e.g., the use of cleansing enemas in preparation for anal sex), is another trend in the microbicides field [126]. Overall, such principles need to be introduced early in the design process of microbicides, including those based on nanotechnology. The development of dual compartment (or dual chamber) microbicide formulations, i.e., for both vaginal and rectal use, has been pursued in the past as a way to meet user's needs and improve adherence [127]. Such strategy would also be interesting for developing nanotechnology-based microbicides, since nanosystems may hold the potential for both vaginal and rectal protection from viral transmission [49]. However, physiological differences between the two anatomical sites often make it difficult to design products that fit dual usage and, thus, the concept has been somewhat left behind over more recent years. A final word for the regulatory concerns of developing nanotechnology-based microbicides is as follows: While still waiting for the first microbicide product to be approved, general guidance from medical agencies on nanomedicines [128–130] and microbicides [131], as well as the particular cases of VivaGel® and the DPV ring, should be considered as relevant standards.

Author Contributions: Conceptualization, J.d.N.; literature search and critical appraisal, L.M., J.G., R.N., B.S., and J.d.N.; writing—original draft preparation, L.M., J.G., and J.d.N.; writing—review and editing, L.M., J.G., R.N., B.S., and J.d.N.

Funding: This work was supported by Programa Gilead GÉNESE, Gilead Portugal (refs. PGG/046/2015 and PGG/002/2016). This article is a result of the project NORTE-01-0145-FEDER-000012, supported by Norte Portugal Regional Operational Programme (NORTE 2020), under the PORTUGAL 2020 Partnership Agreement, through the European Regional Development Fund (ERDF). This work was financed by FEDER-Fundo Europeu de Desenvolvimento Regional funds through the COMPETE 2020-Operacional Programme for Competitiveness and Internationalisation (POCI), Portugal 2020, and by Portuguese funds through FCT-Fundação para a Ciência e a Tecnologia/Ministério da Ciência, Tecnologia e Ensino Superior in the framework of the project "Institute for Research and Innovation in Health Sciences" (POCI-01-0145-FEDER-007274), and also Partnership Agreement PT2020 UID/QUI/50006/2013-POCI/01/0145/FEDER/007265. J.G. gratefully acknowledges FCT for financial support (SFRH/BD/140271/2018 scholarship).

Conflicts of Interest: The authors declare no conflict of interest.

References

1. UNAIDS. *UNAIDS Data 2018*; UNAIDS: Geneva, Switzerland, 2018. Available online: http://www.unaids.org/en/resources/documents/2018/unaids-data-2018 (accessed on 18 March 2019).
2. Baeten, J.; Celum, C. Systemic and topical drugs for the prevention of HIV infection: Antiretroviral pre-exposure prophylaxis. *Annu. Rev. Med.* **2013**, *64*, 219–232. [CrossRef]
3. Turpin, J.A. Considerations and development of topical microbicides to inhibit the sexual transmission of HIV. *Expert Opin. Investig. Drugs* **2002**, *11*, 1077–1097. [CrossRef]
4. Baeten, J.M.; Palanee-Phillips, T.; Brown, E.R.; Schwartz, K.; Soto-Torres, L.E.; Govender, V.; Mgodi, N.M.; Matovu Kiweewa, F.; Nair, G.; Mhlanga, F.; et al. MTN-Aspire Study Team, Use of a vaginal ring containing dapivirine for HIV-1 prevention in women. *N. Engl. J. Med.* **2016**, *375*, 2121–2132. [CrossRef] [PubMed]
5. Nel, A.; van Niekerk, N.; Kapiga, S.; Bekker, L.G.; Gama, C.; Gill, K.; Kamali, A.; Kotze, P.; Louw, C.; Mabude, Z.; et al. Ring Study Team, Safety and efficacy of a dapivirine vaginal ring for HIV prevention in women. *N. Engl. J. Med.* **2016**, *375*, 2133–2143. [CrossRef]
6. International Partnersphip for Microbicides. IPM's Application for Dapivirine Vaginal Ring for Reducing HIV Risk in Women Now Under Review by European Medicines Agency. Available online: https://www.ipmglobal.org/content/ipm%E2%80%99s-application-dapivirine-vaginal-ring-reducing-hiv-risk-women-now-under-review-european (accessed on 10 December 2018).

7. das Neves, J.; Nunes, R.; Rodrigues, F.; Sarmento, B. Nanomedicine in the development of anti-HIV microbicides. *Adv. Drug Deliv. Rev.* **2016**, *103*, 57–75. [CrossRef] [PubMed]
8. Stein, Z.A. HIV prevention: The need for methods women can use. *Am. J. Public Health* **1990**, *80*, 460–462. [CrossRef] [PubMed]
9. Rohan, L.C.; Devlin, B.; Yang, H. Microbicide dosage forms. *Curr. Top. Microbiol. Immunol.* **2014**, *383*, 27–54. [PubMed]
10. Malkovsky, M.; Newell, A.; Dalgleish, A.G. Inactivation of HIV by nonoxynol-9. *Lancet* **1988**, *1*, 645. [CrossRef]
11. Roddy, R.E.; Zekeng, L.; Ryan, K.A.; Tamoufe, U.; Weir, S.S.; Wong, E.L. A controlled trial of nonoxynol 9 film to reduce male-to-female transmission of sexually transmitted diseases. *N. Engl. J. Med.* **1998**, *339*, 504–510. [CrossRef]
12. Van Damme, L.; Ramjee, G.; Alary, M.; Vuylsteke, B.; Chandeying, V.; Rees, H.; Sirivongrangson, P.; Mukenge-Tshibaka, L.; Ettiegne-Traore, V.; Uaheowitchai, C.; et al. Effectiveness of COL-1492, a nonoxynol-9 vaginal gel, on HIV-1 transmission in female sex workers: A randomised controlled trial. *Lancet* **2002**, *360*, 971–977. [CrossRef]
13. Van Damme, L.; Govinden, R.; Mirembe, F.M.; Guedou, F.; Solomon, S.; Becker, M.L.; Pradeep, B.S.; Krishnan, A.K.; Alary, M.; Pande, B.; et al. Lack of effectiveness of cellulose sulfate gel for the prevention of vaginal HIV transmission. *N. Engl. J. Med.* **2008**, *359*, 463–472. [CrossRef] [PubMed]
14. Feldblum, P.J.; Adeiga, A.; Bakare, R.; Wevill, S.; Lendvay, A.; Obadaki, F.; Olayemi, M.O.; Wang, L.; Nanda, K.; Rountree, W. SAVVY vaginal gel (C31G) for prevention of HIV infection: A randomized controlled trial in Nigeria. *PLoS ONE* **2008**, *3*, e1474. [CrossRef] [PubMed]
15. Skoler-Karpoff, S.; Ramjee, G.; Ahmed, K.; Altini, L.; Plagianos, M.G.; Friedland, B.; Govender, S.; De Kock, A.; Cassim, N.; Palanee, T.; et al. Efficacy of Carraguard for prevention of HIV infection in women in South Africa: A randomised, double-blind, placebo-controlled trial. *Lancet* **2008**, *372*, 1977–1987. [CrossRef]
16. Abdool Karim, S.S.; Richardson, B.A.; Ramjee, G.; Hoffman, I.F.; Chirenje, Z.M.; Taha, T.; Kapina, M.; Maslankowski, L.; Coletti, A.; Profy, A.; et al. Safety and effectiveness of BufferGel and 0.5% PRO2000 gel for the prevention of HIV infection in women. *AIDS* **2011**, *25*, 957–966. [CrossRef] [PubMed]
17. Hillier, S.L.; Moench, T.; Shattock, R.; Black, R.; Reichelderfer, P.; Veronese, F. In vitro and in vivo: The story of nonoxynol 9. *J. Acquir. Immune Defic. Syndr.* **2005**, *39*, 1–8. [CrossRef]
18. Mesquita, P.M.; Cheshenko, N.; Wilson, S.S.; Mhatre, M.; Guzman, E.; Fakioglu, E.; Keller, M.J.; Herold, B.C. Disruption of tight junctions by cellulose sulfate facilitates HIV infection: Model of microbicide safety. *J. Infect. Dis.* **2009**, *200*, 599–608. [CrossRef]
19. Turpin, J.A. Topical microbicides to prevent the transmission of HIV: Formulation gaps and challenges. *Drug Deliv. Transl. Res.* **2011**, *1*, 194–200. [CrossRef]
20. Abdool Karim, Q.; Abdool Karim, S.S.; Frohlich, J.A.; Grobler, A.C.; Baxter, C.; Mansoor, L.E.; Kharsany, A.B.; Sibeko, S.; Mlisana, K.P.; Omar, Z.; et al. Effectiveness and safety of tenofovir gel, an antiretroviral microbicide, for the prevention of HIV infection in women. *Science* **2010**, *329*, 1168–1174. [CrossRef]
21. Kashuba, A.D.; Gengiah, T.N.; Werner, L.; Yang, K.H.; White, N.R.; Karim, Q.A.; Abdool Karim, S.S. Genital tenofovir concentrations correlate with protection against HIV infection in the CAPRISA 004 trial: Importance of adherence for microbicide effectiveness. *J. Acquir. Immune Defic. Syndr.* **2015**, *69*, 264–269. [CrossRef]
22. Klatt, N.R.; Cheu, R.; Birse, K.; Zevin, A.S.; Perner, M.; Noel-Romas, L.; Grobler, A.; Westmacott, G.; Xie, I.Y.; Butler, J.; et al. Vaginal bacteria modify HIV tenofovir microbicide efficacy in African women. *Science* **2017**, *356*, 938–945. [CrossRef]
23. Marrazzo, J.M.; Ramjee, G.; Richardson, B.A.; Gomez, K.; Mgodi, N.; Nair, G.; Palanee, T.; Nakabiito, C.; van der Straten, A.; Noguchi, L.; et al. VOICE Study Team, Tenofovir-based preexposure prophylaxis for HIV infection among African women. *N. Engl. J. Med.* **2015**, *372*, 509–518. [CrossRef] [PubMed]
24. Delany-Moretlwe, S.; Lombard, C.; Baron, D.; Bekker, L.G.; Nkala, B.; Ahmed, K.; Sebe, M.; Brumskine, W.; Nchabeleng, M.; Palanee-Philips, T.; et al. Tenofovir 1% vaginal gel for prevention of HIV-1 infection in women in South Africa (FACTS-001): A phase 3, randomised, double-blind, placebo-controlled trial. *Lancet Infect. Dis.* **2018**, *18*, 1241–1250. [CrossRef]
25. van der Straten, A.; Brown, E.R.; Marrazzo, J.M.; Chirenje, M.Z.; Liu, K.; Gomez, K.; Marzinke, M.A.; Piper, J.M.; Hendrix, C.W. Divergent adherence estimates with pharmacokinetic and behavioural measures in the MTN-003 (VOICE) study. *J. Int. AIDS Soc.* **2016**, *19*, 20642. [CrossRef]

26. Nel, A.; Smythe, S.; Young, K.; Malcolm, K.; McCoy, C.; Rosenberg, Z.; Romano, J. Safety and pharmacokinetics of dapivirine delivery from matrix and reservoir intravaginal rings to HIV-negative women. *J. Acquir. Immune Defic. Syndr.* **2009**, *51*, 416–423. [CrossRef]
27. Montgomery, E.T.; Stadler, J.; Naidoo, S.; Katz, A.W.K.; Laborde, N.; Garcia, M.; Reddy, K.; Mansoor, L.E.; Etima, J.; Zimba, C.; et al. Reasons for nonadherence to the dapivirine vaginal ring: Narrative explanations of objective drug-level results. *AIDS* **2018**, *32*, 1517–1525. [CrossRef]
28. Brown, E.; Palanee-Philips, T.; Marzinke, M.; Hendrix, C.; Dezzutti, C.; Soto-Torres, L.; Baeten, J. Residual Dapivirine Ring Levels Indicate Higher Adherence to Vaginal Ring is Associated with HIV-1 Protection. In Proceedings of the AIDS 2016, Durban, South Africa, 18–22 July 2016.
29. Woodsong, C.; Holt, J.D. Acceptability and preferences for vaginal dosage forms intended for prevention of HIV or HIV and pregnancy. *Adv. Drug Deliv. Rev.* **2015**, *15*, 146–154. [CrossRef]
30. Fernández-Romero, J.A.; Deal, C.; Herold, B.C.; Schiller, J.; Patton, D.; Zydowsky, T.; Romano, J.; Petro, C.D.; Narasimhan, M. Multipurpose prevention technologies: The future of HIV and STI protection. *Trends Microbiol.* **2015**, *23*, 429–436. [CrossRef] [PubMed]
31. McGowan, I. The development of rectal microbicides for HIV prevention. *Expert Opin. Drug Deliv.* **2014**, *11*, 69–82. [CrossRef] [PubMed]
32. McGowan, I.; Cranston, R.D.; Duffill, K.; Siegel, A.; Engstrom, J.C.; Nikiforov, A.; Jacobson, C.; Rehman, K.K.; Elliott, J.; Khanukhova, E.; et al. A phase 1 randomized, open label, rectal safety, acceptability, pharmacokinetic, and pharmacodynamic study of three formulations of tenofovir 1% gel (the CHARM-01 study). *PLoS ONE* **2015**, *10*, e0125363. [CrossRef] [PubMed]
33. Cranston, R.D.; Lama, J.R.; Richardson, B.A.; Carballo-Dieguez, A.; Kunjara Na Ayudhya, R.P.; Liu, K.; Patterson, K.B.; Leu, C.S.; Galaska, B.; Jacobson, C.E.; et al. MTN-017: A rectal phase 2 extended safety and acceptability study of tenofovir reduced-glycerin 1% gel. *Clin. Infect. Dis.* **2017**, *64*, 614–620. [CrossRef] [PubMed]
34. Patel, P.; Borkowf, C.B.; Brooks, J.T.; Lasry, A.; Lansky, A.; Mermin, J. Estimating per-act HIV transmission risk: A systematic review. *AIDS* **2014**, *28*, 1509–1519. [CrossRef] [PubMed]
35. Nunes, R.; Sarmento, B.; das Neves, J. Formulation and delivery of anti-HIV rectal microbicides: Advances and challenges. *J. Control. Release* **2014**, *194*, 278–294. [CrossRef]
36. das Neves, J.; Amiji, M.M.; Bahia, M.F.; Sarmento, B. Nanotechnology-based systems for the treatment and prevention of HIV/AIDS. *Adv. Drug Deliv. Rev.* **2010**, *62*, 458–477. [CrossRef] [PubMed]
37. Sanchez-Rodríguez, J.; Vacas-Cordoba, E.; Gomez, R.; De La Mata, F.J.; Muñoz-Fernández, M.Á. Nanotech-derived topical microbicides for HIV prevention: The road to clinical development. *Antivir. Res.* **2015**, *113*, 33–48. [CrossRef]
38. Brako, F.; Mahalingam, S.; Rami-Abraham, B.; Craig, D.Q.; Edirisinghe, M. Application of nanotechnology for the development of microbicides. *Nanotechnology* **2017**, *28*, 052001. [CrossRef] [PubMed]
39. Nunes, R.; Sousa, C.; Sarmento, B.; das Neves, J. Nanotechnology-based systems for microbicide development. In *Drug Delivery and Development of Anti-HIV Microbicides*; das Neves, J., Sarmento, B., Eds.; Pan Stanford: Singapore, 2014; pp. 415–458.
40. Nandy, B.; Saurabh, S.; Sahoo, A.K.; Dixit, N.M.; Maiti, P.K. The SPL7013 dendrimer destabilizes the HIV-1 gp120-CD4 complex. *Nanoscale* **2015**, *7*, 18628–18641. [CrossRef]
41. McGowan, I.; Gomez, K.; Bruder, K.; Febo, I.; Chen, B.A.; Richardson, B.A.; Husnik, M.; Livant, E.; Price, C.; Jacobson, C. Phase 1 randomized trial of the vaginal safety and acceptability of SPL7013 gel (VivaGel) in sexually active young women (MTN-004). *AIDS* **2011**, *25*, 1057–1064. [CrossRef]
42. Moscicki, A.B.; Kaul, R.; Ma, Y.; Scott, M.E.; Daud, I.I.; Bukusi, E.A.; Shiboski, S.; Rebbapragada, A.; Huibner, S.; Cohen, C.R. Measurement of mucosal biomarkers in a phase 1 trial of intravaginal 3% StarPharma LTD 7013 gel (VivaGel) to assess expanded safety. *J. Acquir. Immune Defic. Syndr.* **2012**, *59*, 134–140. [CrossRef] [PubMed]
43. Sepúlveda-Crespo, D.; Ceña-Díez, R.; Jiménez, J.L.; Muñoz-Fernández, M.Á. Mechanistic studies of viral entry: An overview of dendrimer-based microbicides as entry inhibitors against both HIV and HSV-2 overlapped infections. *Med. Res. Rev.* **2017**, *37*, 149–179. [CrossRef] [PubMed]
44. Sepúlveda-Crespo, D.; Serramía, M.J.; Tager, A.M.; Vrbanac, V.; Gómez, R.; De La Mata, F.J.; Jiménez, J.L.; Muñoz-Fernández, M.Á. Prevention vaginally of HIV-1 transmission in humanized BLT mice and mode of antiviral action of polyanionic carbosilane dendrimer G2-S16. *Nanomedicine* **2015**, *11*, 1299–1308. [CrossRef]

45. Ceña-Diez, R.; García-Broncano, P.; Javier de la Mata, F.; Gómez, R.; Resino, S.; Muñoz-Fernández, M. G2-S16 dendrimer as a candidate for a microbicide to prevent HIV-1 infection in women. *Nanoscale* **2017**, *9*, 9732–9742. [CrossRef]
46. Martínez-Ávila, O.; Hijazi, K.; Marradi, M.; Clavel, C.; Campion, C.; Kelly, C.; Penadés, S. Gold manno-glyconanoparticles: Multivalent systems to block HIV-1 gp120 binding to the lectin DC-SIGN. *Chemistry* **2009**, *15*, 9874–9888. [CrossRef]
47. Maisel, K.; Ensign, L.; Reddy, M.; Cone, R.; Hanes, J. Effect of surface chemistry on nanoparticle interaction with gastrointestinal mucus and distribution in the gastrointestinal tract following oral and rectal administration in the mouse. *J. Control. Release* **2015**, *197*, 48–57. [CrossRef]
48. das Neves, J.; Rocha, C.M.; Gonçalves, M.P.; Carrier, R.L.; Amiji, M.; Bahia, M.F.; Sarmento, B. Interactions of microbicide nanoparticles with a simulated vaginal fluid. *Mol. Pharm.* **2012**, *9*, 3347–3356. [CrossRef]
49. das Neves, J.; Araújo, F.; Andrade, F.; Michiels, J.; Ariën, K.K.; Vanham, G.; Amiji, M.; Bahia, M.F.; Sarmento, B. In vitro and ex vivo evaluation of polymeric nanoparticles for vaginal and rectal delivery of the anti-HIV drug dapivirine. *Mol. Pharm.* **2013**, *10*, 2793–2807. [CrossRef]
50. Ariza-Saenz, M.; Espina, M.; Bolanos, N.; Calpena, A.C.; Gomara, M.J.; Haro, I.; Garcia, M.L. Penetration of polymeric nanoparticles loaded with an HIV-1 inhibitor peptide derived from GB virus C in a vaginal mucosa model. *Eur. J. Pharm. Biopharm.* **2017**, *120*, 98–106. [CrossRef]
51. das Neves, J.; Michiels, J.; Ariën, K.K.; Vanham, G.; Amiji, M.; Bahia, M.F.; Sarmento, B. Polymeric nanoparticles affect the intracellular delivery, antiretroviral activity and cytotoxicity of the microbicide drug candidate dapivirine. *Pharm. Res.* **2012**, *29*, 1468–1484. [CrossRef]
52. das Neves, J.; Araújo, F.; Andrade, F.; Amiji, M.; Bahia, M.F.; Sarmento, B. Biodistribution and pharmacokinetics of dapivirine-loaded nanoparticles after vaginal delivery in mice. *Pharm. Res.* **2014**, *31*, 1834–1845. [CrossRef]
53. Nunes, R.; Araújo, F.; Barreiros, L.; Bártolo, I.; Segundo, M.A.; Taveira, N.; Sarmento, B.; das Neves, J. Noncovalent PEG coating of nanoparticle drug carriers improves the local pharmacokinetics of rectal anti-HIV microbicides. *ACS Appl. Mater Interfaces* **2018**, *10*, 34942–34953. [CrossRef]
54. Kovarova, M.; Council, O.D.; Date, A.A.; Long, J.M.; Nochi, T.; Belshan, M.; Shibata, A.; Vincent, H.; Baker, C.E.; Thayer, W.O.; et al. Nanoformulations of rilpivirine for topical pericoital and systemic coitus-independent administration efficiently prevent HIV transmission. *PLoS Pathog.* **2015**, *11*, e1005075.
55. Meng, J.; Zhang, T.; Agrahari, V.; Ezoulin, M.J.; Youan, B.B. Comparative biophysical properties of tenofovir-loaded, thiolated and nonthiolated chitosan nanoparticles intended for HIV prevention. *Nanomedicine (Lond.)* **2014**, *9*, 1595–1612. [CrossRef] [PubMed]
56. das Neves, J.; Sarmento, B. Precise engineering of dapivirine-loaded nanoparticles for the development of anti-HIV vaginal microbicides. *Acta Biomater.* **2015**, *18*, 77–87. [CrossRef] [PubMed]
57. Ariza-Sáenz, M.; Espina, M.; Calpena, A.; Gómara, M.J.; Pérez-Pomeda, I.; Haro, I.; García, M.L. Design, characterization, and biopharmaceutical behavior of nanoparticles loaded with an HIV-1 fusion inhibitor peptide. *Mol. Pharm.* **2018**, *15*, 5005–5018. [CrossRef] [PubMed]
58. Kish-Catalone, T.; Pal, R.; Parrish, J.; Rose, N.; Hocker, L.; Hudacik, L.; Reitz, M.; Gallo, R.; Devico, A. Evaluation of -2 RANTES vaginal microbicide formulations in a nonhuman primate simian/human immunodeficiency virus (SHIV) challenge model. *AIDS Res. Hum. Retrovir.* **2007**, *23*, 33–42. [CrossRef] [PubMed]
59. Ham, A.S.; Cost, M.R.; Sassi, A.B.; Dezzutti, C.S.; Rohan, L.C. Targeted delivery of PSC-RANTES for HIV-1 prevention using biodegradable nanoparticles. *Pharm. Res.* **2009**, *26*, 502–511. [CrossRef]
60. Boyapalle, S.; Xu, W.; Raulji, P.; Mohapatra, S.; Mohapatra, S.S. A multiple siRNA-based anti-HIV/SHIV microbicide shows protection in both in vitro and in vivo models. *PLoS ONE* **2015**, *10*, e0135288. [CrossRef]
61. das Neves, J.; Nunes, R.; Machado, A.; Sarmento, B. Polymer-based nanocarriers for vaginal drug delivery. *Adv. Drug Deliv. Rev.* **2015**, *92*, 53–70. [CrossRef]
62. Alukda, D.; Sturgis, T.; Youan, B.B. Formulation of tenofovir-loaded functionalized solid lipid nanoparticles intended for HIV prevention. *J. Pharm. Sci.* **2011**, *100*, 3345–3356. [CrossRef]
63. Caron, M.; Besson, G.; Etenna, S.L.; Mintsa-Ndong, A.; Mourtas, S.; Radaelli, A.; Morghen Cde, G.; Loddo, R.; La Colla, P.; Antimisiaris, S.G.; et al. Protective properties of non-nucleoside reverse transcriptase inhibitor (MC1220) incorporated into liposome against intravaginal challenge of Rhesus macaques with RT-SHIV. *Virology* **2010**, *405*, 225–233. [CrossRef]

64. Ramanathan, R.; Jiang, Y.; Read, B.; Golan-Paz, S.; Woodrow, K.A. Biophysical characterization of small molecule antiviral-loaded nanolipogels for HIV-1 chemoprophylaxis and topical mucosal application. *Acta Biomater.* **2016**, *36*, 122–131. [CrossRef] [PubMed]
65. Blakney, A.K.; Jiang, Y.; Woodrow, K.A. Application of electrospun fibers for female reproductive health. *Drug Deliv. Transl. Res.* **2017**, *7*, 796–804. [CrossRef] [PubMed]
66. das Neves, J.; Amiji, M.; Sarmento, B. Mucoadhesive nanosystems for vaginal microbicide development: Friend or foe? *Wiley Interdiscip. Rev. Nanomed. Nanobiotechnol.* **2011**, *3*, 389–399. [CrossRef] [PubMed]
67. Maisel, K.; Reddy, M.; Xu, Q.; Chattopadhyay, S.; Cone, R.; Ensign, L.M.; Hanes, J. Nanoparticles coated with high molecular weight PEG penetrate mucus and provide uniform vaginal and colorectal distribution in vivo. *Nanomedicine* **2016**, *11*, 1337–1343. [CrossRef] [PubMed]
68. Nunes, R.; Araújo, F.; Tavares, J.; Sarmento, B.; das Neves, J. Surface modification with polyethylene glycol enhances colorectal distribution and retention of nanoparticles. *Eur. J. Pharm. Biopharm.* **2018**, *130*, 200–206. [CrossRef] [PubMed]
69. Mohideen, M.; Quijano, E.; Song, E.; Deng, Y.; Panse, G.; Zhang, W.; Clark, M.R.; Saltzman, W.M. Degradable bioadhesive nanoparticles for prolonged intravaginal delivery and retention of elvitegravir. *Biomaterials* **2017**, *144*, 144–154. [CrossRef]
70. Ballou, B.; Andreko, S.K.; Osuna-Highley, E.; McRaven, M.; Catalone, T.; Bruchez, M.P.; Hope, T.J.; Labib, M.E. Nanoparticle transport from mouse vagina to adjacent lymph nodes. *PLoS ONE* **2012**, *7*, e51995. [CrossRef]
71. Ramanathan, R.; Park, J.; Hughes, S.M.; Lykins, W.R.; Bennett, H.R.; Hladik, F.; Woodrow, K.A. Effect of mucosal cytokine administration on selective expansion of vaginal dendritic cells to support nanoparticle transport. *Am. J. Reprod. Immunol.* **2015**, *74*, 333–344. [CrossRef]
72. Shattock, R.J.; Moore, J.P. Inhibiting sexual transmission of HIV-1 infection. *Nat. Rev. Microbiol.* **2003**, *1*, 25–34. [CrossRef] [PubMed]
73. Malik, R.; Maikhuri, J.P.; Gupta, G.; Misra, A. Biodegradable nanoparticles in the murine vagina: Trans-cervical retrograde transport and induction of proinflammatory cytokines. *J. Biomed. Nanotechnol.* **2011**, *7*, 45–46. [CrossRef] [PubMed]
74. U.S. Food & Drug Administration. Drugs@FDA Glossary of Terms. Available online: https://www.fda.gov/drugs/informationondrugs/ucm079436.htm (accessed on 16 January 2019).
75. European Directorate for the Quality of Medicines. Standard Terms: Introduction and Guidance for Use. Available online: https://www.edqm.eu/sites/default/files/standard_terms_introduction_and_guidance_for_use.pdf (accessed on 16 January 2019).
76. das Neves, J.; Palmeira-de-Oliveira, R.; Palmeira-de-Oliveira, A.; Rodrigues, F.; Sarmento, B. Vaginal mucosa and drug delivery. In *Mucoadhesive Materials and Drug Delivery Systems*; Khutoryanskiy, V.V., Ed.; Wiley: Chichester, UK, 2014; pp. 99–131.
77. Batchelor, H. Rectal drug delivery. In *Pediatric Formulations: A Roadmap*; Bar-Shalom, D., Rose, K., Eds.; Springer: New York, NY, USA, 2014; pp. 303–310.
78. Fernández-Romero, J.A.; Teleshova, N.; Zydowsky, T.M.; Robbiani, M. Preclinical assessments of vaginal microbicide candidate safety and efficacy. *Adv. Drug Deliv. Rev.* **2015**, *92*, 27–38. [CrossRef] [PubMed]
79. Melo, M.; Nunes, R.; Sarmento, B.; das Neves, J. Rectal administration of nanosystems: From drug delivery to diagnostics. *Mater. Today Chem.* **2018**, *10*, 128–141. [CrossRef]
80. Lakshmi, Y.S.; Kumar, P.; Kishore, G.; Bhaskar, C.; Kondapi, A.K. Triple combination MPT vaginal microbicide using curcumin and efavirenz loaded lactoferrin nanoparticles. *Sci. Rep.* **2016**, *6*, 25479. [CrossRef]
81. Samizadeh, M.; Zhang, X.; Gunaseelan, S.; Nelson, A.G.; Palombo, M.S.; Myers, D.R.; Singh, Y.; Ganapathi, U.; Szekely, Z.; Sinko, P.J. Colorectal delivery and retention of PEG-Amprenavir-Bac7 nanoconjugates-proof of concept for HIV mucosal pre-exposure prophylaxis. *Drug Deliv. Transl. Res.* **2016**, *6*, 1–16. [CrossRef]
82. Dezzutti, C.S.; Brown, E.R.; Moncla, B.; Russo, J.; Cost, M.; Wang, L.; Uranker, K.; Kunjara Na Ayudhya, R.P.; Pryke, K.; Pickett, J.; et al. Is wetter better? An evaluation of over-the-counter personal lubricants for safety and anti-HIV-1 activity. *PLoS ONE* **2012**, *7*, e48328. [CrossRef]
83. Veazey, R.S.; Shattock, R.J.; Klasse, P.J.; Moore, J.P. Animal models for microbicide studies. *Curr. HIV Res.* **2012**, *10*, 79–87. [CrossRef]
84. Ensign, L.M.; Hoen, T.E.; Maisel, K.; Cone, R.A.; Hanes, J.S. Enhanced vaginal drug delivery through the use of hypotonic formulations that induce fluid uptake. *Biomaterials* **2013**, *34*, 6922–6929. [CrossRef]

85. Maisel, K.; Chattopadhyay, S.; Moench, T.; Hendrix, C.; Cone, R.; Ensign, L.M.; Hanes, J. Enema ion compositions for enhancing colorectal drug delivery. *J. Control. Release* **2015**, *209*, 280–287. [CrossRef]
86. Agashe, H.; Hu, M.; Rohan, L. Formulation and delivery of microbicides. *Curr. HIV Res.* **2012**, *10*, 88–96. [CrossRef]
87. Rupp, R.; Rosenthal, S.L.; Stanberry, L.R. VivaGel (SPL7013 Gel): A candidate dendrimer-microbicide for the prevention of HIV and HSV infection. *Int. J. Nanomed.* **2007**, *2*, 561–566.
88. Gaurav, C.; Goutam, R.; Rohan, K.N.; Sweta, K.T.; Abhay, C.S.; Amit, G.K. In situ stabilized AgNPs and (Cu-Cur) CD dispersed gel, a topical contraceptive antiretroviral (ARV) microbicide. *RSC Adv.* **2015**, *5*, 83013–83028. [CrossRef]
89. Schwartz, J.L.; Ballagh, S.A.; Kwok, C.; Mauck, C.K.; Weiner, D.H.; Rencher, W.F.; Callahan, M.M. Fourteen-day safety and acceptability study of the universal placebo gel. *Contraception* **2007**, *75*, 136–141. [CrossRef]
90. Wang, L.; Sassi, A.B.; Patton, D.; Isaacs, C.; Moncla, B.J.; Gupta, P.; Rohan, L.C. Development of a liposome microbicide formulation for vaginal delivery of octylglycerol for HIV prevention. *Drug Dev. Ind. Pharm.* **2012**, *38*, 995–1007. [CrossRef]
91. Lara, H.H.; Ixtepan-Turrent, L.; Garza-Trevino, E.N.; Rodriguez-Padilla, C. PVP-coated silver nanoparticles block the transmission of cell-free and cell-associated HIV-1 in human cervical culture. *J. Nanobiotechnol.* **2010**, *8*, 15. [CrossRef]
92. Di Fabio, S.; Van Roey, J.; Giannini, G.; van den Mooter, G.; Spada, M.; Binelli, A.; Pirillo, M.F.; Germinario, E.; Belardelli, F.; de Bethune, M.P.; et al. Inhibition of vaginal transmission of HIV-1 in hu-SCID mice by the non-nucleoside reverse transcriptase inhibitor TMC120 in a gel formulation. *AIDS* **2003**, *17*, 1597–1604. [CrossRef] [PubMed]
93. Koffi, A.A.; Agnely, F.; Besnard, M.; Kablan Brou, J.; Grossiord, J.L.; Ponchel, G. In vitro and in vivo characteristics of a thermogelling and bioadhesive delivery system intended for rectal administration of quinine in children. *Eur. J. Pharm. Biopharm.* **2008**, *69*, 167–175. [CrossRef]
94. Bouchemal, K.; Frelichowska, J.; Martin, L.; Lievin-Le Moal, V.; Le Grand, R.; Dereuddre-Bosquet, N.; Djabourov, M.; Aka-Any-Grah, A.; Koffi, A.; Ponchel, G. Note on the formulation of thermosensitive and mucoadhesive vaginal hydrogels containing the miniCD4 M48U1 as anti-HIV-1 microbicide. *Int. J. Pharm.* **2013**, *454*, 649–652. [CrossRef] [PubMed]
95. Bodratti, A.M.; Alexandridis, P. Formulation of poloxamers for drug delivery. *J. Funct. Biomater.* **2018**, *9*, 11. [CrossRef] [PubMed]
96. Mandal, S.; Khandalavala, K.; Pham, R.; Bruck, P.; Varghese, M.; Kochvar, A.; Monaco, A.; Prathipati, P.K.; Destache, C.; Shibata, A. Cellulose acetate phthalate and antiretroviral nanoparticle fabrications for HIV pre-exposure prophylaxis. *Polymers (Basel)* **2017**, *9*, 423. [CrossRef]
97. Timur, S.S.; Şahin, A.; Aytekin, E.; Öztürk, N.; Polat, K.H.; Tezel, N.; Gürsoy, R.N.; Çalış, S. Design and in vitro evaluation of tenofovir-loaded vaginal gels for the prevention of HIV infections. *Pharm. Dev. Technol.* **2018**, *23*, 301–310. [CrossRef] [PubMed]
98. Date, A.A.; Shibata, A.; Goede, M.; Sanford, B.; La Bruzzo, K.; Belshan, M.; Destache, C.J. Development and evaluation of a thermosensitive vaginal gel containing raltegravir + efavirenz loaded nanoparticles for HIV prophylaxis. *Antivir. Res.* **2012**, *96*, 430–436. [CrossRef]
99. Date, A.A.; Shibata, A.; McMullen, E.; La Bruzzo, K.; Bruck, P.; Belshan, M.; Zhou, Y.; Destache, C.J. Thermosensitive gel containing cellulose acetate phthalate-efavirenz combination nanoparticles for prevention of HIV-1 infection. *J. Biomed. Nanotechnol.* **2015**, *11*, 416–427. [CrossRef]
100. Destache, C.J.; Mandal, S.; Yuan, Z.; Kang, G.; Date, A.A.; Lu, W.; Shibata, A.; Pham, R.; Bruck, P.; Rezich, M.; et al. Topical tenofovir disoproxil fumarate nanoparticles prevent HIV-1 vaginal transmission in a humanized mouse model. *Antimicrob. Agents Chemother.* **2016**, *60*, 3633–3639. [CrossRef] [PubMed]
101. Melo, M.; Nunes, R.; Sarmento, B.; das Neves, J. Nanoparticles-in-thermosensitive enemas as potential vehicles for microbicide development. *AIDS Res. Hum. Retrovir.* **2018**, *34*, 70.
102. Machado, R.M.; Palmeira-de-Oliveira, A.; Martinez-de-Oliveira, J.; Palmeira-de-Oliveira, R. Vaginal films for drug delivery. *J. Pharm. Sci.* **2013**, *102*, 2069–2081. [CrossRef] [PubMed]
103. Akil, A.; Agashe, H.; Dezzutti, C.S.; Moncla, B.J.; Hillier, S.L.; Devlin, B.; Shi, Y.; Uranker, K.; Rohan, L.C. Formulation and characterization of polymeric films containing combinations of antiretrovirals (ARVs) for HIV prevention. *Pharm. Res.* **2015**, *32*, 458–468. [CrossRef]

104. Zhang, W.; Hu, M.; Shi, Y.; Gong, T.; Dezzutti, C.S.; Moncla, B.; Sarafianos, S.G.; Parniak, M.A.; Rohan, L.C. Vaginal microbicide film combinations of two reverse transcriptase inhibitors, EFdA and CSIC, for the prevention of HIV-1 sexual transmission. *Pharm. Res.* **2015**, *32*, 2960–2972. [CrossRef] [PubMed]
105. Bunge, K.E.; Dezzutti, C.S.; Rohan, L.C.; Hendrix, C.W.; Marzinke, M.A.; Richardson-Harman, N.; Moncla, B.J.; Devlin, B.; Meyn, L.A.; Spiegel, H.M.; et al. A Phase 1 trial to assess the safety, acceptability, pharmacokinetics and pharmacodynamics of a novel dapivirine vaginal film. *J. Acquir. Immune Defic. Syndr.* **2015**, *71*, 498–505. [CrossRef] [PubMed]
106. Castro, P.M.; Baptista, P.; Madureira, A.R.; Sarmento, B.; Pintado, M.E. Combination of PLGA nanoparticles with mucoadhesive guar-gum films for buccal delivery of antihypertensive peptide. *Int. J. Pharm.* **2018**, *547*, 593–601. [CrossRef]
107. das Neves, J.; Sarmento, B. Antiretroviral drug-loaded nanoparticles-in-films: A new option for developing vaginal microbicides? *Expert Opin. Drug Deliv.* **2017**, *14*, 449–452. [CrossRef]
108. das Neves, J.; Amiji, M.; Bahia, M.F.; Sarmento, B. Assessing the physical-chemical properties and stability of dapivirine-loaded polymeric nanoparticles. *Int. J. Pharm.* **2013**, *456*, 307–314. [CrossRef]
109. Cautela, M.P.; Moshe, H.; Sosnik, A.; Sarmento, B.; das Neves, J. Composite films for vaginal delivery of tenofovir disoproxil fumarate and emtricitabine. *Eur. J. Pharm. Biopharm.* **2018**. [CrossRef] [PubMed]
110. Gu, J.; Yang, S.; Ho, E.A. Biodegradable film for the targeted delivery of siRNA-loaded nanoparticles to vaginal immune cells. *Mol. Pharm.* **2015**, *12*, 2889–2903. [CrossRef] [PubMed]
111. Cunha-Reis, C.; Machado, A.; Barreiros, L.; Araújo, F.; Nunes, R.; Seabra, V.; Ferreira, D.; Segundo, M.A.; Sarmento, B.; das Neves, J. Nanoparticles-in-film for the combined vaginal delivery of anti-HIV microbicide drugs. *J. Control. Release* **2016**, *243*, 43–53. [CrossRef]
112. Machado, A.; Cunha-Reis, C.; Araújo, F.; Nunes, R.; Seabra, V.; Ferreira, D.; das Neves, J.; Sarmento, B. Development and in vivo safety assessment of tenofovir-loaded nanoparticles-in-film as a novel vaginal microbicide delivery system. *Acta Biomater.* **2016**, *44*, 332–340. [CrossRef] [PubMed]
113. Srinivasan, P.; Zhang, J.; Martin, A.; Kelley, K.; McNicholl, J.M.; Buckheit, R.W., Jr.; Smith, J.M.; Ham, A.S. Safety and pharmacokinetics of quick dissolving polymeric vaginal films delivering the antiretroviral IQP-0528 for pre-exposure prophylaxis. *Antimicrob. Agents Chemother.* **2016**, *60*, 4140–4150. [CrossRef] [PubMed]
114. Blakney, A.K.; Ball, C.; Krogstad, E.A.; Woodrow, K.A. Electrospun fibers for vaginal anti-HIV drug delivery. *Antivir. Res.* **2013**, *100*, S9–S16. [CrossRef]
115. Carson, D.; Jiang, Y.; Woodrow, K.A. Tunable release of multiclass anti-HIV drugs that are water-soluble and loaded at high drug content in polyester blended electrospun fibers. *Pharm. Res.* **2016**, *33*, 125–136. [CrossRef]
116. Huang, C.; Soenen, S.J.; van Gulck, E.; Vanham, G.; Rejman, J.; Van Calenbergh, S.; Vervaet, C.; Coenye, T.; Verstraelen, H.; Temmerman, M.; et al. Electrospun cellulose acetate phthalate fibers for semen induced anti-HIV vaginal drug delivery. *Biomaterials* **2012**, *33*, 962–969. [CrossRef] [PubMed]
117. Agrahari, V.; Meng, J.; Ezoulin, M.J.; Youm, I.; Dim, D.C.; Molteni, A.; Hung, W.T.; Christenson, L.K.; Youan, B.C. Stimuli-sensitive thiolated hyaluronic acid based nanofibers: Synthesis, preclinical safety and in vitro anti-HIV activity. *Nanomedicine (Lond.)* **2016**, *11*, 2935–2958. [CrossRef]
118. Laborde, N.D.; Leslie, J.; Krogstad, E.; Morar, N.; Mutero, P.; Etima, J.; Woodrow, K.; van der Straten, A. Perceptions of the "Fabric"—An exploratory study of a novel multi-purpose technology among women in Sub Saharan Africa. *PLoS ONE* **2018**, *13*, e0204821. [CrossRef] [PubMed]
119. Krogstad, E.A.; Woodrow, K.A. Manufacturing scale-up of electrospun poly(vinyl alcohol) fibers containing tenofovir for vaginal drug delivery. *Int. J. Pharm.* **2014**, *475*, 282–291. [CrossRef] [PubMed]
120. Krogstad, E.A.; Ramanathan, R.; Nhan, C.; Kraft, J.C.; Blakney, A.K.; Cao, S.; Ho, R.J.Y.; Woodrow, K.A. Nanoparticle-releasing nanofiber composites for enhanced in vivo vaginal retention. *Biomaterials* **2017**, *144*, 1–16. [CrossRef] [PubMed]
121. Kim, S.; Traore, Y.L.; Ho, E.A.; Shafiq, M.; Kim, S.H.; Liu, S. Design and development of pH-responsive polyurethane membranes for intravaginal release of nanomedicines. *Acta Biomater.* **2018**, *82*, 12–23. [CrossRef]
122. Vedha Hari, B.N.; Narayanan, N.; Dhevedaran, K. Efavirenz–eudragit E-100 nanoparticle-loaded aerosol foam for sustained release: In-vitro and ex-vivo evaluation. *Chem. Pap.* **2015**, *69*, 358–367. [CrossRef]
123. Malcolm, R.K.; Boyd, P.J.; McCoy, C.F.; Murphy, D.J. Microbicide vaginal rings: Technological challenges and clinical development. *Adv. Drug Deliv. Rev.* **2016**, *103*, 33–56. [CrossRef]

124. Gunawardana, M.; Baum, M.M.; Smith, T.J.; Moss, J.A. An intravaginal ring for the sustained delivery of antibodies. *J. Pharm. Sci.* **2014**, *103*, 3611–3620. [CrossRef]
125. McBride, J.W.; Boyd, P.; Dias, N.; Cameron, D.; Offord, R.E.; Hartley, O.; Kett, V.L.; Malcolm, R.K. Vaginal rings with exposed cores for sustained delivery of the HIV CCR5 inhibitor 5P12-RANTES. *J. Control. Release* **2019**, *298*, 1–11. [CrossRef]
126. Carballo-Dieguez, A.; Giguere, R.; Lentz, C.; Dolezal, C.; Fuchs, E.J.; Hendrix, C.W. Rectal douching practices associated with anal intercourse: Implications for the development of a behaviorally congruent HIV-prevention rectal microbicide douche. *AIDS Behav.* **2018**. [CrossRef]
127. Dezzutti, C.S.; Rohan, L.C.; Wang, L.; Uranker, K.; Shetler, C.; Cost, M.; Lynam, J.D.; Friend, D. Reformulated tenofovir gel for use as a dual compartment microbicide. *J. Antimicrob. Chemother.* **2012**, *67*, 2139–2142. [CrossRef]
128. U.S. Food & Drug Administration. *Guidance for Industry. Considering Whether an FDA-Regulated Product Involves the Application of Nanotechnology*; U.S. Food & Drug Administration: Silver Spring, MD, USA, 2014. Available online: https://www.fda.gov/downloads/RegulatoryInformation/Guidances/UCM401695.pdf (accessed on 22 February 2019).
129. U.S. Food & Drug Administration. *Guidance for Industry. Drug Products, Including Biological Products, that Contain Nanomaterials*; U.S. Food & Drug Administration: Silver Spring, MD, USA, 2017. Available online: https://www.fda.gov/downloads/Drugs/GuidanceComplianceRegulatoryInformation/Guidances/UCM588857.pdf (accessed on 22 February 2019).
130. Pita, R.; Ehmann, F.; Papaluca, M. Nanomedicines in the EU-Regulatory overview. *AAPS J.* **2016**, *18*, 1576–1582. [CrossRef]
131. U.S. Food & Drug Administration. *Guidance for Industry. Vaginal Microbicides: Development for the Prevention of HIV Infection*; U.S. Food & Drug Administration: Silver Spring, MD, USA, 2014. Available online: https://www.fda.gov/downloads/drugs/guidances/ucm328842.pdf (accessed on 15 February 2019).

pharmaceutics

MDPI

Review

Relating Advanced Electrospun Fiber Architectures to the Temporal Release of Active Agents to Meet the Needs of Next-Generation Intravaginal Delivery Applications

Kevin M. Tyo [1,2,†], Farnaz Minooei [3,†], Keegan C. Curry [4], Sarah M. NeCamp [5], Danielle L. Graves [5], Joel R. Fried [3] and Jill M. Steinbach-Rankins [1,2,5,6,*]

1 Department of Pharmacology and Toxicology, School of Medicine, University of Louisville, Louisville, KY 40202, USA; kevin.tyo@louisville.edu
2 Center for Predictive Medicine, Louisville, KY 40202, USA
3 Department of Chemical Engineering, University of Louisville, Louisville, KY 40292, USA; farnaz.minooei@louisville.edu (F.M.); joel.fried@louisville.edu (J.R.F.)
4 Department of Biology, University of Louisville, Louisville, KY 40292, USA; keegan.curry@louisville.edu
5 Department of Bioengineering, Speed School of Engineering, University of Louisville, Louisville, KY 40292, USA; sarah.necamp@louisville.edu (S.M.N.); danielle.graves@louisville.edu (D.L.G.)
6 Department of Microbiology and Immunology, School of Medicine, University of Louisville, Louisville, KY 40292, USA
* Correspondence: jmstei01@louisville.edu; Tel.: +1-502-852-5486
† Co-first authors.

Received: 7 March 2019; Accepted: 30 March 2019; Published: 3 April 2019

Abstract: Electrospun fibers have emerged as a relatively new delivery platform to improve active agent retention and delivery for intravaginal applications. While uniaxial fibers have been explored in a variety of applications including intravaginal delivery, the consideration of more advanced fiber architectures may offer new options to improve delivery to the female reproductive tract. In this review, we summarize the advancements of electrospun coaxial, multilayered, and nanoparticle-fiber architectures utilized in other applications and discuss how different material combinations within these architectures provide varied durations of release, here categorized as either transient (within 24 h), short-term (24 h to one week), or sustained (beyond one week). We seek to systematically relate material type and fiber architecture to active agent release kinetics. Last, we explore how lessons derived from these architectures may be applied to address the needs of future intravaginal delivery platforms for a given prophylactic or therapeutic application. The overall goal of this review is to provide a summary of different fiber architectures that have been useful for active agent delivery and to provide guidelines for the development of new formulations that exhibit release kinetics relevant to the time frames and the diversity of active agents needed in next-generation multipurpose applications.

Keywords: electrospun fibers; fiber architecture; drug delivery; intravaginal delivery; delivery vehicle

1. Introduction

Intravaginal delivery is an effective strategy to improve the localization of antiviral, antibacterial, antifungal, chemotherapeutic, and contraceptive agents within the female reproductive tract (FRT) [1,2]. Relative to oral administration routes, intravaginal delivery localizes agents to the FRT, avoiding both the harsh gastrointestinal environment and hepatic first pass effect. This results in an increase in drug bioavailability within target tissue and corresponding functional activity by decreasing off-target effects and systemic exposure [3]. The inherent characteristics of the FRT, including its large surface area and low enzymatic activity, additionally make the FRT a favorable site for localized active agent administration and targeting [4,5].

Although intravaginal delivery offers a variety of advantages to enhance the delivery of active agents [6], challenges unique to the FRT must be overcome to provide efficacious prophylaxis and treatment. One of the most important components of the FRT is the mucus layer, which protects the epithelium and lamina propria from incoming pathogens (Figure 1). However, it can also act as a barrier, impeding therapeutic transport to underlying epithelial and immune cells [7,8]. In addition to these challenges, the frequent shedding and production of cervicovaginal mucus can decrease active agent retention, while bacterial flora, enzymes, and the acidic environment created by beneficial bacteria can contribute to metabolization and degradation of active agents, reducing efficaciousness.

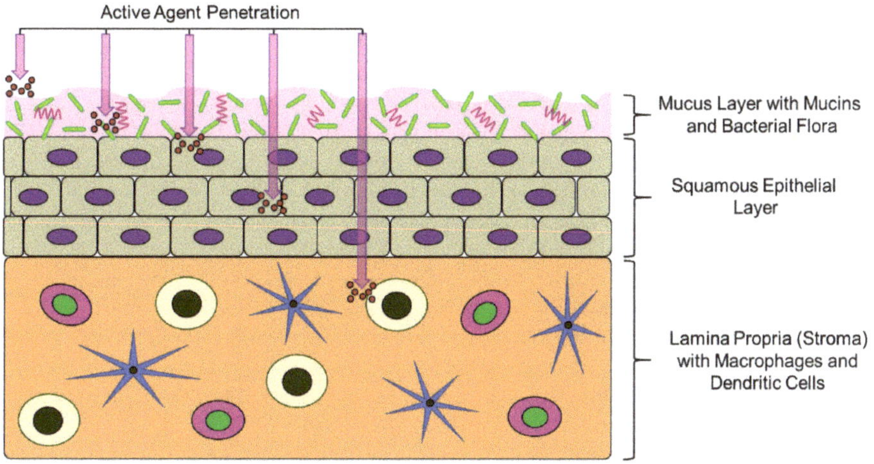

Figure 1. Schematic depicting the structure and specific layers of the vaginal mucosa that can act as a barrier to active agent transport (not to scale). The mucus layer of the female reproductive tract (FRT) frequently sheds and can immobilize active agents (shown in red), leading to decreased efficacy of the administered agents. The bacterial flora normally present within the FRT can also metabolize and degrade agents, further contributing to decreased efficacy. Last, the squamous epithelium can hinder transport to underlying immune cells present near the epithelial surface and/or in the lamina propria.

To address these challenges, intravaginal delivery platforms have been formulated as solid or semi-solid dosage forms that include suppositories, tablets, capsules, gels, rings, and creams to enhance delivery to and retention in the FRT [9–13]. While these dosage forms have enabled high levels of active agent incorporation and localization, these traditionally used delivery platforms still face significant challenges, including difficulty of self-administration, economic feasibility, poor user-compliance, vaginal irritation, the need for frequent administration, and low residence times [14]. Of these platforms, intravaginal rings have provided the "gold standard" for long-term delivery due to their ability to sustain the release of one or multiple active agents for weeks to months, avoid leakage and loss of active agent, and improve drug stability [15–19]. However, some biological agents have difficulty withstanding the high temperature and solvent processes often required for fabrication, limiting their incorporation [20].

More recently, nanoparticles (NPs) have been developed for topical intravaginal delivery due to their ability to encapsulate both hydrophilic and hydrophobic agents and to provide encapsulant stability while enhancing cell specific targeting, transport, and internalization [21–29]. However, NPs can experience low intravaginal retention due to mucus shedding, or conversely experience immobilization within the mucus layer, resulting in inadequate transport to underlying tissue [29]. To improve retention and to maximize transport, NPs have been surface-modified [30,31], while carrier solutions with different osmolarities have been explored to increase retention within and penetration of the vaginal lumen [32–34]. Despite these efforts, hurdles including low encapsulation efficiency and rapid release of hydrophilic agents have hindered the ability to achieve long-term delivery and retention [35]. Given these issues, other delivery platforms have been investigated that may increase the longevity of active agents within the FRT and improve user adherence while also offering a new dosage form alternative to women.

Electrospun fibers have recently gained attention for intravaginal delivery due to their ease of use, ability to be fabricated into various geometries and sizes, and tunable release properties [36,37]. They have been considered for sustained-delivery, a characteristic that is often desirable for intravaginal applications, due to their high surface area-to-volume ratio, degree of interconnected porosity, tunable pore sizes, surface-modification potential, interchangeable polymer options, and diverse fiber architectures that enable finer control over the rate, duration, and site of agent release [38]. Electrospun fibers have the additional advantage that they can be fabricated using a variety of natural or synthetic polymers to tailor release properties [39], and these polymer types are typically selected based on their biocompatibility, hydrophobicity, and related degradation properties.

One of the most significant factors that contributes to active agent release from fibers is the relative hydrophobicity of the selected polymer material [40,41]. In addition to polymer hydrophobicity, the medium (in vitro) or environment (in vivo) surrounding the fiber can impact drug release. Simulated vaginal and seminal fluids, often used to preliminarily assess intravaginal release, may alter the release of agents relative to testing in water or phosphate buffered saline (PBS) (in vitro) or in vivo, due to differences in viscosity, salt, and protein concentrations, as well as pH. Therefore, depending on the degree of polymer hydrophobicity and the environment release it is tested in, the same encapsulated active agent can have distinctly different release profiles, in some cases ranging from hours to months [42,43]. Usually, independent of these conditions, the use of hydrophilic polymers often results in the immediate release of both hydrophilic and hydrophobic active agents due to the high solubility and degradation rate of hydrophilic polymers in aqueous environments [44]. Natural polymers such as collagen, gelatin, chitosan, elastin, and laminin, and synthetic polymers including poly(ethylene oxide) (PEO), polyvinyl alcohol (PVA), and polyvinylpyrrolidone (PVP) are examples of hydrophilic materials that have been fabricated into fibers with micron- and nanometer-scaled properties. In contrast, synthetic hydrophobic polymers including polycaprolactone (PCL), poly(lactic-*co*-glycolic acid) (PLGA), and polyurethane (PU) have demonstrated burst or sustained-release kinetics depending on the hydrophobicity of the incorporated active agent [45–49]. Moreover, synthetic hydrophobic polymers can also serve as a mechanical and structural basis for different fiber architectures in which the release of single or multiple encapsulants may be tailored by using more complex fiber designs or composites. Fiber release rates can also be optimized by adjusting the polymer molecular weight or hydrophilicity, for example, by adding hydrophilic groups such as aliphatic poly(phosphoester) to the polymer structure [50]. Together, these features have enabled the incorporation and release of a variety of antiviral, antimicrobial, and biological agents from fiber scaffolds [51–53].

Active agent release from polymeric fibers typically occurs via diffusion, polymer degradation, and erosion [41]. When fibers are first administered, solvent or solution diffuses through the porous fiber matrix. Once in contact with the solvent or solution, the polymer matrix swells, loosening polymer chains and enabling the diffusion of active agents, dependent in part on molecular size. Concurrently, the fiber surface may undergo bulk erosion at a rate corresponding to polymer hydrophilicity. These features in combination with the large surface-to-volume ratio of the fibers allows for the increased diffusion of encapsulants relative to diffusion from non-porous bulk materials [54]. Traditionally, fibers have been electrospun as uniaxial fibers or fibers that comprise a single polymer or polymer blend and exhibit homogeneous morphology. Diffusion of active agents from more traditional uniaxial fibers is dependent upon the compatibility of the encapsulant, polymer, and surrounding eluant. In contrast with diffusion, polymer degradation is observed when fibers are exposed to aqueous environments, and polymer bonds are cleaved by either passive hydrolysis or enzymatic reaction [55], resulting in slow degradation of the fiber scaffold. This degradation alters the distance between and size of interconnected pores, thereby impacting the diffusion and release of incorporated active agents. For most synthetic polymers, hydrolysis is the most common mechanism of degradation, although hydrolysis-resistant polymers have been utilized [56], which significantly impact active agent release. As the fibers degrade, they can also undergo surface or bulk erosion, which is dependent upon solvent diffusivity into the fiber, polymer solubility, and overall fiber matrix dimensions [57].

As a result of these mechanisms and the materials selected, electrospun fibers can tailor the release of encapsulated agents within different durations to achieve immediate (transient or rapid), short-term, or sustained-release. Within this review, we defined release as transient, when the complete release of active agents occurs within 24 h of administration; short-term, when the release occurs from one day to one week; or sustained, when the release of the active agent occurs over a duration of weeks to months. A schematic showing an example of these different potential release profiles is provided in Figure 2. Factors including the electrospinning parameters, polymer materials, fiber architecture, the resulting structure and morphology, and the distribution and amount of incorporated active agent each contribute to the resulting release kinetics and efficacy of delivery [37].

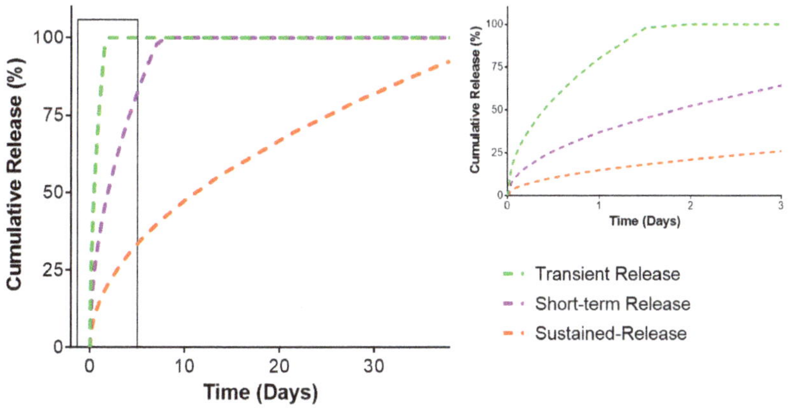

Figure 2. Schematic depicting examples of transient, short-term, and sustained-release profiles.

Traditional uniaxial electrospun fibers in which each individual fiber is composed of a single cohesive polymer layer were the first fiber architectures to be fabricated [58] and have been utilized in a variety of drug delivery applications over the past decade [36,37,59–61]. While uniaxial fibers offer high encapsulation efficiencies, cost-effectiveness, and ease of use, they have suffered from burst release and challenges in tailoring release properties [37,42,62]. These challenges are most evident in achieving the sustained-release of hydrophilic agents, often necessitating hydrophilic polymers to attain high encapsulation efficacy as well as hydrophobic polymers for sustained-release. More complex fiber architectures offer alternative options to address these limitations by combining different polymer types in distinct layers to modulate the release.

While the release characteristics of traditional uniaxial electrospun fibers have been thoroughly reviewed in literature [63–68], to our knowledge, there has not yet been a review of the more advanced fiber architectures used to deliver active agents, nor a review that considers the impact these architectures may have on intravaginal delivery applications. Here, we seek to provide an overview of different polymer architectures including coaxial, multilayered, and nanoparticle-fiber composites (Figure 3) as a function of the materials used to construct these architectures that have been utilized in a diversity of health applications. We seek to present different material combinations in these architectures to systematically relate material type and fiber architecture to active agent release kinetics. Last, we explore how lessons derived from these different architectures might be applied in the context of intravaginal delivery to address the needs of future topical sustained-release platforms for a given prophylactic or therapeutic application. The overall goal of this review is to provide a summary of different fiber architectures that have been useful for active agent delivery and to provide guidelines for the development of new formulations based on the knowledge obtained from previous work across other applications. While some of these more complex architectures have only recently been investigated relative to uniaxial fibers, they have demonstrated promise in enabling greater tunability of release and may be useful to apply as new dosage forms for intravaginal delivery and other similar applications.

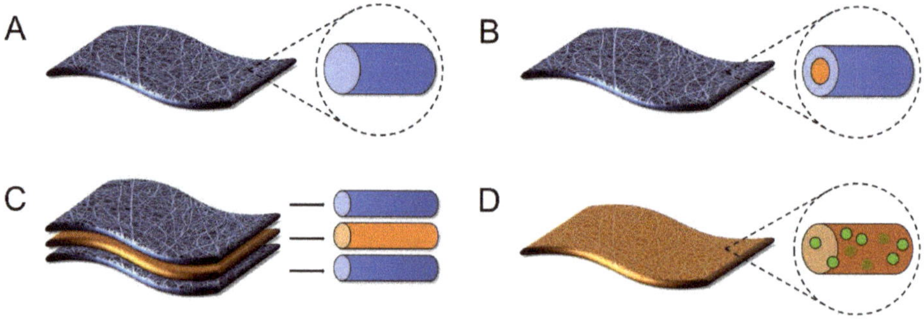

Figure 3. Schematic of different electrospun fiber composites. Diagrams representing (**A**) traditional uniaxial fibers, (**B**) coaxial fibers, (**C**) multilayered fibers, and (**D**) nanoparticle-fiber composites. (**A**) Uniaxial fibers are comprised of a single polymer or polymer blend (shown in blue) that is distributed homogenously throughout the fiber structure. (**B**) In contrast, coaxial fibers contain both core (orange) and shell (blue) layers that are chemically distinct. (**C**) Multilayered fibers result from sequentially electrospinning different fiber layers together or integrating individual layers post-fabrication. (**D**) Finally, nanoparticle-fiber composites consist of hydrophilic or hydrophobic fibers (orange) that encapsulate nanoparticles (green).

2. Coaxial Electrospun Fibers

2.1. Coaxial Architectures and Properties

Coaxial electrospinning, adapted from uniaxial or single axial electrospinning, provides a multicomponent fiber scaffold that easily allows the tunable release of active agents [69,70]. Coaxial fibers are usually comprised of two parts, an outer protective layer or shell and an inner layer or core [71], where encapsulants are typically localized (Figures 3B and 4). Coaxial fibers can provide several advantages relative to uniaxially spun fibers. First, electrospinning the core and shell polymer solutions simultaneously through a coaxial spinneret allows for the design of unique fiber architectures. The thickness and ratios of the core and shell layers can be modulated, providing more reproducible fiber properties with a greater ability to alter encapsulant release relative to other fabrication methods. Additionally, coaxial electrospinning ensures that the active agent in the core phase is protected within harsh physiological environments, such as the female reproductive tract [53]. Furthermore, a variety of materials can be used as either the core or shell to finely regulate encapsulant release (Figure 4) [69,72].

Despite these advantages, the added complexity of simultaneously electrospinning two or more polymer phases and the additional interactions between the core and shell solutions requires additional optimization relative to uniaxial electrospinning in terms of selecting compatible polymers and solvents. In addition to the core-shell architecture itself, the release profiles of active agents from coaxial fibers are impacted by solvent choice, polymer-solvent miscibility, the miscibility between core and shell solvents/solutions, solvent volatility, and layer thicknesses [73,74]. Solvent choice has been shown to alter fiber diameter and structure [75], thereby impacting active agent release [76]. Additionally, miscible core and shell solvents/solutions may lead to the partial dissolution of core encapsulants in the shell, whereas, immiscible core and shell solvents may promote material delamination at the core-shell interface, facilitating burst release of the core encapsulant. Therefore, the polymers and solvents for both core and shell layers must be selected based on their individual properties as well as their anticipated interactions [77,78]. In addition, solvent volatility and evaporation rate can affect the distribution and subsequent release of active agents, while the thickness of the polymer shell, polymer composition, and spinning conditions influence encapsulant diffusion rates [79]. Here, we discuss coaxial fibers as a function of their core-shell design, composition, and incorporated active agents to help relate these considerations to the resulting transient, short-term, or sustained-release characteristics.

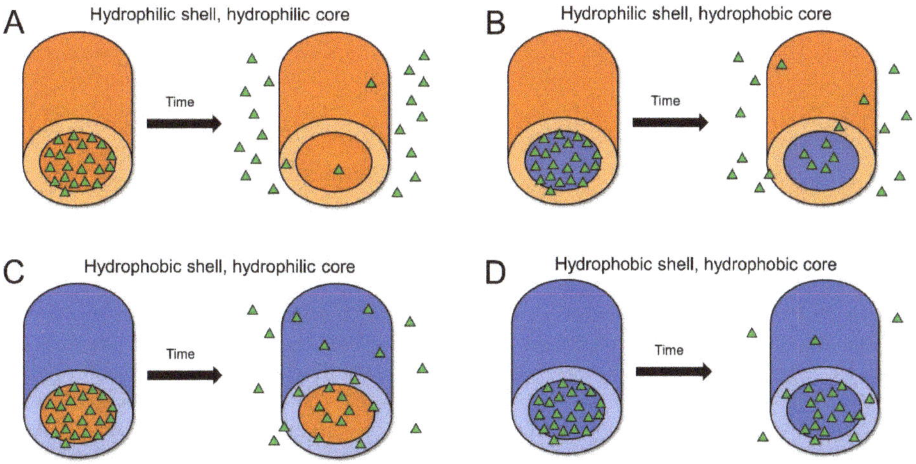

Figure 4. Schematic of anticipated release profiles from different coaxial fiber architectures. Generally, the release of encapsulants from coaxial fibers is dependent on the core and shell hydrophobicity. The release of active agents from coaxial fibers with (**A**) hydrophilic core and shell, (**B**) hydrophobic core and hydrophilic shell, (**C**) hydrophilic core and hydrophobic shell, and (**D**) hydrophobic core and shell are shown. Hydrophilic polymers (shown in orange) typically promote transient release, while more hydrophobic polymers (blue) are typically used to provide short-term or sustained-release.

2.2. *Release Kinetics from Coaxial Fibers*

2.2.1. Transient Release (within 24 h)

Hydrophobic Shell—Hydrophilic Core

Electrospun fibers can be designed to release the active agent immediately or within 24 h of administration if a rapid onset of action is needed for a given application [80]. Moreover, multiple active agents can be incorporated into different layers of a coaxial fiber (core or shell) to provide transient release.

For application to infectious diseases, coaxially spun fibers that demonstrate burst release followed by lower levels of short-term release may provide on-demand protection against incoming pathogens, increasing the immediate efficaciousness of agents by releasing initially high (burst) concentrations. This type of release can be achieved by employing coaxial fibers comprised of hydrophobic shells and hydrophilic cores. In one study, coaxial and triaxial fiber multi-drug delivery platforms that used PCL as the outermost shell released ~15% and ~80% of two different hydrophilic dyes, keyacid blue and keyacid uranine (KAB and KAU), from the PVP core and PCL shell fibers, respectively, within one hour [69]. In both the coaxial and triaxial fibers, the PVP core containing KAB was protected by the surrounding PCL layer containing KAU, which helped to extend the release of the remaining

KAB to 24 h. For the triaxial fibers, a blank PCL layer was electrospun between the outer PCL shell and the inner PEO core. In both the coaxial and triaxial fibers, KAU was released from the shell within 3 h; however, the triaxial fibers better modulated the release of KAB from the core, releasing 50% less during the first hour. The burst release of the KAU dye, observed from both coaxial and triaxial fibers, was attributed to water penetrating the porous fiber shell, allowing transient release. In another example of coaxial fiber design, water-soluble PVP was used as a core with a hydrophobic ethyl cellulose (EC) shell to encapsulate hydrophobic compounds of either quercetin or ketoprofen. Using this architecture, ~75% of both hydrophobic encapsulants were released within 24 h.

Hydrophilic Shell—Hydrophobic or Hydrophilic Cores

Similarly, coaxial fibers that have hydrophilic shells can facilitate the rapid release of encapsulated agents with an initial burst release of 1 to 4 h followed by continued transient release within 24 h of administration. One architecture that has been adopted to achieve rapid- or on-demand release from coaxial fibers is a hydrophilic shell in combination with a hydrophilic or hydrophobic core. In one study, zein-PVP core-shell fibers were developed that incorporated the active agent in both the core (zein) and the shell (PVP) layers [81]. Zein, a natural, moderately hydrophobic polymer was used to achieve immediate and transient release of the hydrophobic drug, ketoprofen. A burst release of 43% was observed within the first hour, followed by transient release of the remaining ketoprofen over 10 h. The initial burst release was correlated with rapid dissolution of the hydrophilic shell, while the more transient 10 h release was attributed to the hydrophobic core. In another study, the release profile of a hydrophobic drug, asiaticoside, was compared between coaxial fibers composed of chitosan cores with either a hydrophilic alginate and PVA-blended polymer shell or a hydrophobic centella triterpenes cream shell [82]. The coaxial fiber with the alginate-PVA shell demonstrated 80% more asiaticoside release relative to the centella control within 10 h, which was attributed to the shell hydrophilicity [82]. Additionally, the trend of burst release followed by more gradual transient release was attributed to rapid degradation of the alginate-PVA shell, followed by subsequent degradation of the chitosan core. While this example incorporated a polymer blend (alginate-PVA) as the hydrophilic shell, to be considered a core-shell structure, it should be noted that the material itself needs to be electrospinnable without other polymers. As this example demonstrates, hydrophilic polymers such as PVP, PVA, or PEO can be electrospun alone or in blends to create hydrophilic core and shell layers.

Core-Shell Architectures with Similar Core-Shell Hydrophobicity

Coaxial fibers comprised of both hydrophilic core and shell layers have also been investigated to provide transient release of active agents. For example, coaxial fibers fabricated with a hydrophilic PVP shell and hydrophilic cellulose acetate core were investigated. These coaxial fibers with both a hydrophilic core and shell released 31% of their hydrophobic encapsulant (epicatechin) within 10 min, followed by 80% release after 4 h [83].

In addition to the utilization of materials with similar hydrophobicities, coaxial fibers consisting of identical core-shell materials have been fabricated to provide the rapid release of active agents. In one study, fibers with PVP shells and cores were investigated to provide rapid release of the hydrophobic drug, quercetin. The PVP shell-PVP core fibers released quercetin within one minute [84], and this burst release was similarly observed in a separate study that used the same fiber formulation to deliver acyclovir [85]. In another study, the hydrophobic antibiotic, allyltriphenylphosphonium bromide, was incorporated within the core of coaxial fibers, and the volumetric ratios of core-shell solutions were varied to study release. Fibers comprised of zein-zein with core-shell volume ratios greater than 1:2 were found to suppress the burst release of the antibiotic, only releasing 15% within the first hour. In contrast, 35% and 45% of the antibiotic were released from fibers with a 1:1 core:shell volumetric ratio or blended fiber controls over the same duration [86]. In a separate study, a triaxial fiber in which all three layers were comprised of ethyl cellulose provided zero-order release of ketoprofen over 20 h due to the gradual increase in the drug content moving from shell to the core [87]. These studies highlight the role of the active agent distribution within the fiber layers, suggesting that encapsulant localization within the fiber core may enhance release.

Finally, the release of fluorescently labeled bovine serum albumin (BSA) from core-shell hydrogel nanofilaments composed of a poly(lactide-*co*-ε-caprolactone) (PLCL) shell and *N*,*N*-isopropylacrylamide (NIPAAm)/*N*,*N*′-methylene bisacrylamide crosslinked core was studied. The crosslinker, *N*,*N*′-methylene bisacrylamide, was used to polymerize NIPAAm during the electrospinning process. This study showed that by changing the NIPAAm-crosslinker (*w*/*w*) ratio from 4:1 to 37:1, the release of BSA increased from 0.15 to 0.7 ug/mg over 24 h. However, in the absence of a hydrogel within the core, BSA showed nearly complete release over the same duration. This study demonstrated that the mechanical and corresponding drug release properties could be more finely tailored by altering the NIPAAm-crosslinker (*w*/*w*) ratio [88].

Stimuli-Responsive Coaxial Architectures

Another method to modulate the release of active agents from coaxial fibers is to integrate stimuli-responsive layers to precisely release agents in response to surrounding physiological conditions [89]. Unlike stimuli-responsive uniaxial fibers, the more complex interactions between the core and shell layers in coaxial fibers can provide increased control of active agent release via pH- or other stimuli-based mechanisms. A variety of natural and synthetic materials have been investigated for their use in pH-responsive applications. In one example, a coaxial fiber comprised of a lecithin-diclofenac sodium core and a Eudragit S100 shell provided the pH-responsive release of ferulic acid for 10 h [90]. Ferulic acid release was facilitated under conditions of neutral pH (pH 7), with minimal release occurring in a more acidic (pH 2) environment. Another pH-sensitive polymethacrylate-based copolymer [90–92], Eudragit EPO, was used to fabricate pH-responsive antibacterial fibers. Here, Eudragit EPO cores, which dissolve below pH 5, were used in combination with Eudragit L100 shells, which dissolve at a pH greater than 6. These coaxial fibers provided pH-responsive release for an hour under slightly acidic conditions (pH 6) while demonstrating attenuated release in very acidic conditions (pH 2) [93]. Additionally, two separate studies investigated coaxial fibers comprised of Eudragit S100 shells and PEO cores to stimulate pH-responsive release within the gastrointestinal tract [94,95]. In both studies, the release of hydrophobic indomethacin and hydrophilic mebeverine hydrochloride agents was minimal (~10%) after 2 h under acidic conditions, followed by rapid release for 6 h when switched to neutral conditions (pH 7.4). Coaxial fibers comprised of cellulose acetate phthalate shells with polyurethane cores, as well as gelatin-sodium bicarbonate shells with PLCL cores have also been used to provide similarly rapid pH-responsive release of ciprofloxacin and rhodamine B (Rhd B). These studies demonstrated the potential of coaxial fibers as pH-sensitive delivery systems [96,97].

Coaxial fibers with other stimuli-responsive properties have been investigated for on-demand, rapid release applications. Although studies with other stimuli-responsive systems have been limited, one study investigated the use of self-immolative polymers, or polymers that depolymerize when exposed to specific external stimuli, for rapid stimuli-responsive release [98]. In this study, self-immolative fibers comprised of dibutyltin dilaurate and phenyl (4-(hydroxymethyl)phenyl) carbamate were blended with polyacrylonitrile and used as shells to surround PVP cores. The fibers provided minimal release of KAB dye when incubated in water; however, the fibers depolymerized when exposed to trifluoroacetic acid, resulting in zero-order release of ~40% dye within a week.

2.2.2. Short-Term Release (One Day to One Week)

Hydrophobic Shell—Hydrophilic Core

A key advantage of short-term release specifically for intravaginal delivery is that the burden of frequent or daily administration may decrease, thereby increasing user adherence of prophylactics and therapeutics. Traditionally, hydrophobic materials have been well-suited to provide longer durations of release (depending on the encapsulant) due to their decreased degradation rates in aqueous environments. For more traditional uniaxial hydrophobic fiber platforms, most hydrophobic small molecule drugs or larger macromolecules achieve release for up to one week due to the similar hydrophobic properties of the polymer and encapsulant [6]. This compatibility allows for hydrophobic encapsulants to partition more evenly within and distribute throughout hydrophobic polymers. However, hydrophilic agents, which have low solubility in nonpolar polymers, often partition to the fiber surface, resulting in burst release and suboptimal short-term and/or sustained-release properties. To address this challenge, coaxial fibers in which hydrophilic agents are encapsulated within a hydrophilic core and surrounded by a protective hydrophobic shell can prolong and adjust the release of hydrophilic molecules.

The use of coaxial fibers with hydrophobic shells and hydrophilic cores has been shown to extend the release of many encapsulants [71,99,100]. In one study, a coaxial fiber comprised of a hydrophobic ethyl cellulose shell with a hydrophilic PVP core was investigated for short-term release. These fibers released maraviroc over a duration of hours to days depending on the thickness of the hydrophobic shell, which was modulated via flow rate and total electrospun volume. The increased thickness of the hydrophobic shell extended encapsulant release from 24 h to five days by increasing the shell-to-core volume ratio from 0.5 to 4 [99]. In another study, a PCL fiber shell surrounding a PVP-graphene oxide blended core was studied. These fibers released 65% of hydrophilic vancomycin hydrochloride within 4 h and attained full release of vancomycin after 96 h [101]. Although this coaxial fiber provided short-term release, the long-term safety of graphene oxide within the FRT is unknown, and further studies are required to assess its safety in intravaginal delivery applications. Finally, a coaxial fiber composed of a synthetic hydrophilic poly-cyclodextrin core and hydrophobic poly(methacrylic acid) shell reduced the burst release of a hydrophilic drug, propranolol hydrochloride, by 50%, and extended release to 180 h relative to the 140 hour release obtained from uniaxial fibers [102].

Hydrophobic Shell—Hydrophobic Core

In addition to the widely used hydrophobic shell-hydrophilic core coaxial architectures, the use of hydrophobic materials in both the core and the shell layers has also been investigated to provide the short-term release of active agents. In one study, a PCL core surrounded by an outer PCL shell was used to prolong the release of the antibiotic ampicillin. Ampicillin, a hydrophilic compound, normally localizes to the surface of PCL when spun as a uniaxial fiber, resulting in burst release [103]. As an alternative, a 4% (*w*/*v*) PCL solution was used to fabricate an ultra-thin shell to delay release. In addition, the parameters for coaxial electrospinning were modified using dilute sheath solutions to improve the control of fiber diameter and morphology. The resulting coaxial fiber efficiently encapsulated ampicillin and provided short-term release for ~80 h [103]. In another study, coaxial fibers comprised of a zein shell with a PCL core reduced the burst release of the hydrophilic antibiotic, metronidazole, achieving short-term release for more than four days [78].

Stimuli-Responsive Coaxial Architectures

Coaxial fibers exhibiting stimuli-responsive properties have also been investigated to provide short-term release of active agents. As one example, poly(*N*-isopropylacrylamide), a thermoresponsive polymer, was used as a core layer in combination with an ethyl cellulose and anhydrous ethanol shell solution. At room temperature, poly(*N*-isopropylacrylamide) exhibits hydrophilic properties; however, at temperatures above 32 °C, the polymer demonstrates more hydrophobic characteristics. At room temperature and after 55 h, the fibers released 65% of ketoprofen in PBS, while only 40% of the same drug was released at 37 °C [104].

Blended Polymers in Coaxial Architectures

Another method of prolonging release is to use blended polymers to formulate coaxial fibers, which can decrease fiber wettability. One study combined gelatin, a natural hydrophilic protein, with the hydrophobic polymer, PCL, to create coaxial fibers with increased hydrophobicity and mechanical stability relative to gelatin alone [105]. In one study, the release of hydrophilic doxycycline was measured from three different fiber architectures—a uniaxial PCL-gelatin blended fiber, coaxial fibers with three different cores (PCL, gelatin, or a PCL-gelatin blend) and a PCL-gelatin blended shell, and a triaxial fiber with both a PCL-gelatin blended core and outer shell and an intermediate gelatin layer. Among these five designs, uniaxial PCL-gelatin blended fibers released the most doxycycline within 24 h (90%), while coaxial fibers with a PCL-gelatin core and shell released the least (50%). Additionally, only coaxial fibers with either a PCL-gelatin or gelatin core prolonged release over five days. Furthermore, the other architectures including the uniaxial PCL-gelatin blend, coaxial fiber with PCL core, and triaxial fibers failed to release doxycycline for more than 30 h. The burst release observed in fibers with PCL cores was attributed to the lack of compatibility between the hydrophobic PCL cores and hydrophilic encapsulant, which caused doxycycline to localize on the core surface. Additionally, the subsequent suboptimal encapsulant release was attributed to low water penetration into the hydrophobic core. These studies demonstrate that utilization of both hydrophobic and hydrophilic polymers alone or as blends can modulate the short-term release of hydrophilic encapsulants due to the variation in the permeability of different layers and core-encapsulant interactions.

2.2.3. Sustained-Release (One Week to Multiple Months)

Hydrophobic Shell—Hydrophilic Core

Similar to fibers that provide short-term release, fibers designed for sustained-release commonly use hydrophobic polymers as the outer shell to prevent the fiber from undergoing rapid hydrolysis. Studies have demonstrated that the most promising coaxial architecture to achieve sustained-delivery utilizes a hydrophobic shell and hydrophilic core [6]. A polymer that is frequently used in coaxial fibers to provide sustained-release is poly(lactic-*co*-glycolic acid) (PLGA). In one study, a coaxial fiber composed of a PLGA shell was used to shield a hydrophilic core consisting of tragacanth gum. The encapsulant, tetracycline hydrochloride, served as a model hydrophilic agent. Investigators observed that PLGA (shell)-tragacanth gum (core) coaxial fibers diminished burst release and provided sustained-release of tetracycline hydrochloride for 75 days, releasing 68% of tetracycline hydrochloride during this period [106]. In another study, a PLGA (shell)-polyethylenimine (PEI, core) architecture was used to prolong the release and stability of bone morphogenetic protein-2 plasmid (pBMP2-2). The hydrophilic PEI core was used to encapsulate and retain the bioactivity of pBMP2-2, while the hydrophobic PLGA shell was used as a protective barrier to prolong release. When compared to uniaxial PLGA-PEI blended fibers, the PLGA (shell)-PEI (core) coaxial fiber exhibited both improved bioactivity and prolonged release of the pBMP2-2 plasmid. The coaxial fiber released 80% of the plasmid over 20 days, while the uniaxial fibers released the same amount over seven days [107].

Polymers other than PLGA have been used as hydrophobic shells to sustain the release of active agents from coaxial fibers. One study formulated coaxial fibers containing a hydrophilic dextran core and hydrophobic PCL shell. The addition of polyethylene glycol (PEG) to the PCL shell increased the release of the encapsulated BSA by forming pores in the shell layer. Although all fibers released ~20% BSA within the first 24 h, increasing the PEG concentration increased the amount of BSA released over extended durations. Interestingly, all fibers demonstrated sustained-release regardless of PEG concentration; coaxial fibers fabricated with 5% PEG shells released ~60% BSA, while fibers containing 40% PEG shells released 90% BSA over 27 days [108]. In another study, the relationship between PEG (core):PCL (shell) molar ratio and the release of BSA or lysozyme was investigated. The thinnest shell layers with a core:shell molar ratio of 1.59 and a core flow rate of 2 mL/h provided complete release of both encapsulants within 24 days, compared to only 50% release from thicker fibers with a core:shell molar ratio of 0.32 and a core flow rate of 0.6 mL/h. Moreover, the fibers preserved the bioactivity of lysozyme and released BSA over 29 days, with no noticeable differences between BSA and lysozyme release rates [109]. In addition to conventional coaxial spinning, the use of emulsion electrospinning has also been investigated to fabricate coaxial fibers, which can be electrospun using a uniaxial spinneret [70]. One study that used emulsion electrospinning fabricated core-shell fibers composed of a PEG-poly(D,L-lactic acid) shell and methyl cellulose core to minimize the burst release of lysozyme [110]. The release of lysozyme from the core was achieved over 15 days and was dependent on the percent of lysozyme loaded, while the structural integrity and bioactivity of lysozyme was protected by the shell. A later study compared these same coaxial fibers to blended uniaxial fibers composed of PCL and PEG and showed that the coaxial fibers improved sustained-release by releasing ~50% of BSA over 35 days relative to blended fibers, which released ~75% BSA [111].

Another study explored the effects of multiple processing parameters, including PEG and PCL concentrations, PEG molecular weight, encapsulant concentration, and fiber diameter, in modulating the release of plasmid DNA (pDNA). Plasmid DNA was encapsulated in a PEI core, and a non-viral gene delivery vector (r-PEI-HA) was incorporated within a PCL shell [112]. An increase in fiber diameter was observed with an increase in all of the three other parameters, while the loading and release of r-PEI-HA were correlated to pDNA concentration in the fiber core and PEG molecular weight. The fibers formulated with high PEG molecular weight and low pDNA concentration exhibited ~30% release of r-PEI-HA over 60 days, while the fibers with high pDNA concentration and low molecular weight PEG completely released pDNA within 60 days.

Core-Shell Architectures with the Same Core-Shell Hydrophobicity

Although coaxial architectures with similar core and shell hydrophobicities have been utilized to obtain transient and short-term release, coaxial fibers that use the same materials have been less frequently investigated to provide sustained-release. In one study, PLGA was utilized in both the core and shell layers to investigate the effect on vancomycin and ceftazidime delivery [113]. Both hydrophilic drugs were encapsulated within the core PLGA layer and exhibited similar burst release kinetics within the first day, followed by a second phase of more gradual release over five to ten days. Ninety percent of the antibiotics were released after 11 days, followed by complete release after 25 days, with the more gradual release attributed to the PLGA barrier layer.

2.3. *Applications for Intravaginal Delivery*

The enhanced tunability and versatility provided by the core and shell layers of coaxial fibers make them excellent candidates for intravaginal delivery applications. While uniaxial fibers have been studied for sustained- and stimuli-responsive release of active agents in the FRT [6,114–119], they have faced challenges in providing the sustained-release of therapeutically relevant concentrations of individual active agents and effectively modulating the release of multiple agents core [6]. Often, compatibility between the polymer and encapsulant can pose challenges to achieving sustained-release with uniaxial fibers, while coaxial fibers may circumvent this issue by integrating two different polymers, enabling the separation of agents within a compatible polymer formulation (core or shell). Moreover, the additional outer shell can help to modulate release. One can envision that with a coaxial architecture, multiple agents may be delivered against a particular infection to provide a synergistic effect or to provide protection against multiple types of viral or bacterial infections. Together, these features allow for enhanced tunability with the option of providing immediate to short-term release for on-demand applications while also providing long-term release that may be particularly useful in prophylactic or contraceptive applications.

A variety of release kinetics can be attained from coaxial fibers by using different combinations of materials in the core and shell layers. Transient or rapid release of active agents is often accomplished with the use of hydrophilic polymers due to their rapid dissolution in aqueous environments. To achieve short-term release extending to one week, a hydrophilic core in combination with a hydrophobic shell is the most frequently used architecture, enabling the slow dissolution of the shell layer, which acts as a barrier to encapsulant diffusion from the core. For sustained-release applications that require delivery on the order of weeks to months, hydrophobic polymers such as

PLGA and PCL are often selected as shell polymers due to their slower degradation kinetics and biocompatibility. Yet, due to the number of parameters involved in the synthesis of coaxial fibers, two similar architectures may still be tailored to perform very differently by altering physical versus chemical properties. An example may be seen in which fibers composed of similar or even the same polymers display very different release rates due to the modulation of shell thickness. In these cases, thinner shells have been shown to provide more transient release, while increasing the shell thickness delays or alters the trend to more gradual release.

Coaxial fibers have been investigated previously for intravaginal delivery [96,99]. In one study, maraviroc release from coaxial fibers was adjusted by varying the drug loading and solution flow rates to provide release over five days [99]. In addition, pH-responsive coaxial fibers have been fabricated to react in the presence of semen by utilizing the pH-sensitive polymer cellulose acetate phthalate as a shell. The outer shell dissolved immediately after exposure to PBS, promoting pH-responsive release of Rhd B [96].

Although coaxial fibers have shown promise in general drug and initial intravaginal delivery applications, further refinements are required to expand their overall utility. First, compatibility between the solvents of the two polymer electrospinning solutions may limit the potential combinations of core-shell materials and encapsulated agents to achieve successful electrospinning. Additionally, residual solvents from the electrospinning process may interact with and inactivate encapsulated active agents in the core layer. Therefore, while research in coaxial fiber design is still ongoing, other fiber architectures such as multilayered fibers may offer additional advantages to advance intravaginal delivery.

3. Multilayered Electrospun Fibers

3.1. Multilayered Fiber Architectures and Properties

Multilayered fibers can provide layer-by-layer delivery platforms that are relatively simple and inexpensive to fabricate while allowing for the encapsulation of different active agents within the individual layers. The topology, thickness, and composition of each individual layer can be easily tuned to provide different release properties based on the envisioned application. Moreover, multilayered fibers have been shown to have increased mechanical stability and flexibility compared to coaxial fibers [120]. While the interactions between two or more polymer solution interfaces must be considered for coaxial fibers, multilayered fibers can be fabricated from normally incompatible polymers due to their sequential versus simultaneous fabrication process.

Electrospun multilayered fibers can be fabricated by sequential layering, stacking, or interweaving fibers [121–123]. In sequential layering, the first layer of polymer is electrospun onto a collector, followed by electrospinning additional polymer layers directly onto the same collector. In comparison, "stacking" fibers refers to individually electrospinning each layer separately and subsequently adhering individual layers together post-spin. Stacked fibers share similar physical properties with sequentially-layered fibers, enabling temporally-programmed or spatially-specific delivery of active agents [124]. Finally, the fabrication of interwoven fibers utilizes dual or multiple-syringes to simultaneously electrospin two or more different polymer solutions (usually one hydrophilic and hydrophobic) onto the same collector. In contrast to fibers produced using the sequential layering and stacking processes, which have distinct, separate layers of polymeric fibers, interwoven fibers result from the blending of these different polymer solutions from syringes placed opposite of or adjacent to each other into one integrated layer [125–127]. This technique seamlessly integrates both hydrophilic and hydrophobic polymers in a way that prevents unwanted interactions between the electrospun polymer solutions [127,128] while enabling the porosity of the hydrophilic fibers to be altered to more finely tune fiber degradation [129]. Although interwoven fibers do not have a shell layer, the interwoven architecture has been beneficial in promoting cell adhesion and growth and has the potential to more finely modulate active agent release via porosity-based mechanisms for drug delivery applications [130,131].

Regardless of fabrication technique, multilayered fibers are beneficial in that they can temporally modulate the release of multiple agents from a single delivery platform and can provide additional tunability by modulating the barrier or discrete layers of the multilayered structure (Figure 5). In addition, the ability to impart spatially-specific release—where specific layers of the multilayered fiber possess distinct release profiles—is a key advantage of this architecture. This advantage may be envisioned for intravaginal delivery applications where one layer provides rapid active agent release to the mucus while another layer enables sustained-delivery specific to underlying epithelial or immune cells [121,123]. For interwoven multilayered fibers, studies have shown that the incorporation of a hydrophilic polymer can alter the overall porosity and wettability [129,132,133], while using a hydrophobic outer layer in multilayered fibers (similar to coaxial fibers) can decrease surface wettability and corresponding active agent release [134].

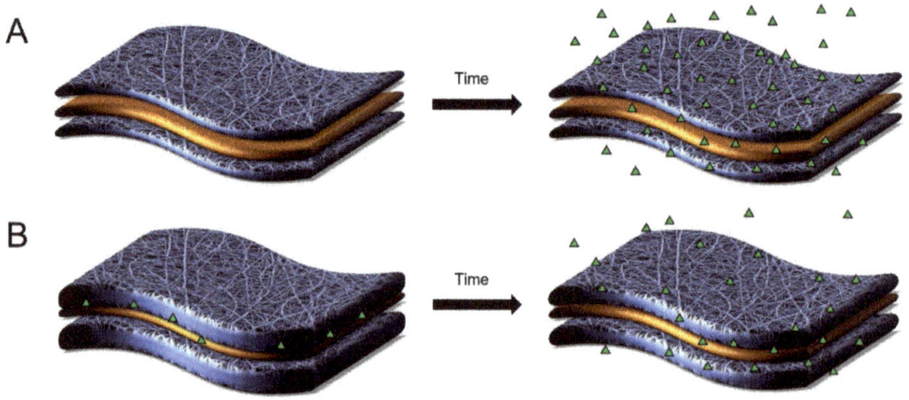

Figure 5. Schematic of anticipated active agent release from multilayered fibers. One method to modulate the release of active agents (shown in green) is to vary the thickness of the outer layer (shown in blue). (**A**) A thin outer layer provides both rapid burst release and limited sustained-release of encapsulants. (**B**) In contrast, increased outer layer thickness can delay the release of some active agents.

While the process of creating multilayered fibers is well established, more work is required to elucidate how each polymer layer impacts release kinetics. Physical properties including the pore size, fiber diameter, and thickness of traditional uniaxial fibers are known to impact the delivery kinetics of active agents from individual layers. Thus, the presence of one or more fiber layers can contribute to the complexity in establishing and predicting the release kinetics of diverse active agents from differently layered architectures. Despite these considerations and complexities, the adoption of different layering techniques to create multilayered fibers can achieve diverse patterns of release for transient, short-term, and sustained-release applications.

3.2. Release Kinetics from Multilayered Fibers

3.2.1. Transient and Short-Term Release

Multilayered fibers have shown promise in providing transient and short-term release of active agents. Conventionally, a hydrophilic layer serves as a reservoir for active agents, while hydrophobic materials provide an outer shell layer to prolong release. One study utilized a multilayered fabrication approach to encapsulate the hydrophobic antibiotic, gentamicin, in a hydrophilic PVA center layer and utilized a PU outer layer to envelop the inner PVA fiber [135]. Three separate fibers were fabricated by altering the thickness of the PU outer layer between 3.4 and 8.1 μm. The release of gentamicin was modulated with the thinnest PU layer (3.4 μm) demonstrating complete release within 1 h, relative to 10% release obtained from the thickest layer (8.1 μm). Furthermore, the thickest PU layer continued to release gentamicin for 24 h. Another study using interwoven electrospun fibers containing PEO and PCL demonstrated that by adjusting the ratio of the two polymers, tunable fiber degradation could be achieved from the resulting changes in pore size and porosity [127]. Although this study investigated interwoven fibers to enhance cell infiltration through the pores, the use of sacrificial fiber layers may be applied to modulate active agent release from the fibers for intravaginal delivery applications [127].

In addition to modulating the outer layer thickness and overall fiber composition, alterations to the number of layers have been shown to impact active agent release. In one study, fibroin-gelatin blended uniaxial fibers exhibited release of trypan blue, fluorescein isothiocyanate (FITC)-inulin, and FITC-BSA within minutes [136]. In contrast, multilayered fibers composed of the same materials extended the release of all three model compounds to 28 days [136]. In another study, dual-release, multilayered electrospun fibers containing the model dyes, 5,10,15,20-tetraphenyl-21*H*,23*H*-porphinetetrasulfonic acid disulfuric acid (TPPS) and chromazurol B, were encapsulated in four-layered PLCL (75:25) fibers. The release rate and duration of the dyes were controlled by the fiber diameter and individual fiber layer thicknesses. Minimal release of both dyes was observed for the first 15 min, followed by a quasi-linear release profile for up to 4 h. However, increasing the thickness of dye-loaded layers resulted in higher quasi-linear release rates due to the reduced density of the fiber surface [137]. In another study, the transient release of ketoprofen was achieved using trilayer fibers composed of two EC outer layers surrounding a center PVP fiber. These fibers provided nearly complete release of ketoprofen within 24 h [121]. Last, asymmetric multilayered polylactide fibers with different designs on each side were fabricated to prevent liver cancer recurrence by promoting one-sided prolonged chemotherapeutic release [138]. The fiber was composed of five poly(lactic acid) (PLA) layers, with each layer serving as either a barrier to release or a drug encapsulating reservoir. In vivo studies in a murine model demonstrated tumor suppression for at least four days, indicating that the multilayered fiber may provide localized chemotherapy for short-term durations [138].

Multilayered fibers with stimuli-responsive properties have also been investigated for transient and short-term release applications. In one of the first studies to investigate multilayered architectures, the pH-responsive polymers, poly(acrylic acid) (PAA) and poly(allylamine hydrochloride) (PAH), were electrospun together to create a blended fiber. These fibers were loaded with a low molecular weight cationic molecule, methylene blue, and demonstrated rapid release of methylene blue (~10 min) at a neutral pH (7.4). However, by gradually adjusting the pH from 6 to 2 in aqueous solutions, the step-wise pH-responsive release of methylene blue was achieved over three and a half days. Building upon this work, the effect of coating the fibers with a thermoresponsive polymer blend, poly(*N*-isopropylacrylamide)-PAA, or perfluorosilane was assessed. The addition of the thermoresponsive poly(*N*-isopropylacrylamide)-PAA coating modulated methylene blue release via temperature. Above a critical temperature, the thermoresponsive polymer became insoluble and formed intramolecular hydrogen bonds, which led to the release of methylene blue within 50 min (PBS, pH 7.4). In comparison, coating with perfluorosilane modulated release for up to 20 h at neutral pH. When both the pH-responsive and multiple layers of thermoresponsive polymers were integrated and evaluated at 25 and 40 °C, dye released for a maximum of 10 h regardless of layer thickness [139].

3.2.2. Sustained-Release

The ability of multilayered fibers to provide long-term release has been demonstrated in a variety of studies [66,67,140]. In one study, the release of a hydrophobic chemotherapeutic agent, 7-ethyl-10-hydroxycamptothecin (SN-38), was prolonged to 30 days by using a triple-layered fiber in which SN-38 was encapsulated in the center layer and surrounded by two superhydrophobic outer layers consisting of PCL and poly(glycerol monostearate-*co*-ε-caprolactone) [134]. Similar to the trends seen for transient and short-term release from multilayered fibers, increasing the thickness of the outer fiber substantially improved the longevity and amount of drug released. In another study, multilayered fibers comprised of a PCL shell and a PEO/Rhd B core were fabricated to assess the effect of increasing the outer layer thicknesses between 46.1, 68.9, and 186.1 μm [141]. While the thinnest 46 μm layers released 85% of Rhd B in one day, the 68.9 and 186.1 μm layers increased release to 15 and 25 days, respectively. Moreover, the release from the two fibers with the thicker outer layers demonstrated zero-order kinetics, producing gradual, even release of drug with respect to time.

3.3. Applications for Intravaginal Delivery

Multilayered fibers have shown promise as a platform to co-deliver or prolong the release of active agents in different environments. The process of creating multilayered fibers is relatively simple, eliminating the more complex set-up and considerations of polymer-solvent interactions between the adjacent, simultaneously spun layers present in coaxial spinning. By removing this complexity of interactions, multilayered fibers can achieve "programmed release" by simply modulating the thickness of each layer.

Multilayered fibers possess other unique features that make them excellent candidates for intravaginal delivery applications. One of the unique strengths of multilayered fibers is that they can provide spatially-specific release in that, unlike other architectures, the individual layers of multilayered fibers can be designed for specific and discrete purposes. For example, one layer may be designed to improve mucoadhesion for enhanced longevity and biocompatibility within the FRT, while another layer may provide release of active agents dependent on its location within the multilayered matrix. Compared to coaxial fibers, the optimization of multilayered fibers is not limited by solvent compatibility, as they can be sequentially spun and assembled post-fabrication. Moreover, multiple individually spun layers can increase the ease of encapsulating multiple types of active agents, which serve mechanistically different roles against a single type of viral infection or as a multipurpose viral-contraceptive or viral-bacterial dosage form. Finally, each fiber layer can be adjusted to have distinct mechanical properties that include tensile strength, porosity, and elasticity, important for comfort and user preference [142].

To date, the use of multilayered fibers for intravaginal delivery has been briefly explored [56,99,123]. In one study, circular sheets of pre-spun PVP and PVP-EC fibers were stacked and annealed via a pressed metal die that was dipped in solvent. The die annealed the edges of the stacked fibers, creating a multilayered fiber with a PVP inner layer surrounded by blended PVP-EC sheaths. Other multilayered fibers were also constructed by folding the outer layers and pressing the seams. Both types of multilayered fibers encapsulated the hydrophilic compound maraviroc and provided biphasic release, exhibiting an initial burst release followed by short-term release for up to five days. Another study from the same group examined tenofovir (TFV) localization within stacked PCL/PLGA fibers. It was found that TFV localization within the multilayered fiber could be predicted by considering the changes in polymer crystalline structure caused by encapsulant-polymer interactions and correlating drug-polymer hydrophilicity [56].

Both multilayered and coaxial fibers have the potential to provide tunable and sustained-release; however, each architecture still faces the challenges surrounding FRT delivery. For example, the interplay between two polymer solutions still needs to be considered for interwoven multilayered (and coaxial) fibers, which may result in challenges to altering active agent release. Additionally, as stated previously, the most significant obstacles to intravaginal delivery are providing a dosage form that can facilitate active agent penetration of mucus and retention and release of therapeutically relevant agent concentrations within the FRT. To improve retention, fibers can be fabricated using polymers or polymer blends that have mucoadhesive properties. However, this longevity is rarely translated to active agents once they have been released from fibers. Thus, new measures may be considered to provide efficacious and sustained-delivery from fibers.

4. Composite Nanoparticle-Fiber Delivery Vehicles

4.1. Nanoparticle-Fiber Architectures and Properties

Over the past two decades, polymeric NPs have been extensively studied as efficacious drug delivery platforms for a variety of applications. Polymeric nanoparticles are an attractive option for intravaginal delivery relative to traditional delivery platforms such as gels and films due to the tunability of active agent release, ability for surface modification, potential for targeted delivery, enhanced distribution potential, and the often resulting enhanced efficacy of encapsulated agents. Additionally, polymeric NPs have been shown to elicit minimal immune response and to improve the delivery and bioactivity of biologics [29,143,144]. Although metallic nanoparticles have also been explored for use in many drug delivery applications, they have been less commonly administered within the FRT, hence, a more comprehensive review of their applications may be found in [145,146].

Many physicochemical characteristics of NPs can be altered, such as particle size, surface charge, and hydrophobicity, which contribute to their success in achieving sustained-release and localization to target sites [147]. Although NPs have proven to be effective delivery platforms, as discussed in previous reviews [148,149], achieving the prolonged release of active agents can be difficult due to the natural clearance mechanisms of the FRT. In particular, NPs are challenged with retention in the vaginal cavity due to mucus clearance and transport through mucus to underlying tissue [28,150,151]. These challenges may be overcome by incorporating NPs into electrospun fibers, thereby creating a composite delivery vehicle that complements the capabilities of both technologies. One might envision that fibers may act as a reservoir for NPs, improving NP and active agent retention, while the innate fiber porosity can help to more finely tune encapsulant release from NPs relative to the release observed from freely administered NPs or fibers.

Nanoparticle-fiber composites are dual-component systems that have the ability to alter the release kinetics of active agents from NPs or NPs themselves [152,153]. Often, the active agent of interest is encapsulated within the NPs, which are then preloaded into polymer solutions for subsequent electrospinning. While a variety of inorganic NPs have been incorporated into fibers [154–156], concerns still persist regarding the safety of their use relative to polymeric NPs, particularly for intravaginal applications. By utilizing biocompatible polymeric materials for both nanoparticles and fibers, composites may provide safe and prolonged release for clinical applications.

4.2. Release Kinetics from Nanoparticle-Fiber Composites

4.2.1. Transient Release

Nanoparticle-fiber composites have been used to rapidly release NPs and their encapsulated agents. A study was conducted with hydrophilic PVA and PEO fibers that incorporated PLGA NPs that contained the dye, Coumarin 6 [157]. PEO fibers released 90% of NPs within 30 min when immersed in a 50:50 ethanol:PBS solution, followed by additional release (5%) after 3.5 h. In comparison, PVA fibers released approximately 70% of PLGA NPs within 30 min, followed by a decrease in NP release (15%) over 8 h. Slightly slower release over 24 h was observed when PVA fibers were crosslinked prior to NP incorporation. This study highlights that nanoparticle-fiber composites can be used to successfully incorporate NPs and to modulate the transient release of NPs from these composites within aqueous solutions [157].

4.2.2. Short-Term Release

Several studies have utilized nanoparticle-fiber composites to provide the short-term release of active agents. One group explored a composite drug delivery system that encapsulated the antibiotic, erythromycin, in gelatin NPs and free lidocaine hydrochloride within PVA-chitosan blended fibers [158]. Eighty percent of the lidocaine hydrochloride was released from the fibers within 54 h, while 70% of the erythromycin was released after 70 h. In contrast, free gelatin NPs released 90% of erythromycin within the same duration. In a separate study, chitosan-PEO blended fibers containing methoxypolyethylene glycol (mPEG)-b-PLA micelles demonstrated a low initial burst release (15%) of 5-fluorouracil (5-FU), followed by prolonged release (91%) for 109 h [159]. In another study, the release of free hydrophobic naproxen and chitosan nanoparticles containing Rhd B was studied from PCL fiber scaffolds [160]. Rhodamine B exhibited low levels (5%) of burst release, while 30–40% of naproxen was released within the first 2 h.

Moreover, after 72 h, only 20% of Rhd B was released, while 60% of naproxen was released. The rapid release of naproxen was achieved via incorporation within the fiber scaffold, while the extended release of Rhd B was obtained and enhanced through nanoparticle-fiber encapsulation. These results demonstrate the utility of nanoparticle-fiber composites in providing the short-term release of multiple agents.

4.2.3. Sustained-Release

Nanoparticle-fiber composites have also demonstrated long-term release capabilities in several studies. In one study, dual-release nanoparticle-fiber composites were used to mend and treat critically sized calvarial defects in rats [161]. These composites, consisting of PCL-*co*-PEG fibers encapsulating dexamethasone and BSA NPs and loaded with bone morphogenic protein-2 (BMP-2), demonstrated sustained-release of both molecules over 35 days. Another study explored the incorporation of siRNA into chitosan NPs and PLGA fiber composites [153]. In these composites, the release of active siRNA was sustained in vitro, with 95% of siRNA released from the fibers over 32 days, while gene silencing activity was maintained. Sustained-release from nanoparticle-fiber composites was also demonstrated in another study with chitosan-PEO electrospun fibers that were loaded with PLGA NPs encapsulating phenytoin. Nearly complete release of phenytoin from the composite scaffold was achieved over nine days [162]. Lastly, PLA fibers encapsulating chitosan particles provided sustained-release of BSA (45%) for 27 days, while chitosan particles alone released 80% BSA in 14 days [163].

In addition to NP incorporation within traditional uniaxial or blended fibers, NPs have been incorporated in more complex fiber architectures to prolong the release of active agents. For instance, the effect of combining a multilayered fiber architecture with nanoparticle-fiber composites was investigated by fabricating alternating layers of poly-L-lactic acid (PLLA) and PCL fibers with layers of PCL fibers encapsulating positively-charged chitosan BSA NPs [164]. The multilayered composite released 80% of the BSA in approximately eight days, whereas the monolayer control released the same concentration of BSA within 24 h.

4.3. *Applications for Intravaginal Delivery*

Composite delivery vehicles containing nanoparticles and fibers have thus far been primarily studied in wound healing and tissue engineering to fabricate scaffolds for tissue regeneration and bone remodeling [86,165–167]. However, these platforms may be promising candidates for intravaginal delivery applications due to their structural stability and ability to sustain the release of active agents. In such systems, the fibers may be utilized as a reservoir for NPs to aid in intravaginal retention by helping to decrease NP clearance during shedding. In addition, it is envisioned that, depending on fiber formulation and, importantly, NP size and charge, NP (and active agent) release may be modulated, enabling NPs to traverse mucus and deliver agents to target cells that reside in the epithelium or underlying lamina propria. Similar to other architectures, fiber parameters such as polymer composition and size can be tailored to impact release in combination with altering NP composition, size, and loading within the fiber.

For intravaginal delivery applications, NPs can impart cell specificity, cell internalization, as well as mucoadhesive or mucopenetrative properties to their encapsulated active agents [14]. Numerous studies have demonstrated the ability of NPs to enhance cell targeting via surface modification [168,169]. Additionally, surface modification can increase cell internalization, which may enhance the transport, subcellular localization, and corresponding efficacy of drugs like tenofovir disoproxil fumarate (TDF), which require cell internalization. Furthermore, the NP surface charge can be modulated to provide either mucoadhesive or mucopenetrative properties that further enhance active agent delivery. Additionally, fibers can be fabricated to encapsulate NPs for sustained-release as well as free agents for rapid release, providing both on-demand and sustained-release in one platform. Finally, nanoparticle-fiber composites, when coupled with coaxial or multilayered fiber architectures, provide an attractive strategy to retain and sustain the release of active agents within the FRT (Figure 6).

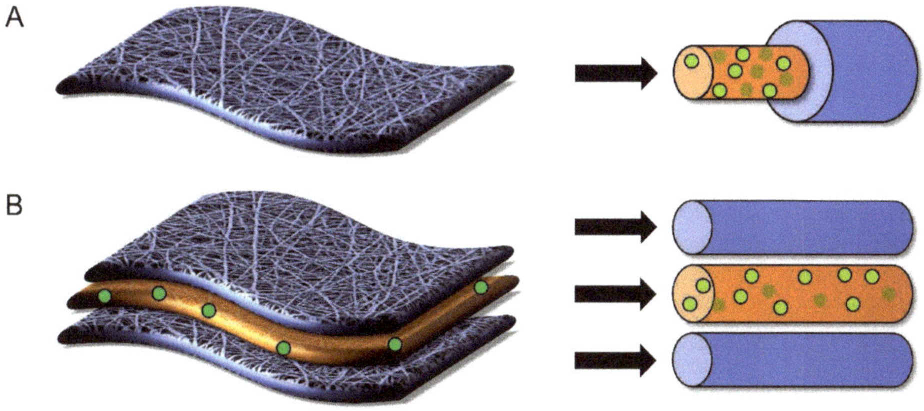

Figure 6. Schematic of electrospun nanoparticle-fiber composites that integrate coaxial and multilayered fiber architectures. (**A**) Coaxial fibers can be fabricated to encapsulate nanoparticles (NPs) within the core fiber, conferring sustained- or delayed-release of active agents that are encapsulated in NPs (shown in green). (**B**) Multilayered fibers that encapsulate NPs can also act as reservoirs for either NP or active agent release.

As with multilayered fibers, the use of nanoparticle-fiber composites has only recently been investigated for intravaginal delivery. In a proof-of-concept study, rapid-release PEO, PVA, or PVP fibers encapsulated PLGA NPs containing C6 dye or etravirine drug [23]. In this study, composites and free NPs were administered within murine FRTs and assessed for retention and release. The encapsulated nanoparticles exhibited a 30-fold increase in retention in the mouse FRTs relative to free NPs. Furthermore, nanoparticles alone provided transient release of etravirine, while all nanoparticle-fiber composites demonstrated release for up to seven days. To date, this is the only investigation of nanoparticle-fiber composites for use in intravaginal delivery. However, the significant difference in retention and release rate achieved with nanoparticle-fiber composites highlights the immense potential of this architecture for sustained-delivery in the FRT.

Although combining nanoparticles and electrospun fibers into one delivery vehicle has demonstrated potential, challenges exist for this platform. The major concern is related to the concentration of nanoparticles that can be effectively encapsulated within fibers without hindering the ability of the polymer solution to be electrospun [170]. Furthermore, the concentration of active agent may decrease with the use of a coaxial or multilayered architecture, as only specific layers of the fiber will encapsulate NPs. Finally, polymeric NPs are often comprised of the same or similar polymers as electrospun fibers, thus care must be taken to prevent polymer solvents from dissolving the NPs prior to or during the electrospinning process [171]. Moreover, the morphology of NPs may also be adversely affected by electrospinning voltage. These factors limit the combinations of fiber and nanoparticle materials available for composite fabrication. Thus, for composite delivery applications to succeed, polymer choice and electrospinning conditions must be taken into consideration.

5. Future Directions and Discussion

Within the past decade, electrospun fibers have been explored as a multipurpose delivery platform to prevent and treat sexually transmitted infections (STIs). For intravaginal applications, fibers have typically been uniaxially electrospun to release active agents targeted to HIV-1/HSV-2 infections and contraceptive applications. However, other electrospun architectures have been developed that may provide more finely-tuned active agent release, the encapsulation of multiple agents, and longer release durations, desirable for next-generation vehicles. Given this, the goal of this review was to summarize the advancements in electrospun fiber architectures including coaxial, multilayered, and nanoparticle-fiber composites, to meet these needs, and to review their use in other drug delivery applications. We sought to relate different temporal regimens of delivery, including transient (occurring within hours), short-term (spanning hours to one week), and sustained (extending from one week to months), to architectural design and materials selection to help guide the design of future platforms that meet the unique temporal needs of intravaginal delivery.

One of the major challenges facing intravaginal delivery is the lack of user adherence surrounding the administration of current delivery platforms. Several clinical trials have highlighted how a lack of user adherence contributes to decreased efficacy in clinical trials. In both the FACTS-001 and VOICE trials, South African women deemed high risk for HIV-1 exposure were given antiretroviral TFV gels to administer prior to intercourse [93,172]. Despite the known efficaciousness of TFV, the gels provided suboptimal protection against HIV-1 infection, which was attributed to low user adherence of the gels prior to intercourse. Another study examined the efficacy of gels that incorporated the antiviral polysaccharide, carrageenan, in women in Thailand. This study demonstrated similarly disappointing clinical outcomes, with low user adherence considered the most significant reason for the lack of clinical efficacy [173]. Negative outcomes in other trials such as PRO-2000 and cellulose sulfate gel studies, which examined the efficacy of anti-HIV gels in female populations, further validated these studies, highlighting that both user preference and adherence regimens must be considered during product design rather than at the clinical trial stage. As a result of these studies, there has been an increased emphasis to design vehicles that decrease the administration frequency by prolonging active agent release after a single topical application.

In conjunction with improving user adherence, the development of multipurpose delivery vehicles that offer long-term protection against the various stages of a single infection or a diversity of different types of infections is highly desirable [174]. For single infections, a delivery platform may administer multiple agents with different mechanisms of action that target different stages of the viral or bacterial life cycle. However, the increased likelihood of viral co-infections, such as HSV-2 and HIV, as well as bacterial and fungal infections will likely require co-administration of antiviral and antimicrobial agents to be successful. Furthermore, applications that seek to meet both antiviral and contraceptive needs in the same dosage form will require the incorporation of multiple types of agents to expand a platform's effectiveness. Therefore, a delivery platform that has the capability to release multiple active agents, each over time frames relevant to the application or active agent, will have greater utility and enable more convenient administration schedules based on specific user needs.

Despite these needs, tailoring the delivery of multiple types of active agents for viral, bacterial, fungal, and contraceptive applications is an ambitious goal given the unique chemical properties of each agent. For example, the antiretroviral TFV and its pro-drug TDF have similar structures and both work as nucleoside reverse transcriptase inhibitors yet possess markedly different hydrophobicities. As such, a delivery platform designed to prolong TFV release may result in different release kinetics of TDF, requiring the formulation of distinct delivery vehicles specific to the selected active agents [56,116,118,175,176]. Furthermore, each active agent may necessitate specific temporal dosing regimens to provide protection or treatment. For example, it may be desirable to administer viral entry inhibitors, which inactivate virions prior to cell entry, over a different time frame than active agents that work inside of cells and need to transport through and localize to target tissue. Several studies have investigated this and have found that more complex and specialized architectures may be useful to achieve temporal delivery goals by tuning the release properties of multiple encapsulants for multiple targets [177,178]. Similarly, for contraceptive applications, on-demand and/or zero-order release with equivalent daily dosing may be desirable for spermicides and hormonal/non-hormonal contraceptives, respectively. Conversely, it may be desirable to deliver active agents such as hormones and small hydrophilic drugs (e.g., etonogestrel and acyclovir) within the same time frame for simultaneous long-term contraception and prevention. However, the drastically different chemical properties of these agents will require more complex solutions to achieve similar release profiles. Given this, multipurpose intravaginal delivery platforms must be tailored to maximize the efficacy of individual active agents, including small molecule drugs, proteins, antibiotics, hormones, and live organisms (e.g., probiotics), to meet the needs of these diverse applications.

While providing distinct release profiles of different active agents is an important criterion for the development of future intravaginal platforms, to date, intravaginal rings (IVRs) are the only platforms that provide delivery over a duration of weeks to months [179–183]. Furthermore, IVR studies indicate that more complex dosage forms, such as rings with drug-encapsulating pods, may more likely succeed, particularly in challenging delivery scenarios, e.g., achieving the sustained-release of small hydrophilic molecules [177]. These and other studies [177,178,184,185] emphasize the need to offer alternative delivery vehicles for women, with the key lesson that platform architecture must be designed to consider the hydrophobicity and chemical compatibility of the encapsulants in combination with its surrounding materials.

In addition to the development of fibers with more complex architectures, active agent release and transport from these platforms must be assessed. Tissue mimetics and ex vivo tissues have been used to assess these parameters within the context of intravaginal delivery applications [116,186–191]. One of the most common ways in which to assess intravaginal delivery is by using human ectocervical tissue explants derived from patients [187–191]. These explants provide a representative environment in which to measure transport by accounting for the three-dimensional structure of patient tissue. However, patient-specific variations and tissue availability can limit the use of vaginal explants. Given this, organotypic three-dimensional vaginal tissue models such as Epivaginal™ tissue have been created to help evaluate the safety, transport, and efficacy of active agents within an FRT mimetic [192,193]. Other in vitro models have also been developed to explore bacteria and host cell interactions in the reproductive environment [194]. Moreover, within the past decade, new biomarkers and assay endpoints have been identified and studied in different models to more fully assess microbicide interactions with the FRT [195]. The use of tissue models promises to streamline the assessment of future fiber platforms as viable intravaginal delivery platforms.

To date, a variety of studies have developed uniaxial electrospun fibers for intravaginal applications, including HIV prevention [56,115–119,175,176,196–199]. In these studies, electrospun fibers have demonstrated promising potential for intravaginal applications due to their mucoadhesive characteristics, mechanical properties, and ability to be fabricated in different shapes and sizes [53]. Depending upon the polymer hydrophilicity, traditional uniaxial fibers have been formulated as transient, short-term, or long-term delivery platforms. For the purposes of on-demand and short-term release, many of these studies use hydrophilic fibers, which dissolve or degrade quickly. In contrast, fibers consisting of more hydrophobic materials are expected to persist within the FRT, acting as reservoirs to sustain the release of active agents. We envision (and have observed) that long-term delivery vehicles maintain their structure during the delivery duration of interest and may require physical removal from the FRT, similar to current IVRs. However, one of the key challenges for intravaginal delivery has been to sustain the release of small hydrophilic antiretrovirals due to their rapid diffusion through the porous fiber matrix, solubility in aqueous solutions, and chemical incompatibility with hydrophobic polymer cores [6]. Many of these uniaxial fibers demonstrated burst release of hydrophilic agents followed by short-term release [59,200], partially attributed to the localization of hydrophilic agents on the fiber surface. Compounding this, concerns exist that the subsequent release of active agents may be insufficient to provide complete protection against future infections. While blended uniaxial fibers have been moderately successful in addressing these challenges, more work is required [119].

The primary parameters that impact release from uniaxial fibers are the choice of solvent and polymer. Other factors such as polymer concentration and electrospinning parameters also play a role in attaining different release profiles; however, it is unlikely that these factors alone are sufficient to overcome the challenge of delivering sustained and therapeutically-relevant concentrations of hydrophilic agents. Furthermore, it is difficult to utilize traditional uniaxial fibers for the encapsulation of multiple diverse agents such as large proteins and small drugs. Due to these issues, other electrospinning architectures may be better suited to meet the diverse challenges of intravaginal delivery.

As discussed previously, coaxial fibers have shown promise for the encapsulation and release of small hydrophilic and hydrophobic molecules, which may be useful for intravaginal delivery applications. The different goals of transient, short-term and long-term release can be achieved by changing the composition and hydrophobicity of core and shell materials as well as by modulating the shell thickness and core:shell ratio. As described, the shell layer can help regulate active agent release, while the core layer is designed to provide optimal compatibility with an encapsulant. For instance, by using pH-responsive polymer shells, an immediate stimuli-responsive release of agents can be achieved when the fiber is in contact with semen. In this scenario, the core layer may be tailored to encapsulate multiple agents, while the shell, comprised of pH-sensitive polymers, retains encapsulants until needed. Another advantage of coaxial fibers is that they can be fabricated to exploit drug-polymer hydrophilicities. For example, a coaxial fiber comprised of a hydrophobic shell and hydrophilic core can be utilized to provide long-term release of hydrophilic compounds. Agent encapsulation into both layers would allow for both transient burst release from the shell due to surface localization and high loading and sustained-release from the core layer. Finally, coaxial fibers can provide release of biological agents such as large proteins. Coaxial cores may be engineered to achieve high protein encapsulation and biocompatibility, while shells can be constructed with porous surfaces, allowing tunable release. This is particularly significant given that many biologics are being investigated as future viral prophylaxes and therapeutics. Although coaxial electrospinning is a more complex process that requires additional optimization, relative to uniaxial spinning, it may enable a versatile platform for transient, short-term, and long-term release [119].

Multilayered fibers combine different polymer layers via sequential or post-spinning to incorporate multiple and chemically distinct drugs within specific layers, thereby tailoring the release kinetics for each encapsulated agent. Multilayered interwoven fibers can be utilized to provide transient release using sacrificial layers to encapsulate agents for on-demand applications. The sacrificial layers comprised of hydrophilic polymers would provide on-demand release of agents based on their immediate degradation when exposed to physiological fluids. Active agent release can be further modulated by the number, thickness, and porosity of each fiber layer [201]. Moreover, blank fibers may be incorporated within the multilayers to either act as a physical barrier for sustained-release or for contraceptive purposes. The layer thickness and level of porosity of blank fibers can be conveniently modulated to delay the release of small hydrophilic molecules from the drug-loaded layers, serving to prolong release. Additionally, multilayered fibers have the potential to deliver biologics and non-hormonal contraceptives. These agents, although efficacious, may degrade when exposed to harsh solvents during the electrospinning process. By incorporating these active agents in distinct layers and integrating barrier layers, multilayered fibers can provide long-term release of drugs and biologics while retaining their activities.

While each of these strategies offers advantages relative to uniaxial spinning, the delivery of active agents may be further enhanced by integrating nanoparticles with fibers. A composite platform may offer a new alternative to address the challenges of intravaginal delivery, such as the maintaining active agent stability, providing cell-specific targeting (via NPs), and enhancing cell internalization. Like electrospun fibers, nanoparticles can be designed to encapsulate virtually any compound. The limitations of nanoparticle-fiber composites mentioned earlier may be overcome by utilizing fibers as a reservoir for both active agents and nanoparticles to release multiple therapeutics. Furthermore, the release rates of encapsulants from both nanoparticles and fibers may be modulated by adjusting the composition of the polymeric scaffold. For on-demand transient release, hydrophilic polymers may be used to enable rapid release of NPs for immediate distribution through and enhanced retention within tissue. In contrast, more hydrophobic fibers may be used to delay the release of NPs or NP-encapsulated agents. Although drug-polymer hydrophobicity is a major contributor to release, other factors such as polymer choice, molecular weight, and crystallinity, as well as solvent choice and electrospinning parameters, also affect the release of agents from fibers.

The application of advanced fiber architectures has only recently been explored in the context of intravaginal delivery. Advanced fiber architectures demonstrate the potential to provide the sustained-release of individual active agents in addition to concurrently providing both transient and sustained-delivery of multiple active agents. These are key advantages over traditional uniaxial fibers, which are challenged with the long-term delivery of small hydrophilic molecules, in addition to providing transient and sustained-release simultaneously. We envision that future fiber architectures will localize active agents within specific sections of the fiber to tailor the release of individual agents independent of other encapsulants. Moreover, we anticipate that future platforms will combine architectures to maximize or complement the advantages of individual platforms. As previous clinical trials have shown, effective protection will be dependent upon fulfilling user preferences, offering convenience, and providing necessary release profiles from one vehicle, which fibers have the potential to realize.

Funding: This research was funded by the Jewish Heritage Fund for Excellence grant and by the NIH P20 COBRE, grant number GM125504.

Conflicts of Interest: The authors declare no conflict of interest. The funders had no role in the writing of the manuscript.

References

1. Wira, C.R.; Patel, M.V.; Ghosh, M.; Mukura, L.; Fahey, J.V. Innate immunity in the human female reproductive tract: Endocrine regulation of endogenous antimicrobial protection against HIV and other sexually transmitted infections. *Am. J. Reprod. Immunol.* **2011**, *65*, 196–211. [CrossRef]
2. Unnithan, A.R.; Barakat, N.A.; Pichiah, P.T.; Gnanasekaran, G.; Nirmala, R.; Cha, Y.-S.; Jung, C.-H.; El-Newehy, M.; Kim, H.Y. Wound-dressing materials with antibacterial activity from electrospun polyurethane–dextran nanofiber mats containing ciprofloxacin HCl. *Carbohydr. Polym.* **2012**, *90*, 1786–1793. [CrossRef] [PubMed]
3. Tourgeman, D.E.; Gentzchein, E.; Stanczyk, F.Z.; Paulson, R.J. Serum and tissue hormone levels of vaginally and orally administered estradiol. *Am. J. Obstet. Gynecol.* **1999**, *180*, 1480–1483. [CrossRef]
4. Steinbach, J.M. Protein and oligonucleotide delivery systems for vaginal microbicides against viral STIs. *Cell. Mol. Life Sci.* **2015**, *72*, 469–503. [CrossRef] [PubMed]
5. Bernkop-Schnürch, A.; Hornof, M. Intravaginal drug delivery systems. *Am. J. Drug Deliv.* **2003**, *1*, 241–254. [CrossRef]
6. Chou, S.F.; Carson, D.; Woodrow, K.A. Current strategies for sustaining drug release from electrospun nanofibers. *J. Control. Release* **2015**, *220 Pt B*, 584–591. [CrossRef]
7. Hickey, D.K.; Patel, M.V.; Fahey, J.V.; Wira, C.R. Innate and adaptive immunity at mucosal surfaces of the female reproductive tract: Stratification and integration of immune protection against the transmission of sexually transmitted infections. *J. Reprod. Immunol.* **2011**, *88*, 185–194. [CrossRef]
8. Wiggins, R.; Hicks, S.; Soothill, P.; Millar, M.; Corfield, A. Mucinases and sialidases: Their role in the pathogenesis of sexually transmitted infections in the female genital tract. *Sex. Transm. Infect.* **2001**, *77*, 402–408. [CrossRef]
9. Das Neves, J.; Bahia, M.F. Gels as vaginal drug delivery systems. *Int. J. Pharm.* **2006**, *318*, 1–14. [CrossRef]
10. Andrews, G.P.; Donnelly, L.; Jones, D.S.; Curran, R.M.; Morrow, R.J.; Woolfson, A.D.; Malcolm, R.K. Characterization of the rheological, mucoadhesive, and drug release properties of highly structured gel platforms for intravaginal drug delivery. *Biomacromolecules* **2009**, *10*, 2427–2435. [CrossRef]
11. Devlin, B.; Nuttall, J.; Wilder, S.; Woodsong, C.; Rosenberg, Z. Development of dapivirine vaginal ring for HIV prevention. *Antivir. Res.* **2013**, *100*, S3–S8. [CrossRef] [PubMed]
12. Derby, N.; Zydowsky, T.; Robbiani, M. In search of the optimal delivery method for anti-HIV microbicides: Are intravaginal rings the way forward? *Expert Rev. Anti-Infect. Ther.* **2013**, *11*, 5–8. [CrossRef]
13. Ho, E.A. Intravaginal rings as a novel platform for mucosal vaccination. *J. Mol. Pharm. Org. Process Res.* **2013**, *1*, e103.
14. Mallipeddi, R.; Rohan, L.C. Nanoparticle-based vaginal drug delivery systems for HIV prevention. *Expert Opin. Drug Deliv.* **2010**, *7*, 37–48. [CrossRef] [PubMed]

15. Kiser, P.F.; Johnson, T.J.; Clark, J.T. State of the art in intravaginal ring technology for topical prophylaxis of HIV infection. *Aids Rev.* **2012**, *14*, 62–77.
16. Dieben, T.O.; Roumen, F.J.; Apter, D. Efficacy, cycle control, and user acceptability of a novel combined contraceptive vaginal ring. *Obstet. Gynecol.* **2002**, *100*, 585–593.
17. Malcolm, R.K.; Edwards, K.-L.; Kiser, P.; Romano, J.; Smith, T.J. Advances in microbicide vaginal rings. *Antivir. Res.* **2010**, *88*, S30–S39. [CrossRef] [PubMed]
18. Roumen, F.; Apter, D.; Mulders, T.; Dieben, T. Efficacy, tolerability and acceptability of a novel contraceptive vaginal ring releasing etonogestrel and ethinyl oestradiol. *Hum. Reprod.* **2001**, *16*, 469–475. [CrossRef]
19. Nel, A.; van Niekerk, N.; Kapiga, S.; Bekker, L.-G.; Gama, C.; Gill, K.; Kamali, A.; Kotze, P.; Louw, C.; Mabude, Z. Safety and efficacy of a dapivirine vaginal ring for HIV prevention in women. *N. Engl. J. Med.* **2016**, *375*, 2133–2143. [CrossRef]
20. Kim, S.; Traore, Y.L.; Chen, Y.; Ho, E.A.; Liu, S. Switchable On-Demand Release of a Nanocarrier from a Segmented Reservoir Type Intravaginal Ring Filled with a pH-Responsive Supramolecular Polyurethane Hydrogel. *ACS Appl. Bio Mater.* **2018**, *1*, 652–662. [CrossRef]
21. Vanić, Ž.; Škalko-Basnet, N. Nanopharmaceuticals for improved topical vaginal therapy: Can they deliver? *Eur. J. Pharm. Sci.* **2013**, *50*, 29–41. [CrossRef] [PubMed]
22. Zhang, T.; Sturgis, T.F.; Youan, B.-B.C. pH-responsive nanoparticles releasing tenofovir intended for the prevention of HIV transmission. *Eur. J. Pharm. Biopharm.* **2011**, *79*, 526–536. [CrossRef] [PubMed]
23. Krogstad, E.A.; Ramanathan, R.; Nhan, C.; Kraft, J.C.; Blakney, A.K.; Cao, S.; Ho, R.J.; Woodrow, K.A. Nanoparticle-releasing nanofiber composites for enhanced in vivo vaginal retention. *Biomaterials* **2017**, *144*, 1–16. [CrossRef]
24. Martínez-Pérez, B.; Quintanar-Guerrero, D.; Tapia-Tapia, M.; Cisneros-Tamayo, R.; Zambrano-Zaragoza, M.L.; Alcalá-Alcalá, S.; Mendoza-Muñoz, N.; Piñón-Segundo, E. Controlled-release biodegradable nanoparticles: From preparation to vaginal applications. *Eur. J. Pharm. Sci.* **2018**, *115*, 185–195. [CrossRef] [PubMed]
25. Marciello, M.; Rossi, S.; Caramella, C.; Remuñán-López, C. Freeze-dried cylinders carrying chitosan nanoparticles for vaginal peptide delivery. *Carbohydr. Polym.* **2017**, *170*, 43–51. [CrossRef]
26. Leyva-Gómez, G.; Piñón-Segundo, E.; Mendoza-Muñoz, N.; Zambrano-Zaragoza, M.; Mendoza-Elvira, S.; Quintanar-Guerrero, D. Approaches in Polymeric Nanoparticles for Vaginal Drug Delivery: A Review of the State of the Art. *Int. J. Mol. Sci.* **2018**, *19*, 1549. [CrossRef]
27. Sims, L.B.; Frieboes, H.B.; Steinbach-Rankins, J.M. Nanoparticle-mediated drug delivery to treat infections in the female reproductive tract: Evaluation of experimental systems and the potential for mathematical modeling. *Int. J. Nanomed.* **2018**, *13*, 2709–2727. [CrossRef]
28. El-Hammadi, M.M.; Arias, J.L. Nanotechnology for Vaginal Drug Delivery and Targeting. In *Nanotechnology and Drug Delivery, Volume Two: Nano-Engineering Strategies and Nanomedicines against Severe Diseases*; CRC Press: Boca Raton, FL, USA, 2016; p. 191.
29. Ensign, L.M.; Tang, B.C.; Wang, Y.Y.; Tse, T.A.; Hoen, T.; Cone, R.; Hanes, J. Mucus-penetrating nanoparticles for vaginal drug delivery protect against herpes simplex virus. *Sci. Transl. Med.* **2012**, *4*, 138ra79. [CrossRef] [PubMed]
30. Maisel, K.; Reddy, M.; Xu, Q.; Chattopadhyay, S.; Cone, R.; Ensign, L.M.; Hanes, J. Nanoparticles coated with high molecular weight PEG penetrate mucus and provide uniform vaginal and colorectal distribution in vivo. *Nanomedicine* **2016**, *11*, 1337–1343. [CrossRef] [PubMed]
31. Henry, C.E.; Wang, Y.-Y.; Yang, Q.; Hoang, T.; Chattopadhyay, S.; Hoen, T.; Ensign, L.M.; Nunn, K.L.; Schroeder, H.; McCallen, J. Anti-PEG antibodies alter the mobility and biodistribution of densely PEGylated nanoparticles in mucus. *Acta Biomater.* **2016**, *43*, 61–70. [CrossRef] [PubMed]
32. Mohideen, M.; Quijano, E.; Song, E.; Deng, Y.; Panse, G.; Zhang, W.; Clark, M.R.; Saltzman, W.M. Degradable bioadhesive nanoparticles for prolonged intravaginal delivery and retention of elvitegravir. *Biomaterials* **2017**, *144*, 144–154. [CrossRef] [PubMed]
33. Ensign, L.; Cone, R.; Hanes, J. Nanoparticle Formulations with Enhanced Mucosal Penetration. U.S. Patent 9,415,020, 16 August 2016.
34. Lai, S.K.; O'Hanlon, E.D.; Man, S.T.; Cone, R.; Hanes, J. Real-time transport of polymer nanoparticles in cervical vaginal mucus. In Proceedings of the 05AIChE: 2005 AIChE Annual Meeting and Fall Showcase, Cincinnati, OH, USA, 30 October–4 November 2005.

35. Meng, J.; Agrahari, V.; Ezoulin, M.J.; Zhang, C.; Purohit, S.S.; Molteni, A.; Dim, D.; Oyler, N.A.; Youan, B.-B.C. Tenofovir containing thiolated chitosan core/shell nanofibers: In vitro and in vivo evaluations. *Mol. Pharm.* **2016**, *13*, 4129–4140. [CrossRef]
36. Zamani, M.; Prabhakaran, M.P.; Ramakrishna, S. Advances in drug delivery via electrospun and electrosprayed nanomaterials. *Int. J. Nanomed.* **2013**, *8*, 2997–3017.
37. Hu, X.; Liu, S.; Zhou, G.; Huang, Y.; Xie, Z.; Jing, X. Electrospinning of polymeric nanofibers for drug delivery applications. *J. Control. Release* **2014**, *185*, 12–21. [CrossRef]
38. Jain, K.K. *Drug Delivery Systems*; Springer Science & Business Media: Berlin, Germany, 2008.
39. Sharma, R.; Singh, H.; Joshi, M.; Sharma, A.; Garg, T.; Goyal, A.K.; Rath, G. Recent advances in polymeric electrospun nanofibers for drug delivery. *Crit. Rev. Ther. Drug Carr. Syst.* **2014**, *31*, 187–217. [CrossRef]
40. Repanas, A.; Andriopoulou, S.; Glasmacher, B. The significance of electrospinning as a method to create fibrous scaffolds for biomedical engineering and drug delivery applications. *J. Drug Deliv. Sci. Technol.* **2016**, *31*, 137–146. [CrossRef]
41. Fu, Y.; Kao, W.J. Drug release kinetics and transport mechanisms of non-degradable and degradable polymeric delivery systems. *Expert Opin. Drug Deliv.* **2010**, *7*, 429–444. [CrossRef] [PubMed]
42. Kim, T.G.; Lee, D.S.; Park, T.G. Controlled protein release from electrospun biodegradable fiber mesh composed of poly(ε-caprolactone) and poly(ethylene oxide). *Int. J. Pharm.* **2007**, *338*, 276–283. [CrossRef]
43. Qi, H.; Hu, P.; Xu, J.; Wang, A. Encapsulation of drug reservoirs in fibers by emulsion electrospinning: Morphology characterization and preliminary release assessment. *Biomacromolecules* **2006**, *7*, 2327–2330. [CrossRef] [PubMed]
44. Nair, L.S.; Laurencin, C.T. Biodegradable polymers as biomaterials. *Prog. Polym. Sci.* **2007**, *32*, 762–798. [CrossRef]
45. Liu, H.; Leonas, K.K.; Zhao, Y. Antimicrobial properties and release profile of ampicillin from electrospun poly (ε-caprolactone) nanofiber yarns. *J. Eng. Fabr. Fiber.* **2010**, *5*, 10–19. [CrossRef]
46. Yoshimoto, H.; Shin, Y.; Terai, H.; Vacanti, J. A biodegradable nanofiber scaffold by electrospinning and its potential for bone tissue engineering. *Biomaterials* **2003**, *24*, 2077–2082. [CrossRef]
47. Luu, Y.; Kim, K.; Hsiao, B.; Chu, B.; Hadjiargyrou, M. Development of a nanostructured DNA delivery scaffold via electrospinning of PLGA and PLA–PEG block copolymers. *J. Control. Release* **2003**, *89*, 341–353. [CrossRef]
48. Puppi, D.; Zhang, X.; Yang, L.; Chiellini, F.; Sun, X.; Chiellini, E. Nano/microfibrous polymeric constructs loaded with bioactive agents and designed for tissue engineering applications: A review. *J. Biomed. Mater. Res. B* **2014**, *102*, 1562–1579. [CrossRef]
49. Cipitria, A.; Skelton, A.; Dargaville, T.; Dalton, P.; Hutmacher, D. Design, fabrication and characterization of PCL electrospun scaffolds—A review. *J. Mater. Chem. A* **2011**, *21*, 9419–9453. [CrossRef]
50. Uhrich, K.E.; Cannizzaro, S.M.; Langer, R.S.; Shakesheff, K.M. Polymeric systems for controlled drug release. *Chem. Rev.* **1999**, *99*, 3181–3198. [CrossRef]
51. Ulery, B.D.; Nair, L.S.; Laurencin, C.T. Biomedical applications of biodegradable polymers. *J. Polym. Sci. Part B Polym. Phys.* **2011**, *49*, 832–864. [CrossRef] [PubMed]
52. Chen, D.W.-C.; Liu, S.-J. Nanofibers used for delivery of antimicrobial agents. *Nanomedicine* **2015**, *10*, 1959–1971. [CrossRef] [PubMed]
53. Blakney, A.K.; Ball, C.; Krogstad, E.A.; Woodrow, K.A. Electrospun fibers for vaginal anti-HIV drug delivery. *Antivir. Res.* **2013**, *100*, S9–S16. [CrossRef] [PubMed]
54. Ramakrishna, S.; Fujihara, K.; Teo, W.-E.; Yong, T.; Ma, Z.; Ramaseshan, R. Electrospun nanofibers: Solving global issues. *Mater. Today* **2006**, *9*, 40–50. [CrossRef]
55. Göpferich, A. Mechanisms of polymer degradation and erosion. *Biomaterials* **1996**, *17*, 103–114. [CrossRef]
56. Chou, S.F.; Woodrow, K.A. Relationships between mechanical properties and drug release from electrospun fibers of PCL and PLGA blends. *J. Mech. Behav. Biomed. Mater.* **2017**, *65*, 724–733. [CrossRef] [PubMed]
57. von Burkersroda, F.; Schedl, L.; Göpferich, A. Why degradable polymers undergo surface erosion or bulk erosion. *Biomaterials* **2002**, *23*, 4221–4231. [CrossRef]
58. Doshi, J.; Reneker, D.H. Electrospinning process and applications of electrospun fibers. *J. Electrostat.* **1995**, *35*, 151–160. [CrossRef]
59. Zeng, J.; Xu, X.; Chen, X.; Liang, Q.; Bian, X.; Yang, L.; Jing, X. Biodegradable electrospun fibers for drug delivery. *J. Control Release* **2003**, *92*, 227–231. [CrossRef]

60. Sill, T.J.; von Recum, H.A. Electrospinning: Applications in drug delivery and tissue engineering. *Biomaterials* **2008**, *29*, 1989–2006. [CrossRef]
61. Kenawy, E.-R.; Abdel-Hay, F.I.; El-Newehy, M.H.; Wnek, G.E. Processing of polymer nanofibers through electrospinning as drug delivery systems. In *Nanomaterials: Risks and Benefits*; Springer: Dordrecht, The Netherlands, 2009; pp. 247–263.
62. Ji, W.; Sun, Y.; Yang, F.; van den Beucken, J.J.; Fan, M.; Chen, Z.; Jansen, J.A. Bioactive electrospun scaffolds delivering growth factors and genes for tissue engineering applications. *Pharm. Res.* **2011**, *28*, 1259–1272. [CrossRef]
63. Pillay, V.; Dott, C.; Choonara, Y.E.; Tyagi, C.; Tomar, L.; Kumar, P.; du Toit, L.C.; Ndesendo, V.M. A review of the effect of processing variables on the fabrication of electrospun nanofibers for drug delivery applications. *J. Nanomater.* **2013**, *2013*, 789289. [CrossRef]
64. Xie, J.; Li, X.; Xia, Y. Putting electrospun nanofibers to work for biomedical research. *Macromol. Rapid Commun.* **2008**, *29*, 1775–1792. [CrossRef]
65. Subbiah, T.; Bhat, G.; Tock, R.; Parameswaran, S.; Ramkumar, S. Electrospinning of nanofibers. *J. Appl. Polym. Sci.* **2005**, *96*, 557–569. [CrossRef]
66. Hadjiargyrou, M.; Chiu, J.B. Enhanced composite electrospun nanofiber scaffolds for use in drug delivery. *Expert Opin. Drug Deliv.* **2008**, *5*, 1093–1106. [CrossRef]
67. Zhang, Y.; Lim, C.T.; Ramakrishna, S.; Huang, Z.-M. Recent development of polymer nanofibers for biomedical and biotechnological applications. *J. Mater. Sci. Mater. Med.* **2005**, *16*, 933–946. [CrossRef]
68. Verreck, G.; Chun, I.; Rosenblatt, J.; Peeters, J.; van Dijck, A.; Mensch, J.; Noppe, M.; Brewster, M.E. Incorporation of drugs in an amorphous state into electrospun nanofibers composed of a water-insoluble, nonbiodegradable polymer. *J. Control. Release* **2003**, *92*, 349–360. [CrossRef]
69. Han, D.; Steckl, A.J. Triaxial electrospun nanofiber membranes for controlled dual release of functional molecules. *ACS Appl. Mater. Interfaces* **2013**, *5*, 8241–8245. [CrossRef]
70. Yarin, A. Coaxial electrospinning and emulsion electrospinning of core–shell fibers. *Polym. Adv. Technol.* **2011**, *22*, 310–317. [CrossRef]
71. He, C.L.; Huang, Z.M.; Han, X.J.; Liu, L.; Zhang, H.S.; Chen, L.S. Coaxial electrospun poly (L-lactic acid) ultrafine fibers for sustained drug delivery. *J. Macromol. Sci. B* **2006**, *45*, 515–524. [CrossRef]
72. Lu, Y.; Huang, J.; Yu, G.; Cardenas, R.; Wei, S.; Wujcik, E.K.; Guo, Z. Coaxial electrospun fibers: Applications in drug delivery and tissue engineering. *Wiley Interdiscip. Rev. Nanomed. Nanobiotechnol.* **2016**, *8*, 654–677. [CrossRef]
73. Yu, D.G.; Branford-White, C.; Bligh, S.A.; White, K.; Chatterton, N.P.; Zhu, L.M. Improving Polymer Nanofiber Quality Using a Modified Co-axial Electrospinning Process. *Macromol. Rapid Commun.* **2011**, *32*, 744–750. [CrossRef]
74. Nezarati, R.M.; Eifert, M.B.; Cosgriff-Hernandez, E. Effects of humidity and solution viscosity on electrospun fiber morphology. *Tissue Eng. Part C Methods* **2013**, *19*, 810–819. [CrossRef]
75. Wang, J.; Jákli, A.; West, J.L. Morphology tuning of electrospun liquid crystal/polymer fibers. *ChemPhysChem* **2016**, *17*, 3080–3085. [CrossRef]
76. Yang, J.-M.; Zha, L.-S.; Yu, D.-G.; Liu, J. Coaxial electrospinning with acetic acid for preparing ferulic acid/zein composite fibers with improved drug release profiles. *Colloids Surf. B Biointerfaces* **2013**, *102*, 737–743. [CrossRef]
77. Tang, C.; Ozcam, A.E.; Stout, B.; Khan, S.A. Effect of pH on protein distribution in electrospun PVA/BSA composite nanofibers. *Biomacromolecules* **2012**, *13*, 1269–1278. [CrossRef]
78. He, M.; Jiang, H.; Wang, R.; Xie, Y.; Zhao, C. Fabrication of metronidazole loaded poly (ε-caprolactone)/zein core/shell nanofiber membranes via coaxial electrospinning for guided tissue regeneration. *J. Colloid Interface Sci.* **2017**, *490*, 270–278. [CrossRef]
79. Wang, C.; Yan, K.-W.; Lin, Y.-D.; Hsieh, P.C. Biodegradable core/shell fibers by coaxial electrospinning: Processing, fiber characterization, and its application in sustained drug release. *Macromolecules* **2010**, *43*, 6389–6397. [CrossRef]
80. Perrie, Y.; Rades, T. *FASTtrack Pharmaceutics: Drug Delivery and Targeting*; Pharmaceutical Press: London, UK, 2012.
81. Jiang, Y.-N.; Mo, H.-Y.; Yu, D. Electrospun drug-loaded core-sheath PVP/zein nanofibers for biphasic drug release. *Int. J. Pharm.* **2012**, *438*, 232–239. [CrossRef]

82. Zhu, L.; Liu, X.; Du, L.; Jin, Y. Preparation of asiaticoside-loaded coaxially electrospinning nanofibers and their effect on deep partial-thickness burn injury. *Biomed. Pharmacother.* **2016**, *83*, 33–40. [CrossRef]
83. Castillo-Ortega, M.; Montaño-Figueroa, A.; Rodríguez-Félix, D.; Prado-Villegas, G.; Pino-Ocaño, K.; Valencia-Córdova, M.; Quiroz-Castillo, J.; Herrera-Franco, P. Preparation by coaxial electrospinning and characterization of membranes releasing (−) epicatechin as scaffold for tissue engineering. *Mater. Sci. Eng. C* **2015**, *46*, 184–189. [CrossRef] [PubMed]
84. Li, X.-Y.; Li, Y.-C.; Yu, D.-G.; Liao, Y.-Z.; Wang, X. Fast disintegrating quercetin-loaded drug delivery systems fabricated using coaxial electrospinning. *Int. J. Mol. Sci.* **2013**, *14*, 21647–21659. [CrossRef]
85. Yu, D.-G.; Zhu, L.-M.; Branford-White, C.J.; Yang, J.-H.; Wang, X.; Li, Y.; Qian, W. Solid dispersions in the form of electrospun core-sheath nanofibers. *Int. J. Nanomed.* **2011**, *6*, 3271–3280. [CrossRef]
86. Fu, L.; Zhang, J.; Yang, G. Present status and applications of bacterial cellulose-based materials for skin tissue repair. *Carbohydr. Polym.* **2013**, *92*, 1432–1442. [CrossRef]
87. Yu, D.-G.; Li, X.-Y.; Wang, X.; Yang, J.-H.; Bligh, S.A.; Williams, G.R. Nanofibers fabricated using triaxial electrospinning as zero order drug delivery systems. *ACS Appl. Mater. Interfaces* **2015**, *7*, 18891–18897. [CrossRef]
88. Nakielski, P.; Pawłowska, S.; Pierini, F.; Liwińska, W.; Hejduk, P.; Zembrzycki, K.; Zabost, E.; Kowalewski, T.A. Hydrogel nanofilaments via core-shell electrospinning. *PLoS ONE* **2015**, *10*, e0129816.
89. Zhu, Y.J.; Chen, F. pH-Responsive Drug-Delivery Systems. *Chem.—Asian J.* **2015**, *10*, 284–305. [CrossRef]
90. Yang, C.; Yu, D.-G.; Pan, D.; Liu, X.-K.; Wang, X.; Bligh, S.A.; Williams, G.R. Electrospun pH-sensitive core–shell polymer nanocomposites fabricated using a tri-axial process. *Acta Biomater.* **2016**, *35*, 77–86. [CrossRef]
91. Thakral, S.; Thakral, N.K.; Majumdar, D.K. Eudragit®: A technology evaluation. *Expert Opin. Drug Deliv.* **2013**, *10*, 131–149. [CrossRef]
92. Yoshida, T.; Lai, T.C.; Kwon, G.S.; Sako, K. pH-and ion-sensitive polymers for drug delivery. *Expert Opin. Drug Deliv.* **2013**, *10*, 1497–1513. [CrossRef] [PubMed]
93. Marrazzo, J.M.; Ramjee, G.; Richardson, B.A.; Gomez, K.; Mgodi, N.; Nair, G.; Palanee, T.; Nakabiito, C.; van der Straten, A.; Noguchi, L.; et al. Tenofovir-based preexposure prophylaxis for HIV infection among African women. *N. Engl. J. Med.* **2015**, *372*, 509–518. [CrossRef] [PubMed]
94. Jin, M.; Yu, D.-G.; Geraldes, C.F.; Williams, G.R.; Bligh, S.A. Theranostic fibers for simultaneous imaging and drug delivery. *Mol. Pharm.* **2016**, *13*, 2457–2465. [CrossRef] [PubMed]
95. Jia, D.; Gao, Y.; Williams, G.R. Core/shell poly (ethylene oxide)/Eudragit fibers for site-specific release. *Int. J. Pharm.* **2017**, *523*, 376–385. [CrossRef]
96. Hua, D.; Liu, Z.; Wang, F.; Gao, B.; Chen, F.; Zhang, Q.; Xiong, R.; Han, J.; Samal, S.K.; de Smedt, S.C.; et al. pH responsive polyurethane (core) and cellulose acetate phthalate (shell) electrospun fibers for intravaginal drug delivery. *Carbohydr. Polym.* **2016**, *151*, 1240–1244. [CrossRef]
97. Sang, Q.; Li, H.; Williams, G.; Wu, H.; Zhu, L.-M. Core-shell poly (lactide-co-ε-caprolactone)-gelatin fiber scaffolds as pH-sensitive drug delivery systems. *J. Biomater. Appl.* **2018**, *32*, 1105–1118. [CrossRef] [PubMed]
98. Han, D.; Yu, X.; Chai, Q.; Ayres, N.; Steckl, A.J. Stimuli-responsive self-immolative polymer nanofiber membranes formed by coaxial electrospinning. *ACS Appl. Mater. Interfaces* **2017**, *9*, 11858–11865. [CrossRef] [PubMed]
99. Ball, C.; Chou, S.-F.; Jiang, Y.; Woodrow, K.A. Coaxially electrospun fiber-based microbicides facilitate broadly tunable release of maraviroc. *Mater. Sci. Eng. C* **2016**, *63*, 117–124. [CrossRef] [PubMed]
100. Zhang, Y.; Wang, X.; Feng, Y.; Li, J.; Lim, C.; Ramakrishna, S. Coaxial electrospinning of (fluorescein isothiocyanate-conjugated bovine serum albumin)-encapsulated poly (ε-caprolactone) nanofibers for sustained release. *Biomacromolecules* **2006**, *7*, 1049–1057. [CrossRef]
101. Yu, H.; Yang, P.; Jia, Y.; Zhang, Y.; Ye, Q.; Zeng, S. Regulation of biphasic drug release behavior by graphene oxide in polyvinyl pyrrolidone/poly (ε-caprolactone) core/sheath nanofiber mats. *Colloids Surf. B Biointerfaces* **2016**, *146*, 63–69. [CrossRef] [PubMed]
102. Oliveira, M.F.; Suarez, D.; Rocha, J.C.B.; de Carvalho Teixeira, A.V.N.; Cortés, M.E.; De Sousa, F.B.; Sinisterra, R.D. Electrospun nanofibers of polyCD/PMAA polymers and their potential application as drug delivery system. *Mater. Sci. Eng. C* **2015**, *54*, 252–261. [CrossRef]
103. Sultanova, Z.; Kaleli, G.; Kabay, G.; Mutlu, M. Controlled release of a hydrophilic drug from coaxially electrospun polycaprolactone nanofibers. *Int. J. Pharm.* **2016**, *505*, 133–138. [CrossRef]

104. Lv, Y.; Pan, Q.; Bligh, S.W.; Li, H.; Wu, H.; Sang, Q.; Zhu, L.M. Core-Sheath Nanofibers as Drug Delivery System for Thermoresponsive Controlled Release. *J. Pharm. Sci.* **2017**, *106*, 1258–1265. [CrossRef] [PubMed]
105. Khalf, A.; Madihally, S.V. Modeling the permeability of multiaxial electrospun poly (ε-caprolactone)-gelatin hybrid fibers for controlled doxycycline release. *Mater. Sci. Eng. C* **2017**, *76*, 161–170. [CrossRef] [PubMed]
106. Ranjbar-Mohammadi, M.; Zamani, M.; Prabhakaran, M.P.; Bahrami, S.H.; Ramakrishna, S. Electrospinning of PLGA/gum tragacanth nanofibers containing tetracycline hydrochloride for periodontal regeneration. *Mater. Sci. Eng. C Mater. Biol. Appl.* **2016**, *58*, 521–531. [CrossRef]
107. Xie, Q.; Jia, L.-N.; Xu, H.-Y.; Hu, X.-G.; Wang, W.; Jia, J. Fabrication of core-shell PEI/pBMP2-PLGA electrospun scaffold for gene delivery to periodontal ligament stem cells. *Stem Cells Int.* **2016**, *2016*, 5385137. [CrossRef]
108. Jiang, H.; Hu, Y.; Zhao, P.; Li, Y.; Zhu, K. Modulation of protein release from biodegradable core-shell structured fibers prepared by coaxial electrospinning. *J. Biomed. Mater. Res. B Appl. Biomater.* **2006**, *79*, 50–57. [CrossRef]
109. Jiang, H.; Hu, Y.; Li, Y.; Zhao, P.; Zhu, K.; Chen, W. A facile technique to prepare biodegradable coaxial electrospun nanofibers for controlled release of bioactive agents. *J. Control. Release* **2005**, *108*, 237–243. [CrossRef]
110. Yang, Y.; Li, X.; Qi, M.; Zhou, S.; Weng, J. Release pattern and structural integrity of lysozyme encapsulated in core–sheath structured poly (DL-lactide) ultrafine fibers prepared by emulsion electrospinning. *Eur. J. Pharm. Biopharm.* **2008**, *69*, 106–116. [CrossRef]
111. Ji, W.; Yang, F.; van den Beucken, J.J.; Bian, Z.; Fan, M.; Chen, Z.; Jansen, J.A. Fibrous scaffolds loaded with protein prepared by blend or coaxial electrospinning. *Acta Biomater.* **2010**, *6*, 4199–4207. [CrossRef]
112. Saraf, A.; Baggett, L.S.; Raphael, R.M.; Kasper, F.K.; Mikos, A.G. Regulated non-viral gene delivery from coaxial electrospun fiber mesh scaffolds. *J. Control. Release* **2010**, *143*, 95–103. [CrossRef]
113. Hsu, Y.-H.; Lin, C.-T.; Yu, Y.-H.; Chou, Y.-C.; Liu, S.-J.; Chan, E.-C. Dual delivery of active antibactericidal agents and bone morphogenetic protein at sustainable high concentrations using biodegradable sheath-core-structured drug-eluting nanofibers. *Int. J. Nanomed.* **2016**, *11*, 3927–3937. [CrossRef]
114. Aniagyei, S.E.; Sims, L.B.; Malik, D.A.; Tyo, K.M.; Curry, K.C.; Kim, W.; Hodge, D.A.; Duan, J.; Steinbach-Rankins, J.M. Evaluation of poly(lactic-co-glycolic acid) and poly(dl-lactide-co-ε-caprolactone) electrospun fibers for the treatment of HSV-2 infection. *Mater. Sci. Eng. C* **2017**, *72*, 238–251. [CrossRef]
115. Ball, C.; Woodrow, K.A. Electrospun Solid Dispersions of Maraviroc for Rapid Intravaginal Preexposure Prophylaxis of HIV. *Antimicrob. Agents Chemother.* **2014**, *58*, 4855–4865. [CrossRef]
116. Tyo, K.M.; Vuong, H.R.; Malik, D.A.; Sims, L.B.; Alatassi, H.; Duan, J.; Watson, W.H.; Steinbach-Rankins, J.M. Multipurpose tenofovir disoproxil fumarate electrospun fibers for the prevention of HIV-1 and HSV-2 infections in vitro. *Int. J. Pharm.* **2017**, *531*, 118–133. [CrossRef]
117. Grooms, T.N.; Vuong, H.R.; Tyo, K.M.; Malik, D.A.; Sims, L.B.; Whittington, C.P.; Palmer, K.E.; Matoba, N.; Steinbach-Rankins, J.M. Griffithsin-Modified Electrospun Fibers as a Delivery Scaffold To Prevent HIV Infection. *Antimicrob. Agents Chemother.* **2016**, *60*, 6518–6531. [CrossRef]
118. Tyo, K.M.; Duan, J.; Kollipara, P.; Cerna, M.V.C.D.; Lee, D.; Palmer, K.E.; Steinbach-Rankins, J.M. pH-responsive delivery of Griffithsin from electrospun fibers. *Eur. J. Pharm. Biopharm.* **2018**. [CrossRef] [PubMed]
119. Carson, D.; Jiang, Y.H.; Woodrow, K.A. Tunable Release of Multiclass Anti-HIV Drugs that are Water-Soluble and Loaded at High Drug Content in Polyester Blended Electrospun Fibers. *Pharm. Res.* **2016**, *33*, 125–136. [CrossRef] [PubMed]
120. Liu, L.; Kamei, K.-i.; Yoshioka, M.; Nakajima, M.; Li, J.; Fujimoto, N.; Terada, S.; Tokunaga, Y.; Koyama, Y.; Sato, H. Nano-on-micro fibrous extracellular matrices for scalable expansion of human ES/iPS cells. *Biomaterials* **2017**, *124*, 47–54. [CrossRef] [PubMed]
121. Huang, L.-Y.; Branford-White, C.; Shen, X.-X.; Yu, D.-G.; Zhu, L.-M. Time-engineeringed biphasic drug release by electrospun nanofiber meshes. *Int. J. Pharm.* **2012**, *436*, 88–96. [CrossRef]
122. Meinel, A.J.; Germershaus, O.; Luhmann, T.; Merkle, H.P.; Meinel, L. Electrospun matrices for localized drug delivery: Current technologies and selected biomedical applications. *Eur. J. Pharm. Biopharm.* **2012**, *81*, 1–13. [CrossRef]
123. Blakney, A.K.; Krogstad, E.A.; Jiang, Y.H.; Woodrow, K.A. Delivery of multipurpose prevention drug combinations from electrospun nanofibers using composite microarchitectures. *Int. J. Nanomed.* **2014**, *9*, 2967–2978. [CrossRef]

124. Mehrotra, S.; Lynam, D.; Maloney, R.; Pawelec, K.M.; Tuszynski, M.H.; Lee, I.; Chan, C.; Sakamoto, J. Time controlled protein release from layer-by-layer assembled multilayer functionalized agarose hydrogels. *Adv. Funct. Mater.* **2010**, *20*, 247–258. [CrossRef]
125. Pan, H.; Li, L.; Hu, L.; Cui, X. Continuous aligned polymer fibers produced by a modified electrospinning method. *Polymer* **2006**, *47*, 4901–4904. [CrossRef]
126. Shin, J.-W.; Shin, H.; Heo, S.; Lee, Y.; Hwang, Y.; Kim, D.; Kim, J.; Shin, J. Hybrid nanofiber scaffolds of polyurethane and poly (ethylene oxide) using dual-electrospinning for vascular tissue engineering. In *3rd Kuala Lumpur International Conference on Biomedical Engineering 2006*; Springer: Berlin/Heidelberg, Germany, 2007; pp. 692–695.
127. Baker, B.M.; Gee, A.O.; Metter, R.B.; Nathan, A.S.; Marklein, R.A.; Burdick, J.A.; Mauck, R.L. The potential to improve cell infiltration in composite fiber-aligned electrospun scaffolds by the selective removal of sacrificial fibers. *Biomaterials* **2008**, *29*, 2348–2358. [CrossRef]
128. Tijing, L.D.; Ruelo, M.T.G.; Amarjargal, A.; Pant, H.R.; Park, C.-H.; Kim, C.S. One-step fabrication of antibacterial (silver nanoparticles/poly(ethylene oxide))—Polyurethane bicomponent hybrid nanofibrous mat by dual-spinneret electrospinning. *Mater. Chem. Phys.* **2012**, *134*, 557–561. [CrossRef]
129. Wulkersdorfer, B.; Kao, K.; Agopian, V.; Ahn, A.; Dunn, J.; Wu, B.; Stelzner, M. Bimodal porous scaffolds by sequential electrospinning of poly (glycolic acid) with sucrose particles. *Int. J. Polym. Sci.* **2010**, *2010*, 436178. [CrossRef]
130. Wan, A.C.; Ying, J.Y. Nanomaterials for in situ cell delivery and tissue regeneration. *Adv. Drug Deliv. Rev.* **2010**, *62*, 731–740. [CrossRef]
131. Dvir, T.; Timko, B.P.; Kohane, D.S.; Langer, R. Nanotechnological strategies for engineering complex tissues. *Nat. Nanotechnol.* **2011**, *6*, 13–22. [CrossRef]
132. Kharaziha, M.; Fathi, M.; Edris, H. Tunable cellular interactions and physical properties of nanofibrous PCL-forsterite: Gelatin scaffold through sequential electrospinning. *Compos. Sci. Technol.* **2013**, *87*, 182–188. [CrossRef]
133. Tan, L.; Hu, J.; Zhao, H. Design of bilayered nanofibrous mats for wound dressing using an electrospinning technique. *Mater. Lett.* **2015**, *156*, 46–49. [CrossRef]
134. Falde, E.J.; Freedman, J.D.; Herrera, V.L.M.; Yohe, S.T.; Colson, Y.L.; Grinstaff, M.W. Layered superhydrophobic meshes for controlled drug release. *J. Control. Release* **2015**, *214*, 23–29. [CrossRef]
135. Sirc, J.; Kubinova, S.; Hobzova, R.; Stranska, D.; Kozlik, P.; Bosakova, Z.; Marekova, D.; Holan, V.; Sykova, E.; Michalek, J. Controlled gentamicin release from multi-layered electrospun nanofibrous structures of various thicknesses. *Int. J. Nanomed.* **2012**, *7*, 5315–5325. [CrossRef] [PubMed]
136. Mandal, B.B.; Mann, J.K.; Kundu, S. Silk fibroin/gelatin multilayered films as a model system for controlled drug release. *Eur. J. Pharm. Sci.* **2009**, *37*, 160–171. [CrossRef] [PubMed]
137. Okuda, T.; Tominaga, K.; Kidoaki, S. Time-programmed dual release formulation by multilayered drug-loaded nanofiber meshes. *J. Control. Release* **2010**, *143*, 258–264. [CrossRef] [PubMed]
138. Liu, S.; Wang, X.; Zhang, Z.; Zhang, Y.; Zhou, G.; Huang, Y.; Xie, Z.; Jing, X. Use of asymmetric multilayer polylactide nanofiber mats in controlled release of drugs and prevention of liver cancer recurrence after surgery in mice. *Nanomed. Nanotechnol. Biol. Med.* **2015**, *11*, 1047–1056. [CrossRef]
139. Chunder, A.; Sarkar, S.; Yu, Y.; Zhai, L. Fabrication of ultrathin polyelectrolyte fibers and their controlled release properties. *Colloids Surf. B Biointerfaces* **2007**, *58*, 172–179. [CrossRef] [PubMed]
140. Son, Y.J.; Kim, W.J.; Yoo, H.S. Therapeutic applications of electrospun nanofibers for drug delivery systems. *Arch. Pharmacal Res.* **2014**, *37*, 69–78. [CrossRef] [PubMed]
141. Yoon, H.; Kim, G.H. Layer-by-layered electrospun micro/nanofibrous mats for drug delivery system. *Macromol. Res.* **2012**, *20*, 402–406. [CrossRef]
142. Park, J.H.; Kim, B.S.; Yoo, Y.C.; Khil, M.S.; Kim, H.Y. Enhanced mechanical properties of multilayer nano-coated electrospun nylon 6 fibers via a layer-by-layer self-assembly. *J. Appl. Polym. Sci.* **2008**, *107*, 2211–2216. [CrossRef]
143. Woodrow, K.A.; Cu, Y.; Booth, C.J.; Saucier-Sawyer, J.K.; Wood, M.J.; Saltzman, W.M. Intravaginal gene silencing using biodegradable polymer nanoparticles densely loaded with small-interfering RNA. *Nat. Mater.* **2009**, *8*, 526–533. [CrossRef] [PubMed]
144. Ensign, L.M.; Cone, R.; Hanes, J. Nanoparticle-based drug delivery to the vagina: A review. *J. Control. Release* **2014**, *190*, 500–514. [CrossRef]

145. Ahmad, M.Z.; Akhter, S.; Jain, G.K.; Rahman, M.; Pathan, S.A.; Ahmad, F.J.; Khar, R.K. Metallic nanoparticles: Technology overview & drug delivery applications in oncology. *Expert Opin. Drug Deliv.* **2010**, *7*, 927–942. [PubMed]
146. Mody, V.V.; Siwale, R.; Singh, A.; Mody, H.R. Introduction to metallic nanoparticles. *J. Pharm. Bioallied Sci.* **2010**, *2*, 282–289. [CrossRef] [PubMed]
147. Singh, R.; Lillard, J.W., Jr. Nanoparticle-based targeted drug delivery. *Exp. Mol. Pathol.* **2009**, *86*, 215–223. [CrossRef]
148. Blanco, E.; Shen, H.; Ferrari, M. Principles of nanoparticle design for overcoming biological barriers to drug delivery. *Nat. Biotechnol.* **2015**, *33*, 941–951. [CrossRef]
149. Peer, D.; Karp, J.M.; Hong, S.; Farokhzad, O.C.; Margalit, R.; Langer, R. Nanocarriers as an emerging platform for cancer therapy. *Nat. Nanotechnol.* **2007**, *2*, 751–760. [CrossRef]
150. Gu, J.; Yang, S.; Ho, E.A. Biodegradable film for the targeted delivery of siRNA-loaded nanoparticles to vaginal immune cells. *Mol. Pharm.* **2015**, *12*, 2889–2903. [CrossRef] [PubMed]
151. Wang, Y.Y.; Lai, S.K.; Suk, J.S.; Pace, A.; Cone, R.; Hanes, J. Addressing the PEG mucoadhesivity paradox to engineer nanoparticles that "slip" through the human mucus barrier. *Angew. Chem. Int. Ed.* **2008**, *47*, 9726–9729. [CrossRef] [PubMed]
152. Wang, S.; Zhao, Y.; Shen, M.; Shi, X. Electrospun hybrid nanofibers doped with nanoparticles or nanotubes for biomedical applications. *Ther. Deliv.* **2012**, *3*, 1155–1169. [CrossRef] [PubMed]
153. Chen, M.; Gao, S.; Dong, M.; Song, J.; Yang, C.; Howard, K.A.; Kjems, J.; Besenbacher, F. Chitosan/siRNA nanoparticles encapsulated in PLGA nanofibers for siRNA delivery. *ACS Nano* **2012**, *6*, 4835–4844. [CrossRef] [PubMed]
154. Sridhar, R.; Lakshminarayanan, R.; Madhaiyan, K.; Barathi, V.A.; Lim, K.H.C.; Ramakrishna, S. Electrosprayed nanoparticles and electrospun nanofibers based on natural materials: Applications in tissue regeneration, drug delivery and pharmaceuticals. *Chem. Soc. Rev.* **2015**, *44*, 790–814. [CrossRef]
155. Mehrasa, M.; Asadollahi, M.A.; Nasri-Nasrabadi, B.; Ghaedi, K.; Salehi, H.; Dolatshahi-Pirouz, A.; Arpanaei, A. Incorporation of mesoporous silica nanoparticles into random electrospun PLGA and PLGA/gelatin nanofibrous scaffolds enhances mechanical and cell proliferation properties. *Mater. Sci. Eng. C* **2016**, *66*, 25–32. [CrossRef]
156. Song, B.; Wu, C.; Chang, J. Controllable delivery of hydrophilic and hydrophobic drugs from electrospun poly (lactic-co-glycolic acid)/mesoporous silica nanoparticles composite mats. *J. Biomed. Mater. Res. B* **2012**, *100*, 2178–2186. [CrossRef]
157. Beck-Broichsitter, M.; Thieme, M.; Nguyen, J.; Schmehl, T.; Gessler, T.; Seeger, W.; Agarwal, S.; Greiner, A.; Kissel, T. Novel 'Nano in Nano'Composites for Sustained Drug Delivery: Biodegradable Nanoparticles Encapsulated into Nanofiber Non-Wovens. *Macromol. Biosci.* **2010**, *10*, 1527–1535. [CrossRef] [PubMed]
158. Fathollahipour, S.; Mehrizi, A.A.; Ghaee, A.; Koosha, M. Electrospinning of PVA/chitosan nanocomposite nanofibers containing gelatin nanoparticles as a dual drug delivery system. *J. Biomed. Mater. Res. Part A* **2015**, *103*, 3852–3862. [CrossRef]
159. Hu, J.; Zeng, F.; Wei, J.; Chen, Y.; Chen, Y. Novel controlled drug delivery system for multiple drugs based on electrospun nanofibers containing nanomicelles. *J. Biomater. Sci. Polym. Ed.* **2014**, *25*, 257–268. [CrossRef]
160. Wang, Y.; Wang, B.; Qiao, W.; Yin, T. A novel controlled release drug delivery system for multiple drugs based on electrospun nanofibers containing nanoparticles. *J. Pharm. Sci.* **2010**, *99*, 4805–4811. [CrossRef]
161. Li, L.; Zhou, G.; Wang, Y.; Yang, G.; Ding, S.; Zhou, S. Controlled dual delivery of BMP-2 and dexamethasone by nanoparticle-embedded electrospun nanofibers for the efficient repair of critical-sized rat calvarial defect. *Biomaterials* **2015**, *37*, 218–229. [CrossRef]
162. Ali, I.H.; Khalil, I.A.; El-Sherbiny, I.M. Single-Dose Electrospun Nanoparticles-in-Nanofibers Wound Dressings with Enhanced Epithelialization, Collagen Deposition, and Granulation Properties. *ACS Appl. Mater. Interfaces* **2016**, *8*, 14453–14469. [CrossRef]
163. Sun, X.; Li, K.; Chen, S.; Yao, B.; Zhou, Y.; Cui, S.; Hu, J.; Liu, Y. Rationally designed particle preloading method to improve protein delivery performance of electrospun polyester nanofibers. *Int. J. Pharm.* **2016**, *512*, 204–212. [CrossRef]
164. Vakilian, S.; Mashayekhan, S.; Shabani, I.; Khorashadizadeh, M.; Fallah, A.; Soleimani, M. Structural stability and sustained release of protein from a multilayer nanofiber/nanoparticle composite. *Int. J. Biol. Macromol.* **2015**, *75*, 248–257. [CrossRef]

165. Nie, H.; Wang, C.-H. Fabrication and characterization of PLGA/HAp composite scaffolds for delivery of BMP-2 plasmid DNA. *J. Control. Release* **2007**, *120*, 111–121. [CrossRef]
166. Cui, W.; Zhou, Y.; Chang, J. Electrospun nanofibrous materials for tissue engineering and drug delivery. *Sci. Technol. Adv. Mater.* **2010**, *11*, 014108. [CrossRef]
167. Shao, W.; He, J.; Sang, F.; Ding, B.; Chen, L.; Cui, S.; Li, K.; Han, Q.; Tan, W. Coaxial electrospun aligned tussah silk fibroin nanostructured fiber scaffolds embedded with hydroxyapatite–tussah silk fibroin nanoparticles for bone tissue engineering. *Mater. Sci. Eng. C* **2016**, *58*, 342–351. [CrossRef]
168. Weissleder, R.; Kelly, K.; Sun, E.Y.; Shtatland, T.; Josephson, L. Cell-specific targeting of nanoparticles by multivalent attachment of small molecules. *Nat. Biotechnol.* **2005**, *23*, 1418–1423. [CrossRef]
169. Kohler, N.; Fryxell, G.E.; Zhang, M. A bifunctional poly (ethylene glycol) silane immobilized on metallic oxide-based nanoparticles for conjugation with cell targeting agents. *JACS* **2004**, *126*, 7206–7211. [CrossRef]
170. Yao, L.; Lin, Y.; Watkins, J.J. Ultrahigh loading of nanoparticles into ordered block copolymer composites. *Macromolecules* **2014**, *47*, 1844–1849. [CrossRef]
171. Zhu, J.; Wei, S.; Chen, X.; Karki, A.B.; Rutman, D.; Young, D.P.; Guo, Z. Electrospun polyimide nanocomposite fibers reinforced with core-shell Fe-FeO nanoparticles. *J. Phys. Chem. C* **2010**, *114*, 8844–8850. [CrossRef]
172. Delany-Moretlwe, S.; Lombard, C.; Baron, D.; Bekker, L.-G.; Nkala, B.; Ahmed, K.; Sebe, M.; Brumskine, W.; Nchabeleng, M.; Palanee-Philips, T.; et al. Tenofovir 1% vaginal gel for prevention of HIV-1 infection in women in South Africa (FACTS-001): A phase 3, randomised, double-blind, placebo-controlled trial. *Lancet Infect. Dis.* **2018**, *18*, 1241–1250. [CrossRef]
173. Skoler-Karpoff, S.; Ramjee, G.; Ahmed, K.; Altini, L.; Plagianos, M.G.; Friedland, B.; Govender, S.; de Kock, A.; Cassim, N.; Palanee, T.; et al. Efficacy of Carraguard for prevention of HIV infection in women in South Africa: A randomised, double-blind, placebo-controlled trial. *Lancet* **2008**, *372*, 1977–1987. [CrossRef]
174. Thurman, A.R.; Clark, M.R.; Doncel, G.F. Multipurpose prevention technologies: Biomedical tools to prevent HIV-1, HSV-2, and unintended pregnancies. *Infect. Dis. Obstet. Gynecol.* **2011**, *2011*, 429403. [CrossRef]
175. Blakney, A.K.; Simonovsky, F.I.; Suydam, I.T.; Ratner, B.D.; Woodrow, K.A. Rapidly Biodegrading PLGA-Polyurethane Fibers for Sustained Release of Physicochemically Diverse Drugs. *ACS Biomater. Sci. Eng.* **2016**, *2*, 1595–1607. [CrossRef]
176. Halwes, M.E.; Tyo, K.M.; Steinbach-Rankins, J.M.; Frieboes, H.B. Computational Modeling of Antiviral Drug Diffusion from Poly(lactic- co-glycolic-acid) Fibers and Multicompartment Pharmacokinetics for Application to the Female Reproductive Tract. *Mol. Pharm.* **2018**, *15*, 1534–1547. [CrossRef]
177. Moss, J.A.; Malone, A.M.; Smith, T.J.; Kennedy, S.; Nguyen, C.; Vincent, K.L.; Motamedi, M.; Baum, M.M. Pharmacokinetics of a Multipurpose Pod-Intravaginal Ring Simultaneously Delivering Five Drugs in an Ovine Model. *Antimicrob. Agents Chemother.* **2013**, *57*, 3994–3997. [CrossRef]
178. Smith, J.M.; Moss, J.A.; Srinivasan, P.; Butkyavichene, I.; Gunawardana, M.; Fanter, R.; Miller, C.S.; Sanchez, D.; Yang, F.; Ellis, S.; et al. Novel multipurpose pod-intravaginal ring for the prevention of HIV, HSV, and unintended pregnancy: Pharmacokinetic evaluation in a macaque model. *PLoS ONE* **2017**, *12*, e0185946. [CrossRef]
179. Morrow, R.J.; Woolfson, A.D.; Donnelly, L.; Curran, R.; Andrews, G.; Katinger, D.; Malcolm, R.K. Sustained release of proteins from a modified vaginal ring device. *Eur. J. Pharm. Biopharm.* **2011**, *77*, 3–10. [CrossRef]
180. Han, Y.A.; Singh, M.; Saxena, B.B. Development of vaginal rings for sustained release of nonhormonal contraceptives and anti-HIV agents. *Contraception* **2007**, *76*, 132–138. [CrossRef]
181. Malcolm, R.K.; Veazey, R.S.; Geer, L.; Lowry, D.; Fetherston, S.M.; Murphy, D.J.; Boyd, P.; Major, I.; Shattock, R.J.; Klasse, P.J.; et al. Sustained Release of the CCR5 Inhibitors CMPD167 and Maraviroc from Vaginal Rings in Rhesus Macaques. *Antimicrob. Agents Chemother.* **2012**, *56*, 2251. [CrossRef]
182. Johnson, T.J.; Gupta, K.M.; Fabian, J.; Albright, T.H.; Kiser, P.F. Segmented polyurethane intravaginal rings for the sustained combined delivery of antiretroviral agents dapivirine and tenofovir. *Eur. J. Pharm. Sci.* **2010**, *39*, 203–212. [CrossRef] [PubMed]
183. Woolfson, A.D.; Toner, C.F.; Malcolm, R.K.; Morrow, R.J.; McCullagh, S.D. Long-term, controlled release of the HIV microbicide TMC120 from silicone elastomer vaginal rings. *J. Antimicrob. Chemother.* **2005**, *56*, 954–956.
184. Baum, M.M.; Butkyavichene, I.; Gilman, J.; Kennedy, S.; Kopin, E.; Malone, A.M.; Nguyen, C.; Smith, T.J.; Friend, D.R.; Clark, M.R.; et al. An Intravaginal Ring for the Simultaneous Delivery of Multiple Drugs. *J. Pharm. Sci.* **2012**, *101*, 2833–2843. [CrossRef]

185. Johnson, T.J.; Clark, M.R.; Albright, T.H.; Nebeker, J.S.; Tuitupou, A.L.; Clark, J.T.; Fabian, J.; McCabe, R.T.; Chandra, N.; Doncel, G.F.; et al. A 90-Day Tenofovir Reservoir Intravaginal Ring for Mucosal HIV Prophylaxis. *Antimicrob. Agents Chemother.* **2012**, *56*, 6272–6283. [CrossRef]
186. Blakney, A.K.; Little, A.B.; Jiang, Y.; Woodrow, K.A. In vitro–ex vivo correlations between a cell-laden hydrogel and mucosal tissue for screening composite delivery systems. *Drug Deliv.* **2017**, *24*, 582–590. [CrossRef]
187. Rohan, L.C.; Moncla, B.J.; Ayudhya, R.P.K.N.; Cost, M.; Huang, Y.; Gai, F.; Billitto, N.; Lynam, J.; Pryke, K.; Graebing, P. In vitro and ex vivo testing of tenofovir shows it is effective as an HIV-1 microbicide. *PLoS ONE* **2010**, *5*, e9310. [CrossRef]
188. Patton, D.; Sweeney, Y.C.; Balkus, J.; Rohan, L.; Moncla, B.; Parniak, M.; Hillier, S. Preclinical safety assessments of UC781 anti-human immunodeficiency virus topical microbicide formulations. *Antimicrob. Agents Chemother.* **2007**, *51*, 1608–1615. [CrossRef]
189. Robinson, J.A.; Marzinke, M.A.; Fuchs, E.J.; Bakshi, R.P.; Spiegel, H.M.L.; Coleman, J.S.; Rohan, L.C.; Hendrix, C.W. Comparison of the Pharmacokinetics and Pharmacodynamics of Single-Dose Tenofovir Vaginal Film and Gel Formulation (FAME 05). *J. Acquir. Immune Defic. Syndr.* **2018**, *77*, 175–182. [CrossRef]
190. Hu, M.; Zhou, T.; Dezzutti, C.S.; Rohan, L.C. The effect of commonly used excipients on the epithelial integrity of human cervicovaginal tissue. *Aids Res. Hum. Retrovir.* **2016**, *32*, 992–1004. [CrossRef]
191. Merbah, M.; Introini, A.; Fitzgerald, W.; Grivel, J.C.; Lisco, A.; Vanpouille, C.; Margolis, L. Cervico-vaginal tissue ex vivo as a model to study early events in HIV-1 infection. *Am. J. Reprod. Immunol.* **2011**, *65*, 268–278. [CrossRef] [PubMed]
192. Ayehunie, S.; Cannon, C.; Lamore, S.; Kubilus, J.; Anderson, D.J.; Pudney, J.; Klausner, M. Organotypic human vaginal-ectocervical tissue model for irritation studies of spermicides, microbicides, and feminine-care products. *Toxicol. Vitr.* **2006**, *20*, 689–698. [CrossRef] [PubMed]
193. Ayehunie, S.; Cannon, C.; LaRosa, K.; Pudney, J.; Anderson, D.J.; Klausner, M. Development of an in vitro alternative assay method for vaginal irritation. *Toxicology* **2011**, *279*, 130–138. [CrossRef]
194. aniewski, P.; Gomez, A.; Hire, G.; So, M.; Herbst-Kralovetz, M.M. Human three-dimensional endometrial epithelial cell model to study host interactions with vaginal bacteria and Neisseria gonorrhoeae. *Infect. Immun.* **2017**, *85*, e01049-16. [CrossRef] [PubMed]
195. Doncel, G.F.; Clark, M.R. Preclinical evaluation of anti-HIV microbicide products: New models and biomarkers. *Antivir. Res.* **2010**, *88* (Suppl. 1), S10–S18. [CrossRef]
196. Huang, C.; Soenen, S.J.; van Gulck, E.; Vanham, G.; Rejman, J.; van Calenbergh, S.; Vervaet, C.; Coenye, T.; Verstraelen, H.; Temmerman, M.; et al. Electrospun cellulose acetate phthalate fibers for semen induced anti-HIV vaginal drug delivery. *Biomaterials* **2012**, *33*, 962–969. [CrossRef]
197. Ball, C.; Krogstad, E.; Chaowanachan, T.; Woodrow, K.A. Drug-eluting fibers for HIV-1 inhibition and contraception. *PLoS ONE* **2012**, *7*, e49792. [CrossRef]
198. Krogstad, E.A.; Woodrow, K.A. Manufacturing scale-up of electrospun poly(vinyl alcohol) fibers containing tenofovir for vaginal drug delivery. *Int. J. Pharm.* **2014**, *475*, 282–291. [CrossRef]
199. Jiang, J.; Xie, J.; Ma, B.; Bartlett, D.E.; Xu, A.; Wang, C.H. Mussel-inspired protein-mediated surface functionalization of electrospun nanofibers for pH-responsive drug delivery. *Acta Biomater.* **2014**, *10*, 1324–1332. [CrossRef] [PubMed]
200. Sun, X.-Z.; Williams, G.R.; Hou, X.-X.; Zhu, L.-M. Electrospun curcumin-loaded fibers with potential biomedical applications. *Carbohydr. Polym.* **2013**, *94*, 147–153. [CrossRef] [PubMed]
201. Berg, M.C.; Zhai, L.; Cohen, R.E.; Rubner, M.F. Controlled drug release from porous polyelectrolyte multilayers. *Biomacromolecules* **2006**, *7*, 357–364. [CrossRef] [PubMed]

pharmaceutics

MDPI

Article

Development of a Transdermal Delivery System for Tenofovir Alafenamide, a Prodrug of Tenofovir with Potent Antiviral Activity Against HIV and HBV

Ashana Puri [1], Sonalika A. Bhattacharjee [1], Wei Zhang [2], Meredith Clark [2], Onkar N. Singh [2], Gustavo F. Doncel [2] and Ajay K. Banga [1,*]

[1] Center for Drug Delivery Research, Department of Pharmaceutical Sciences, College of Pharmacy, Mercer University, Atlanta, GA 30341, USA; ashana.puri@live.mercer.edu (A.P.); sonalika.arup.bhattacharjee@live.mercer.edu (S.A.B.)

[2] CONRAD, Department of Obstetrics and Gynecology, Eastern Virginia Medical School, Arlington, VA 22209, USA; wzhang@conrad.org (W.Z.); mclark@conrad.org (M.C.); osingh@conrad.org (O.S.); gfdoncel@conrad.org (G.F.D.)

* Correspondence: banga_ak@mercer.edu; Tel.: +1-678-547-6243; Fax: +1-678-547-6423

Received: 27 February 2019; Accepted: 3 April 2019; Published: 9 April 2019

Abstract: Tenofovir alafenamide (TAF) is an effective nucleotide reverse transcriptase inhibitor that is used in the treatment of HIV-1 and HBV. Currently, it is being investigated for HIV prophylaxis. Oral TAF regimens require daily intake, which hampers adherence and increases the possibility of viral resistance. Long-acting formulations would significantly reduce this problem. Therefore, the aim of this study was to develop a transdermal patch containing TAF and investigate its performance in vitro through human epidermis. Two types of TAF patches were manufactured. Transparent patches were prepared using acrylate adhesive (DURO-TAK 87-2516), and suspension patches were prepared using silicone (BIO-PSA 7-4301) and polyisobutylene (DURO-TAK 87-6908) adhesives. In vitro permeation studies were performed while using vertical Franz diffusion cells for seven days. An optimized silicone-based patch was characterized for its adhesive properties and tested for skin irritation. The acrylate-based patches, comprising 2% w/w TAF and a combination of chemical enhancers, showed a maximum flux of 0.60 ± 0.09 μg/cm^2/h. However, the silicone-based patch comprising of 15% w/w TAF showed the highest permeation (7.24 ± 0.47 μg/cm^2/h). This study demonstrates the feasibility of developing silicone-based transdermal patches that can deliver a therapeutically relevant dose of TAF for the control of HIV and HBV infections.

Keywords: transdermal patch; tenofovir alafenamide; acrylate adhesive; in vitro permeation; silicone adhesive; suspension patch

1. Introduction

Tenofovir (TFV) represents the cornerstone of prophylactic and therapeutic approaches to control HIV infection [1]. Tenofovir alafenamide fumarate (TAF), a prodrug, is currently replacing TFV disoproxil fumarate, which shows better affinity for lymphoid tissue, yielding higher levels of TFV-diphosphate (active metabolite) and lower levels of plasma TFV, and displaying a better renal and bone safety profile [2]. Due to these advantages, TAF is also being considered in the treatment of chronic HBV infection [1]. For HIV, TAF is typically administered, in combination with other antiretrovirals (ARVs), in daily single tablet regimens (STRs). Although STRs represent a significant advantage over the need to take multiple tablets three times a day, which characterized the beginnings of highly active antiretroviral therapy, it still remains a high burden to the patient, leading to poor adherence, lower efficacy, and increased risk of viral resistance. Therefore, sustained release of ARVs, which can reduce these problems, is needed. The development of sustained release delivery technologies in this space would be of high clinical relevance.

A variety of delivery systems, such as tablets (oral), tablet-like inserts (vaginal/rectal), gels (rectal/vaginal), vaginal films, ointments, rings, diaphragm based devices and electrospun fibers, oral/parenteral liposomes and nanoparticles, injectables, and subdermal implants for prophylactic or therapeutic anti-HIV agents have been explored [3–9]. Drug delivery systems, such as daily oral tablets and coitally-dependent vaginal gels, are typically short-acting and they have been associated with poorer user adherence. Long-acting coitally-independent systems, such as implants and injectables, do not present these problems, but are riskier if safety or idiosyncratic issues arise [7,10–13]. An intermediate duration drug delivery system, such as a transdermal patch, exerting its pharmacological effect for a few days to a week, would thus be beneficial to the users, providing a more sophisticated and convenient therapy option.

Transdermal drug delivery offers an attractive alternative to oral and other systemic routes of delivery, allowing for the drug substances to reach the systemic circulation across the skin barrier. Transdermal drug delivery systems can be beneficial in preventing the hepatic first pass effect of drugs, eliminating peak and valley plasma drug concentrations, which are usually associated with the oral and injectable drug delivery, avoiding the degradation of drugs in GI tract, and, in general, being non-invasive and convenient to administer [14]. ARV drug substances may potentially be delivered through the transdermal route in a consistent and sustained manner (e.g., zero order release kinetics), which would provide several advantages over the other delivery systems that have been explored so far [15]. Being non-invasive and much easier to apply, without the intervention of a health care provider, transdermal patches may provide more user adherence than the injectable or implantable delivery systems that are currently under investigation. Additionally, transdermal patches can be easily discontinued if the need arises, making them safer to use. The aim of this study was to develop and characterize transdermal patches for the delivery of ARV agents, in particular TAF, and investigate its permeation through human epidermis.

There are generally two types of passive transdermal patches, namely, matrix (drug-in-adhesive) and reservoir. The matrix-based transdermal patch, which was the system of choice in this study, consists of a release liner, an adhesive matrix, and an impermeable backing membrane. In addition to the ease of use and manufacturability, as well as the acceptable cost of goods, one of the significant advantages of matrix type transdermal patches when compared to reservoir patches is the absence of dose dumping [16]. The matrix transdermal patches are usually prepared using organic solvent based pressure sensitive adhesives (PSAs), such as acrylate copolymer, silicone, polyisobutylene (PIB), either alone or in combination with each other [17]. Drug substance can be either dissolved or dispersed in the adhesive matrix, resulting in clear/transparent/translucent or suspension patches, respectively.

From a transdermal patch development perspective, drugs with low molecular weight (<500 Da), suitable melting point (<250 °C), and moderate log P (1–3) are ideal for passive permeation through skin. With a molecular weight of 476.47 g/mol, melting point of 279 °C, and having a logP =1.8, as calculated using chemicalize software (MarvinSketch: version 6.2.2, ChemAxon, Hungary, Europe), TAF, the newest prodrug of TFV in the market, is a promising ARV candidate for the development of transdermal patch formulation. TAF is commercially available with oral daily doses of 10 mg and 25 mg [18]. When considering the higher dose of 25 mg and low bioavailability of orally administered TAF (25%) vs transdermal, 6 mg of it would be needed in the systemic circulation [19,20]. However, 8 mg/day was established as the target dose to be on the safer side and keeping a window of deviations.

The aim of this study was to develop a transdermal patch for sustained release of TAF for approximately one-week duration. To our knowledge, transdermal patches comprising of ARV agents for HIV therapy or prophylaxis have not yet been developed. Thus, this is the first study to report such a system. In the present study, transparent acrylate based patches, as well as silicone and PIB based suspension patches of various compositions, were prepared and evaluated for in vitro permeation across human epidermis for seven days. In order to release a TAF dose of 8 mg/day from a 50 cm^2 transdermal patch, the targeted permeation flux was about 7 $\mu g/cm^2/h$. Different chemical enhancers, plasticizers, as well as crystallization inhibiting agents, were included in the patch formulations. Patches were evaluated for stability, skin irritation, and physical characteristics, such as tack, peel, and shear adhesion.

2. Materials and Methods

2.1. Materials

TAF (free base, CAS number: 379270-37-8) was purchased from Pharmacodia Co., LTD (Beijing, China). Backing membranes (Scotchpak™ 9733-2.05 mil polyester film; CoTran™ 9702, 9706, 9728- ethylene vinyl acetate copolymer (EVAC); CoTran™ 9718, 9720 – polyethylene film); and, release liner (Scotchpak™ 1022- 3 mil fluoropolymer coated polyester film) were gifted by 3M (St. Paul, MN, USA). Silicone coated poly-ethylene terephthalate (PET) films (1 mil): 48101-4400B/000, 44916-7300AM/000, and 40987-U4162/000 were procured from Loparex (Cary, NC, USA). Acrylate PSA (DURO-TAK 87-2516) as well as PIB adhesive (DURO-TAK 87-6908) were obtained as gift samples from Henkel Corporation (Dusseldorf, Germany). The Dow Corning Corporation provided Silicone adhesive (BIO-PSA 7-4301) provided as a gift sample (Washington, DC, USA). Mineral oil, oleyl alcohol, polyethylene glycol (PEG), and polyvinyl pyrrolidone (PVP 360) were obtained from Sigma–Aldrich (St. Louis, MO, USA). Propylene glycol (PG) and oleic acid (OA) were purchased from Ekichem (Joliet, IL, USA) and Croda Inc. (Edison, NJ, USA), respectively. Triacetin, octisalate, gentamycin sulfate, tetra hydrogen furan (THF), sodium dihydrogen phosphate, sodium hydroxide, and phosphate buffered saline, pH 7.4 (PBS) were obtained from Fisher Scientific (Pittsburgh, PA, USA). Kollidon® VA64, Kollidon® 90F, and Kollidon® 30LP were purchased from BASF (Florham Park, NJ, USA). Methanol, ethanol, acetonitrile, and trifluoroacetic acid were purchased from Pharmco-aaper (Brookfield, CT, USA). Dermatomed human cadaver skin was obtained from New York Fire Fighters (New York, NY, USA).

2.2. Methods

2.2.1. Slide Crystallization Studies

TAF was dissolved in methanol (10 mg/mL) and a drop of this solution was placed on a glass slide. Methanol was allowed to evaporate at room temperature (RT) under a fume hood. The drug crystals that were obtained were then observed under Leica DM 750 optical microscope (Leica Microsystems Inc., Buffalo Grove, IL, USA). The DFC-295 camera, which was attached to the microscope, was used to capture images at magnifications of 10× or 20× (as specified). Similar procedure was used to determine the saturation solubility of TAF in different adhesives (DURO-TAK 87-2516, DURO-TAK 87-6908, and BIO-PSA 7-4301), solution of 5% (*w*/*w*) OA in DURO-TAK 87-2516, as well as in the additives: PVP 360, Kollidon® VA64, Kollidon® 90F, and Kollidon® 30LP. For solubility studies in adhesives, blends of different concentrations of TAF in adhesives, ranging from 1–12% *w*/*w* (dry weight) were prepared and a drop of the transparent blends (with completely dissolved drug) was individually placed on glass slides and kept in a flameproof oven at 100 °C for 30 min. for the evaporation of the solvents in the adhesives. Similarly, different blends with TAF concentration, ranging from 3–5% (*w*/*w*) in a mixture of 5% (*w*/*w*) OA in DURO-TAK 87-2516 (dry weight basis), were prepared and a drop of each of the solutions was placed on the glass slides and then exposed to the same drying conditions, as mentioned above. In order to determine the solubility of TAF in different additives, the two components were weighed in the following ratios (% *w*/*w*): 1:9, 2:8, 3:7, 4:6, 5:5, 6:4, 7:3, 8:2, 9:1, and then dissolved in methanol. A drop of each of these solutions was individually mounted on glass slides and observed for crystals after allowing for the evaporation of methanol at RT. The appearance of crystals in all of the test samples was observed for a week before concluding the saturation solubility of TAF in different components. The latter was indicated by the highest concentration at which no crystals were observed [21,22]. The percentages of solid content that were used to calculate the wet weight of adhesives were 41.5% for DURO-TAK 87-2516, 38% for DURO-TAK 87-6908, and 60% for BIO-PSA 7-4301.

2.2.2. Preparation of Drug in Adhesive Patches of TAF

Formulation of Acrylate-Based Patches

Pre-determined amounts of TAF, adhesives, and additives (OA, oleyl alcohol, PG, octisalate, triacetin, as specified in Table 1) were weighed into glass jars with airtight lids to minimize the loss of organic solvents. For the patch formulations AP4 to AP11, excipients, such as PVP 360, Kollidon® VA64, Kollidon® 90F, and Kollidon® 30LP were separately dissolved in methanol and the solutions were then added to the mixture of adhesive and other excipients. The blends were kept overnight on the rotary mixer (Preiser Scientific Inc., St. Albans, WV, USA). Table 1 shows the compositions of the different acrylate based TAF patches. Solid content % of 41.5 was used for the calculation of the wet weight of DURO-TAK 87-2516 for the formulations. The homogenous mixtures were then casted on the release liner (fluoropolymer coated side of Scotchpak™ 1022) using a Gardner film casting knife (BYK-AG-4300 series, Columbia, MD, USA). The target coat weight of different patches varied from 100–400 gsm (grams per square meter). The casted sheets were dried in a flameproof oven at 95 °C for 40 min. After drying, they were laminated using Scotchpak™ 9733 (backing membrane) with the help of a roller, ensuring that no air pockets were formed. The 200 gsm and 400 gsm patches were made by preparing one 100 gsm patch and then stacking 1 and 3, 100 gsm cast films (coated on release liner), respectively, on that one by one, after removing the liner of the previous laminate.

Table 1. Composition of acrylate based Tenofovir alafenamide fumarate (TAF) transdermal patches.

Properties	Patch Formulation Codes														
	AP1	**AP2**	**AP3**	**AP4**	**AP5**	**AP6**	**AP7**	**AP8**	**AP9**	**AP10**	**AP11**	**AP12**	**AP13**	**AP14**	**AP15**
Coat weight (gsm)	200	200	200	100	100	100	100	100	100	100	100	400	400	400	400
Composition	**Components (% Dry Weight, *w/w*)**														
TAF	2	3	4	5	10	15	10	10	10	7.5	7.5	2	2	2	2
OA	5	5	5	5	5	5	5	5	5	5	5	5	5	5	-
DURO-TAK 87-2516	93	92	91	87.86	80.71	73.57	83.89	83.89	83.89	77.50	79.29	83	83	79.5	83
PVP 360	-	-	-	2.14	4.29	6.43	-	-	-	5	-	-	-	-	-
Kollidon® VA64	-	-	-	-	-	-	1.11	-	-	-	-	-	-	-	-
Kollidon® 90 F	-	-	-	-	-	-	-	1.11	-	-	-	-	-	-	-
Kollidon® 30 LP	-	-	-	-	-	-	-	-	1.11	-	3.21	-	-	-	-
Oleyl alcohol	-	-	-	-	-	-	-	-	-	-	-	5	-	-	10
PG	-	-	-	-	-	-	-	-	-	5	5	5	5	5	5
Triacetin	-	-	-	-	-	-	-	-	-	-	-	-	5	-	-
Octisalate	-	-	-	-	-	-	-	-	-	-	-	-	-	8.5	-

Table 2. Composition and manufacturing parameters of the silicone based TAF suspension transdermal patches.

Properties	Patch Formulation Codes													
	SP1	**SP2**	**SP3**	**SP4**	**SP5**	**SP6**	**SP7**	**SP8**	**SP9**	**SP10**	**SP11**	**SP12**	**SP13**	**SP14**
Composition of matrix	Components (% Dry Weight, *w/w*)													
TAF	15	15	15	15	15	15	15	15	15	5	10	20	25	15
OA	5	5	5	5	5	5	5	5	5	5	5	5	5	5
Oleyl alcohol	-	10	10	10	10	10	10	10	10	10	10	10	10	-
Mineral oil	14	14	14	14	14	14	14	14	14	14	14	14	14	14
BIO-PSA 7-4301	66	56	56	56	56	56	56	56	56	66	61	51	46	-
DURO-TAK 87-6908	-	-	-	-	-	-	-	-	-	-	-	-	-	66
Coat weight (gsm)	~100	~250	~250	~250	~250	~250	~300	~300	~300	~350	~350	~200	~200	~50
Release liner	Fluoro-polymer coated side of Scotchpak™ 1022	Uncoated side of Scotchpak™ 1022												Fluoro-polymer coated side of Scotchpak™ 1022
Backing membrane	Scotchpak™ 9733 (polyester)	Scotchpak™ 9733 (polyester)	CoTran™ 9702, 9706, 9728 (EVAC)	CoTran™ 9720 (poly-ethylene)	CoTran™ 9718 (poly-ethylene)	CoTran™ 9718 (poly-ethylene)	Silicone coated PET film (48101)	Silicone coated PET film (44916)	Silicone coated PET film (40987)	Silicone coated PET film (48101)				Scotchpak™ 9733 (polyester)
Homogenization speed (rpm) and time (min) before addition of adhesive	5000; 3	30,000; 20				32,000; 30	32,000; 5							5000; 3
Homogenization speed (rpm) and time (min) after addition of adhesive	5000; 1	30,000; 1				32,000; 1	15,000; 1							5000; 1
Drying conditions	78 °C for 10 min	Air drying for 5 min. followed by drying at 78 °C for 15 min					Air drying, 15 min							78 °C for 10 min

Microscopy of Acrylate-Based Patches

The prepared acrylate patches were punched and stored at 40 °C, RT, and −20 °C, and then observed for the occurrence of crystallization under an optical microscope (Leica DM 750) at predetermined time points for up to six months [23]. In case of patches that showed the appearance of crystals in the first week of storage at RT or 40 °C, microscopic observations after storage at −20 °C were not performed.

Formulation of TAF Suspension Patches

The suspension patches of TAF were prepared using silicone (BIO-PSA 7-4301) and PIB (DURO-TAK 87-6908) adhesive. TAF was first levigated with mineral oil and other additives (OA and oleyl alcohol) were then added and mixed using Benchmixer^{TM} (Benchmark Scientific Inc., Edison, NJ, USA). Heptane was then added to the blend and the drug suspension was homogenized using a high-speed homogenizer (OmniTHQ, Omni International, NW, GA, USA). Heptane was then allowed to evaporate at 90 °C for 5 min. and the adhesive was added to the remaining mixture. The final blend was allowed to mix overnight, then homogenized for 1 min, and then casted on the release liner. It was dried in a flameproof oven and the casted film was then laminated with the backing membrane. Table 2 elaborates on the compositions, homogenization conditions, release liners, backing membranes, and drying conditions employed for the formulation of different TAF suspension patches.

Visual Observations of the TAF Suspension Patches

The suspension patches with varying composition and material components were prepared and observed for the following visual changes for about two weeks: phase separation, contraction/shrinkage of the film, residue on release liner after peeling, ease of peeling off the patches applied on human skin, as well as any residue on skin after removal of the patches.

Effect of Homogenization on Particle Size of TAF

The particle size of TAF, before and after homogenization, was determined using optical microscopy. For the former, pure TAF was dispersed in heptane and a drop of the same was placed on a glass slide and observed under optical microscope. For the size of TAF particles after homogenization, a drop of the suspension, before the addition of the adhesive, was mounted on glass slide and observed under the optical microscope. Size of particles (n = 50) was measured using ImageJ 1.41o software (National Institutes of Health, USA) and it has been reported as average ± SD.

2.2.3. Coat Weight and TAF Content of the Patches

Punching and weighing 4.91 cm^2 or 1 cm^2 laminates (from different areas of the patch, n = 3) using analytical balance (Mettler Toledo, Columbus, OH, USA) and subtracting the weight of equal sized respective release liner and backing membrane from the same determined the coat weight of the prepared patches. For the determination of drug content in the patches, the punched patches were placed in 10 mL of THF after removal of the release liner and allowed to shake overnight. The solutions were then diluted ten times with methanol and centrifuged at 13,400 rpm for 10 min. The supernatants were then analyzed while using high performance liquid chromatography (HPLC) to quantitate the amount of TAF in the same. Comparing the experimental and theoretical drug content values used for processing indicated the stability of TAF after exposure to high temperature conditions. The latter were calculated based on the targeted coat weight for each laminate and the percentage of drug loaded (dry weight% when considering the solvent loss upon drying) in the formulation blend. Results have been reported as average ± SD.

2.2.4. In Vitro Skin Permeation Studies

In vitro permeation studies of TAF from transdermal patches through human epidermis were performed for seven days using in vitro vertical Franz diffusion cells (PermeGear, Inc., Hellertown, PA, USA), providing an effective diffusion area of 0.64 cm^2 (n = 6).

Separation of Epidermis

For each permeation study, human epidermis was freshly isolated from dermatomed human skin by the heat-separation method. Skin was immersed in 10 mM PBS, pH 7.4 for 2 min. at 60 °C. Epidermis was then carefully manually peeled off, with the help of forceps and spatula, and thereafter cut into pieces of suitable size for mounting on the Franz cells [24].

Epidermal Integrity and Thickness Assessment

Resistance of human epidermis was measured in order to select the pieces with acceptable initial barrier integrity for the in vitro permeation study [25]. This was carried out with the use of silver-silver chloride electrodes that were attached to a digital multimeter: 34410A 6 $\frac{1}{2}$ (Agilent Technologies, CA, USA) as well as an arbitrary waveform generator (Agilent 33220A, 20 MHz Function). Epidermis was mounted on the Franz diffusion cells and left for equilibration for 15 min. Phosphate buffer, pH 6.0, was then added in the donor (300 μL) and receptor (5 mL), respectively. Following equilibration, silver and silver chloride electrodes were placed in the receptor and donor compartment, respectively. Load resistor (R_L) attachment was in series with skin, and the voltage drop across the entire circuit and skin (V_S) was displayed on the multimeter (V_O). The following formula was used to measure the skin resistance (R_S):

$$Rs = Vs\,R_L / \,(Vo\ -\ Vs)$$

where, V_O and R_L were 100 mV and 100 kΩ, respectively. Epidermis pieces with electrical resistance of more than 10 kΩ were selected for the permeation study. A thickness gauge measured the thickness of the selected epidermis pieces (Cedarhurst, NY, USA).

Selection of the Receptor Solution

The stability of TAF in different solvent systems (10 mM PBS, pH 7.4; phosphate buffer, pH 6.0; 10 mM PBS, pH 7.4: PEG 400 - 1:1; PEG 400) at 37 °C was assessed in order to select a suitable receptor solution for the seven day permeation studies. For this study, different concentrations of TAF were prepared (0.5, 5, 50 μg/mL) in the above specified solvents and kept at 37 °C. The solutions were analyzed for drug content at 0, and after 24 h using HPLC. The solubility of TAF in the selected receptor media (phosphate buffer, pH 6.0) was determined to ensure the maintenance of sink conditions. For this purpose, excess drug was added to 1.0 mL of the buffer and it was allowed to shake for 24 h at RT. Thereafter, the solution was filtered using 0.22 μm syringe filters (Cell treat Scientific Products, Shirley, MA, USA) and analyzed using HPLC after suitable dilution.

In Vitro Skin Permeation Set Up

The permeation of TAF from different transdermal patch formulations (donor), through human epidermis was investigated using the Franz diffusion cell set up as mentioned in Table 3. While taking the stability of TAF, as well as the maintenance of sink conditions, into consideration, phosphate buffer, pH 6.0 having gentamycin sulfate (80 mg/L) as an anti-microbial agent for the seven day permeation study [22], was selected as the receptor compartment. The temperature of the receptor phase was maintained at 37 °C and it was constantly stirred at 600 rpm. The freshly isolated epidermis pieces were clamped between the donor and receptor compartments. After measuring the skin resistance through the procedure described above, phosphate buffer from the donor was pipetted out, epidermis was removed, placed flat on a glass plate, and then dried with the help of Kimwipes. The release liners of the patches (of size enough to cover the diffusion area) were removed and the latter were applied carefully on the dried epidermis, such that the adhesive layer of the patch was adhered to the stratum corneum side of the skin. The glass rod was rolled on the patches to ensure their adherence to the skin. The epidermis pieces with the applied patches were then mounted between the donor and receptor compartment and the entire set up was secured in place using a clamp. Receptor (0.3 mL) was sampled at 0, 2, 4, 6, and 8 h with replacement with fresh receptor solution and entire receptor (5 mL) was removed and replaced with fresh buffer at 24, 48, 72, 96, 120, 144, and 168 h. All of the samples were analyzed for TAF content using HPLC. As human skin from different donors was employed for the different permeation studies, the in vitro permeation data was normalized using the flux that was obtained by repeating permeation of a previously evaluated patch, within every new permeation study. Permeation flux was calculated from the slope of the linear profile of amount of TAF permeated/ cm^2 and time. The average of all the replicates ± SE has been reported. The calculated flux was extrapolated to patch size of 50 cm^2 and time duration of 24 h, in order to calculate the dose administered/day.

Table 3. Coat weight and drug content of TAF patches.

Patch Formulation Code	COAT WEIGHT (mg/cm^2), mean ± SD		DRUG CONTENT (mg/0.64 cm^2), mean ± SD	
	Targeted	Experimental	Theoretical	Experimental
AP1	20	18.20 ± 1.88	0.256	0.25 ± 0.02
AP5	10	7.10 ± 0.14	0.640	0.53 ± 0.05
AP12	40	37.37 ± 6.94	0.512	0.478 ± 0.09
AP13	40	40.80 ± 1.47	0.512	0.522 ± 0.02
AP14	40	44.17 ± 4.96	0.512	0.565 ± 0.06
AP15	40	40.43 ± 2.35	0.512	0.643 ± 0.106
SP1	10	12.30 ± 3.15	0.96	1.35 ± 0.28
SP5	25	24.23 ± 2.88	2.40	2.52 ± 0.19
SP6	25	20.99 ± 5.80	2.40	2.17 ± 0.55
SP7	30	29.60 ± 1.94	2.88	2.94 ± 0.65
SP10	35	34.15 ± 3.75	1.12	1.11 ± 0.06
SP11	35	36.05 ± 1.29	2.24	2.53 ± 0.11
SP12	20	16.43 ± 0.39	2.56	2.31 ± 0.16
SP13	20	18.29 ± 1.82	3.20	3.02 ± 0.14
SP14	5	4.54 ± 0.55	0.48	0.80 ± 0.06

2.2.5. In Vitro Drug Release Studies

The release profile of TAF from the optimized patch (SP7) was evaluated using Paddle over Disk, USP V dissolution bath apparatus (Sotax AT7 Smart, Sotax AG, Switzerland) [26,27]. This study was performed during week 1 and after 1.5 and 3 months of patch preparation and storage at RT and 40 °C. The patches (0.712 cm^2) were applied on the teflon mesh and then placed on the glass discs, such that the release surface was facing upwards (n = 3). The distance between the paddle and the surface of the disk was about 25 mm. The disks were placed at the bottom of the vessels containing 500 mL of phosphate buffer (pH 6) as the receptor media. Temperature of the media was set at 32 ± 0.5 °C and the paddle speed was 100 rpm. The samples (1 mL) were drawn at 0, 2, 4, 6, 8 h with replacement with fresh media and the entire receptor (500 mL) was removed and replaced with fresh buffer at 24, 48, 72, 96, 120, 144, 168 h. All the samples were analyzed for TAF content using HPLC. Results have been presented as average cumulative amount of TAF released ± SE. The average percentage of TAF released after 7 days from the tested patches has been reported as well and was calculated for each replicate using the following equation:

$$\%TAF\ released = \left(\frac{Cumulative\ amount\ of\ TAF\ released\ over\ 7\ days}{Total\ amount\ of\ TAF\ in\ patch\ tested}\right) \times 100\%$$

2.2.6. Coat Thickness, Coat Weight, and Drug Content of Optimized Patch

Coat thickness of the 1 cm^2 laminates of SP7 (from different areas of the patch) was determined by measuring the patch thickness using a thickness gauge (Cedarhurst, NY, USA), and then subtracting the thickness of the release liner and backing membrane from the former. The coat weight and drug content of the optimized patch were determined using the procedure described in Section 2.2.3. These parameters were determined at day 1 and after 1.5 and 3 months of storage of patch at RT and 40 °C (n = 4–6).

2.2.7. Quantitative Analysis

A UV detection based reverse phase HPLC method was used for quantitative analysis of TAF. Waters Alliance 2695 separation module (Milford, MA, USA) that was attached to a 2996 photodiode array detector was used. Isocratic elution was performed on Phenomenex Luna 5µ C8 (2) 100A, 250 × 4.6 mm (Phenomenex, CA, USA), at a flow rate of 1.0 mL/min. and column temperature of 35 °C, after injecting 50 µL of sample. The mobile phase consisted of acetonitrile (phase A) and 0.1% v/v trifluoroacetic acid in DI water (phase B) in the ratio of 30:70. The run time was 12 min. and the retention time of TAF was around 7.4 min. The drug standards were prepared in the receptor solution and were detected at wavelength of 262 nm. The precision limit of detection and quantification were 0.01 µg/mL and 0.03 µg/mL, respectively, and linearity was observed in the concentration range of 0.1–50 µg/mL (R^2 = 0.9999).

2.2.8. Physical Characterizations of Optimized Patch

Peel Adhesion

The 180° peel adhesion tester (ChemInstruments, Fairfield, OH, USA) was used to evaluate the peel adhesion force (force that is required to peel away the transdermal patch from dermatomed human cadaver skin) [28]. The instrument was calibrated for experimental parameters, such as tension, speed, and peel length with a load cell weighing 50 g prior to running the test patches. Transdermal patches with the length and width of 3.0 inches and 1.0 inch, respectively, were cut. One end of the adhesive coated backing membrane was adhered to the human skin affixed on the stainless steel plate and the other end was attached to load cell grip. The average force that is required to peel off the adhesive film was measured and recorded during the first week after patch preparation at RT and after three months of its storage at RT and 40 °C (n = 3) as well. The results have been shown as average ± SD.

Tack Properties

A texture analyzer (Texture Technologies Corp, Marietta, GA, USA) consisting of a 7 mm probe was used to determine the adhesion efficiency of the TAF suspension patch [29]. Prior to running the test samples, the instrument was calibrated and parameters, such as target force, approach speed, return speed, as well as distance and hold time, were optimized. A transdermal patch (width and length of about 1.0 inch × 1.0 inch, respectively) was cut and adhered on to the sample holder after removal of release liner. As the test run was initiated, the probe was allowed to touch the adhesive film with a target force and hold time of 50 g and 10 s, respectively. This resulted in the creation of a bond between the probe surface and transdermal patch. Further, as the probe was pulled off, it resulted in debonding between the two surfaces. Parameters, such as work of adhesion, positive area, and separation distance were hence recorded during the first week after patch preparation at RT and after three months of storage at RT and 40 °C (n = 3). The results have been shown as average ± SD.

2.2.9. Evaluation of Skin Irritation Potential of Optimized Patch

An in vitro EpiDerm™ skin irritation test (EPI-200-SIT) with a three-dimensional (3D) in vitro reconstructed epidermis (RhE) model (MatTek Corporation 200 Homer Ave, Ashland, MA 01721) was employed to assess the cell viability of skin while using a methyl thiazolyl tetrazolium viability assay [30] after the application of patch SP7. Three replicates of each of the patches were tested and compared to a negative control (Dulbecco's phosphate buffered saline) and positive control (5% aqueous sodium dodecyl sulfate solution). After the removal of the release liners, patches were adhered to the tissues and kept in an incubator for 1 h at 37 °C. Afterwards, the patches were removed from the surface of tissue inserts, the inserts were washed using PBS, and then transferred to a fresh assay medium for 24 h incubation. The media for the inserts was exchanged and incubated again for 18 h. This was followed by transferring the inserts into yellow methyl thiazolyl tetrazolium solution and incubation for 3 h. During the 3 h incubation, mitochondrial metabolism was expected to occur and it was detected by the formation of a purple-blue formazan salt. The plate with the inserts was filled with isopropyl alcohol (2 mL) and kept on a shaker at 120 rpm for 2 -3 h. The aliquots were then transferred to a 96 well enzyme linked immunosorbent assay (ELISA) plate and the optical density of the extracted formazan salt was measured at 560 nm while using a Synergy HT plate reader (BioTek Instruments, Inc, Winooski, VT, USA).

2.2.10. Data Analysis

Microsoft Excel and SPSS software package version 21.0 (IBM, USA) were used for analyzing data. Single factor analysis of variance (ANOVA) and Student's *t*-test were used for statistical analysis and a p value of less than 0.05 was considered for concluding significant difference between the test groups.

3. Results and Discussion

3.1. Slide Crystallization Studies

Adhesiveness is a fundamental property of transdermal patches that is essentially required to ensure the complete contact between the entire surface area of the patch and skin during the wear period, for the efficient delivery of drugs. PSAs deform upon the application of slight pressure, provide intimate contact with surfaces by establishing inter-atomic and molecular forces at the interface, and are thus used for the preparation of transdermal patches [31]. A good adhesive is one that does not leave any residue upon removal, is easy to use, stable to environmental changes, non-irritant and non-sensitive to skin, compatible with other formulation components, allows sufficient drug solubility, and possesses the necessary adhesive properties, such as tack, shear, and skin adhesion [22,32]. Acrylic, silicone, and PIB-based adhesives are most commonly used in the design of transdermal patches [33]. Therefore, all three types of adhesives (acrylate: DURO-TAK 87-2516, silicone: BIO-PSA 7-4301, and PIB: DURO-TAK 87-6908) were explored for formulation development of TAF transdermal patches.

The determination of saturation solubility of TAF in different adhesives was required to determine the maximum amount of dissolvable drug that could be incorporated in the adhesives. Slide crystallization is a preliminary and relatively fast method that is employed as an alternate to preparing complete patches for estimating the solubility of drugs in adhesives [21]. Images of pure TAF, as observed under the microscope at a magnification of 10×, are shown in Figure 1A. TAF, at a concentration of 1% w/w (dry weight), was not soluble in silicone (BIO-PSA 7-4301) as well as PIB (DURO-TAK 87-6908) adhesive, even before the evaporation of solvents. Therefore, due to poor solubility in these adhesives, further slide crystallization studies were not performed with them. However, TAF blends at a concentration of 1–12% w/w (dry weight) in DURO-TAK 87 2516 were successfully prepared and drops of these were allowed to dry on glass slides and images of dried drug-adhesive blends, as observed under the microscope, are shown in Figure 1B–F. TAF at concentrations of 3–12% w/w was observed to immediately crystallize after drying (Figure 1B–D). However, 2% w/w TAF blend crystallized after three days (Figure 1E) and 1% w/w blend did not crystallize, even after seven days at RT (Figure 1F). Therefore, the saturation solubility of TAF in DURO-TAK 87-2516 was observed to be between 1–2% w/w (dry weight). Moreover, the addition of 5% w/w OA to DURO-TAK 87-2516 adhesive enhanced the solubility of TAF to about 4% w/w.

The blend containing 5% *w*/*w* TAF in adhesive-OA blend showed crystals after seven days, whereas the 4% *w*/*w* TAF solution did not, as shown in Figure 1H,G, respectively. Further, in order to select the composition ratios of different crystallization inhibitors and TAF for patch formulations with 5%, 10%, and 15% *w*/*w* TAF (higher than its solubility in adhesive alone), different ratios of the two components were tested in slide crystallization studies. The mixtures of TAF and PVP 360 in the ratios of 9:1 and 8:2 showed the appearance of crystals after seven days, but the compositions with TAF:PVP ratios from 7:3 to 1:9 did not show any crystals (Figure 1I–K). All of the other crystallization inhibitors (Kollidon® VA64, Kollidon® 90F, and Kollidon® 30LP) with TAF, each in the ratio of 1:9, did not show appearance of any crystals after seven days. Therefore, ratios of 7:3 (TAF:PVP 360) and 9:1 (TAF: Kollidon® VA64 or Kollidon® 90F or Kollidon® 30LP), comprising of the lowest amount of PVP with the potential to inhibit crystallization, were selected for the preparation of patches containing 5% *w*/*w* or higher concentrations of TAF.

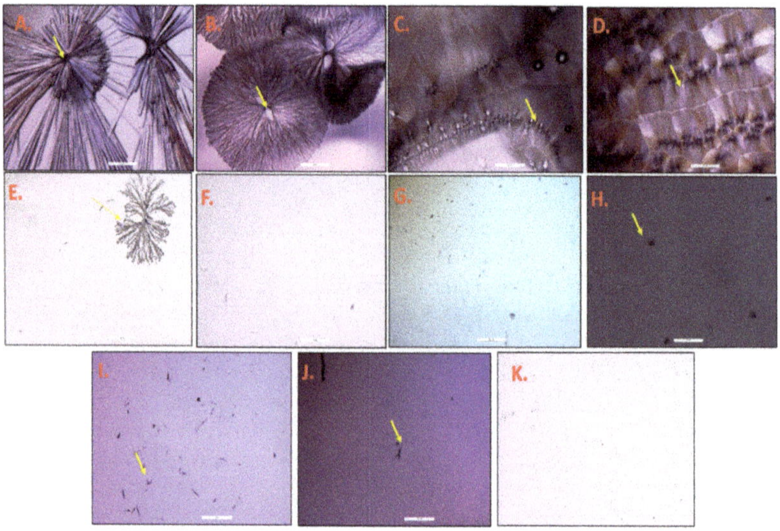

Figure 1. Images of slide crystallization studies, under optical microscope. (**A**). Pure TAF (at 10×). (**B–D**). TAF crystals in DURO-TAK 87-2516 immediately after evaporation of organic solvent at 10×: (**B**) 3% *w*/*w* TAF (**C**). 5% *w*/*w* TAF (**D**). 12% *w*/*w* TAF. (**E**). 2% *w*/*w* TAF in DURO-TAK 87-2516, 3 days after evaporation of organic solvent (10×). (**F**). 1% *w*/*w* TAF in DURO-TAK 87-2516, 7 days after evaporation of organic solvent (20×). (**G**). 4% *w*/*w* TAF in 5% *w*/*w* OA in DURO-TAK 87-2516 (10×). (**H**). 5% *w*/*w* TAF in 5% *w*/*w* OA in DURO-TAK 87-2516 (20×). (**I**). TAF:PVP 360 (9:1) under 20×. (**J**). TAF:PVP 360 (8:2) under 10×. (**K**).TAF:PVP 360 (7:3) under 20× (arrows depicting TAF crystals).

3.2. Formulation of TAF Acrylate Patches

Slide crystallization is a faster means for estimating the saturation solubility of drugs in adhesives. However, the thickness of the actual patches, as well as processing conditions and scale up procedures, may affect the phenomenon of crystallization. Therefore, based on the observations of slide crystallization studies, different transparent TAF patches were prepared in DURO-TAK 87-2516 adhesive, as shown in Table 1 and observed for crystallization for six months under the microscope. The solubility of TAF in 5% *w*/*w* OA in DURO-TAK 87-2516 blend (dry weight basis), as determined by slide crystallization studies, was found to be about 4%. Therefore, 2%, 3%, and 4% TAF patches (200 gsm) were prepared in this matrix (AP1, AP2, AP3, as specified in Table 1). Due to the solubility limitations of TAF in the adhesive, it was not possible to prepare patches with higher drug concentrations in the absence of crystallization inhibiting agents that act as solubilizers and anti-nucleants. The use of different grades of PVP as crystallization inhibitors has been well-reported in literature and some of them were, hence, investigated in the current study [22,34,35]. In order to prepare 100 gsm patches containing higher percentage of TAF (5%, 10%, and 15% *w*/*w*—AP4, AP5, AP6, respectively), PVP 360 was added as a solubilizer and crystallization inhibitor, with drug and PVP in the ratio of 7:3, in addition to OA. Patch 'AP10' was prepared to investigate the effect of enhancing the percentage of PVP 360, as well as the addition of 5% *w*/*w* PG in the 7.5% TAF patch on drug crystallization (100 gsm). Drug and PVP were in the ratio of 6:4 (*w*/*w*) in this formulation. Formulations AP7 to AP9 included the addition of other grades of PVP (Kollidon® VA64, Kollidon® 90F, and Kollidon® 30LP) as solubilizers and crystallization inhibitors with TAF (10% *w*/*w*), in the ratio of 1:9 (PVP: TAF) in OA and DURO-TAK 87-2516 blend. Patch AP11 (100 gsm) was prepared to investigate the effect of enhancing the percentage of Kollidon® 30LP as well as the addition of 5% *w*/*w* PG in the 7.5% TAF patch on drug crystallization. Drug and Kollidon® 30LP were in the ratio of 7:3 (*w*/*w*) in this formulation. Further, patches AP12 to AP14 (400 gsm) containing 2% *w*/*w* TAF were prepared to evaluate the effect of combination of permeation enhancers (5% *w*/*w* OA+ 5% *w*/*w* PG + 5% *w*/*w* oleyl alcohol, 5% *w*/*w* OA+ 5% *w*/*w* PG + 5% *w*/*w* triacetin, 5% *w*/*w* OA+ 5% *w*/*w* PG + 8.5% *w*/*w* octisalate, respectively), as well as increasing the coat weight on the skin permeation of TAF and comparing it with that of AP1. Further, the AP15 (400 gsm) patch was prepared by replacing 5% *w*/*w* OA in AP12 with 5% oleyl alcohol, such that the total oleyl alcohol content was 10% *w*/*w* in this formulation. The transdermal permeation enhancing effects of OA [36–38], oleyl alcohol [39,40], triacetin [41–43], and octisalate [44,45] are well-known and they have been reported in literature, and thus these chemical enhancers were selected for the formulation development of TAF patches.

Microscopy and Stability Assessment of TAF Acrylate Patches

During the storage period, changes in temperature can alter the thermodynamic activity of the transdermal patch formulation, and they result in precipitation or crystallization or changes in the crystal habit of the active ingredient. Therefore, during stability testing, patches are exposed to stress and real world storage conditions that are representative of the product's proposed marketing conditions [46]. The stability of the acrylate based TAF patches was assessed in terms of the occurrence of crystallization over a period of time at different temperature conditions. Table 4 shows the microscopic observations of the same after six months. Figure 2 shows the representative images of TAF crystals in transdermal patches.

Table 4. Microscopy observations of acrylate based TAF transdermal patch formulations up to 6 months.

Patch Codes	Temperature		
	−20 °C	RT	40 °C
AP1	NO	NO	NO
AP2	NO	Crystals after three weeks	Crystals after three weeks
AP3	NO	Crystals after two weeks	Crystals after two weeks
AP4	NO	Crystals after three weeks	Crystals after three weeks
AP5	NO	Crystals after two weeks	Crystals after two weeks
AP6	Crystals after 2 months	Crystals after 9 days	Crystals after one week
AP7	Not observed	Crystals in first week	Crystals in first week
AP8	Not observed	Crystals in first week	Crystals in first week
AP9	Not observed	Smaller crystals than AP7 and AP8 after 12 days	Crystals in first week (2–3 crystals)
AP10	NO	Crystals after three weeks	Crystals after two weeks
AP11	NO	Crystals after three weeks	Crystals after two weeks
AP12	NO	NO	NO
AP13	NO	NO	NO
AP14	NO	NO	NO
AP15	NO	NO	NO

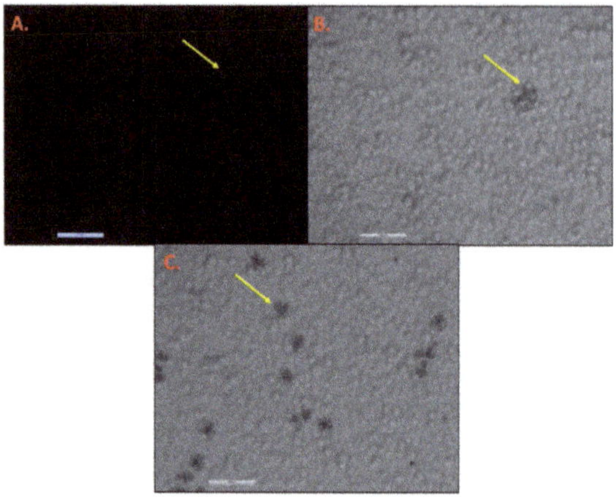

Figure 2. TAF crystals in acrylate based transdermal patches (at 10×). (**A**) Patch AP5 (crystals at 40 °C after 2 weeks). (**B**) Patch AP9 (crystals at RT in 12 days). (**C**) Patch AP8 (crystals at RT in first week). Arrows depict TAF crystals

Patches with 2% *w*/*w* TAF were found to be stable and they did not show crystals at all of the tested temperature conditions, even after six months, as the drug concentrations were within the saturation solubility limits in these matrices. However, in order to incorporate drugs in amounts that are higher than its saturation solubility in the adhesive alone, crystallization inhibitors were included in the patches. Mechanistically, these additives stabilize the systems against crystallization, either due to enhancing the solubility of the drug or by aiding adsorption of drug crystals in the adhesive matrix. PVP has been previously demonstrated as one of the most effective additives in inhibiting drug crystallization in patch formulations acting by the formation of amorphous co-precipitates with the drugs [35,47–49]. Additionally, it has been previously shown that PVP inhibits the crystallization of drugs by being adsorbed on the growing crystal surfaces [35,50]. Therefore, different grades of PVP were incorporated in patches containing higher concentrations of TAF. However, patches with 15% *w*/*w* TAF (AP6) and PVP 360, crystallized in less than ten days at RT and 40 °C, depicting the inability of PVP 360 to inhibit crystallization at high drug levels. At lower drug concentrations (5% and 10% *w*/*w* TAF, AP4, and AP5, respectively), the addition of PVP 360 in the same ratio with TAF (3:7, *w*/*w*), as in AP6, slowed down the process of crystallization, where it was observed after three and two weeks, respectively, at RT as well as 40 °C. Further, increasing the PVP:TAF ratio to 4:6 (*w*/*w*) and adding PG as additional solubilizer in patch AP10 formulation did not delay the process of crystallization further and crystals were still observed in less than a month as well at RT and 40 °C. Interestingly, these patches did not show the presence of any crystals, even after six months at −20 ° C. Further, other crystallization inhibitors (Kollidon® VA64, Kollidon® 90F) were found to be ineffective, as crystals were visible during the first week after storage of all these patches at RT as well as 40 °C. However, the crystals that were observed in patch AP9 (containing Kollidon® 30LP) were smaller than AP7 and AP8 and they were observed after about two weeks at RT and after first week at 40 °C. Therefore, patch AP11 was formulated with reduced drug concentration (7.5% *w*/*w* TAF), a higher amount of Kollidon® 30LP than AP9, and the addition of PG, and showed a slower appearance of crystals (observed after three weeks at RT). However, the inability of PVP to prevent the appearance of TAF crystals in the transdermal patch formulation for longer time duration may be attributed to the difficulty in the molecular interactions between PVP and growing TAF crystals, because of the higher viscosity of the patch formulation as compared to a single drop of blend on the slides [35]. Overall, although most of the studies showed PVP to be an effective drug crystallization inhibitor, there are a few findings regarding ineffectiveness of some grades of PVP. For example, Weng et al. reported that PVP K30 was found to be ineffective in inhibiting the crystallization of risperidone in drug-in-acrylate adhesive transdermal patches. However, in the same study, OA was found to be an effective crystallization inhibitor [34]. In this study, all the compositions of 2% *w*/*w* TAF transdermal patches were found to be stable at all of the temperature conditions for more than six months, which may be attributed to drug solubility in the adhesive matrix being lower than the saturation level, as well as the crystallization inhibiting effect that is conferred by OA. Additionally, the transdermal patches formulated with higher drug concentrations (AP2, AP3, AP4, AP5, AP10, AP11) were only stable at −20 °C.

3.3. Formulation of TAF Suspension Patches

Due to low solubility of TAF in silicone and PIB adhesives, the drug was suspended in the blends of these adhesives with different additives, as elaborated in Table 2. Mineral oil was added as a plasticizer that increases diffusivity of the drug by decreasing the resistance that is offered by the patch matrix [22]. In addition to the plasticizing effect, mineral oil has also been reported as a transdermal permeation enhancer [51]. OA and oleyl alcohol are also well-known chemical penetration enhancers [36–40] that were included in the TAF suspension patches. A number of different silicone based TAF patches were prepared in order to select the optimal type of the release liner and the backing membrane. Homogenization speed and time were also optimized to reduce the particle size of the drug.

Patch SP1 (~100 gsm) was prepared while using the fluoropolymer coated side of Scotchpak™ 1022 and Scotchpak™ 9733 (polyester backing). For TAF suspension patch formulations, SP2-SP13, a higher coat weight (~200–350 gsm) was applied and oleyl alcohol was included as an additional enhancer. For the preparation of SP2 patch, the same release liner and backing membrane as SP1 were initially selected. However, in terms of the TAF suspension patch formulation with oleyl alcohol, it was found that, when the formulation was coated on the fluoropolymer coated side of the liner, it could not come off the release liner, depicting the affinity of the formulation binding to the fluoropolymer coated release liner was more than the polyester backing membrane. Therefore, the formulation was casted on the uncoated side of the release liner (polyester only) and then laminated while using the polyester (Scotchpak™ 9733) backing film. This was an interesting experimental observation, as silicone based transdermal patches are usually casted on the fluoropolymer coated side of release liner in order to have easy release/transfer of the formulation from the liner to backing. Further, when the patch formulation SP2 was peeled off, it was observed that the adhesive layer transiently formed a film on the skin, depicting less affinity of the formulation for polyester backing material.

Therefore, other materials, such as EVAC based (Corona-treated - CoTran™ 9702, 9706, 9728) and polyethylene based (CoTran™ 9720 and CoTran™ 9718), were tried for the optimization of the backing membrane, while the uncoated side of Scotchpak™ 1022 was used as the release liner for the patch formulations, SP3-SP6. The formulation SP5 and SP6, using CoTran™ 9718 as the backing membrane, showed comparatively better peeling characteristics than the SP4 patch. The SP3 patch could not be successfully prepared, as the formulation film did not come off from the liner onto the EVAC backing membranes. Higher speed and duration of homogenization was employed for formulation SP6 when compared to that SP5 in an attempt to investigate the effect of homogenization with different parameters on the drug particle size and skin permeation. However, the issue of the residual film of the formulation on skin was not completely resolved in SP5 and SP6, indicating the requirement of a more compatible backing membrane material to facilitate the peeling off process of the patch formulation. In addition, upon long term storage (after a month), patches SP5 and SP6 showed the contraction of the film and the appearance of streaks.

As was evident from patches SP2-SP6, the silicone-based TAF suspension patch formulation did not have sufficient affinity to the backing membrane materials, including polyester and polyethylene. Therefore, silicone coated PET films (48101-4400B/000, 44916-7300AM/000, and or 40987-U4162/000) were explored as the backing membrane for SP7, SP8, and SP9, respectively. Patch SP7 (containing 15% *w*/*w* TAF) showed the best efficiency in terms of least resistance and minimal formulation residue, while peeling the patch formulation off from the skin. Hence, patches SP10-13 were prepared while using the same material components and processing parameters as SP7, except that different concentrations of TAF (5, 10, 20, 25% *w*/*w*) were included in these patches. For the preparation of PIB-based TAF suspension patch formulation (SP14), Scotchpak^{TM} 1022 (Fluoropolymer coated side) and Scotchpak^{TM} 9733 were used as the release liner and the backing membrane, respectively.

3.3.1. Visual Observation of TAF Suspension Patches

Table 5 summarizes the details of the visual observations of the different TAF suspension patches. Patches SP1, SP7, SP10-13, and SP14 were observed to be acceptable in terms of the properties that are specified in the table. The visual observations depicted that the addition of oleyl alcohol in the silicone suspension formulation rendered it more lipophilic, and thus provided more affinity or binding with skin. Using a silicone-coated film as a backing membrane resolved the issue of the residual film after peeling off. The PIB patch formulation (SP14) did not have any issues during the peeling off process as compared to the silicone patch formulations.

Table 5. Visual observations of TAF suspension patches.

Formulation Code	Properties				
	Phase Separation	Contraction/Shrinkage of Films	Residue on Release Liner after Peeling	Ease of Peeling Patches off the Skin	Residue on the Glove after Patch Removal
SP1	No	No	No	Yes	No
SP2	No	No	No	Yes	Yes
SP4	No	No	No	Yes	Yes
SP5	No	Yes (after a month)	No	Yes	Not when applied afresh, but if applied after storing for few days
SP6	No	Yes (after a month)	No	Yes	Not when applied afresh, but if applied after storing for few days
SP7	No	No	No	Yes	No
SP8	No	No	No	No	No
SP9	No	No	No	No	No
SP10	No	No	No	Yes	No
SP11	No	No	No	Yes	No
SP12	No	No	No	Yes	No
SP13	No	No	No	Yes	No
SP14	No	No	No	Yes	No

3.3.2. Effect of Homogenization on the Particle Size of TAF

The effect of homogenization parameters on the particle size of TAF was also investigated, in order to eventually evaluate the effect of the particle size of API in the adhesive matrix on the skin permeation. Homogenization is a process that consists of micronizing or reducing the particle size of dispersions by the application of high sheer, pressure, turbulence, as well as acceleration and impact [52]. Decreased particle size would aid in enhancing the solubility of the drug in the adhesive formulation, and therefore improving its permeation rate [53]. The effect of homogenization speed and duration on the particle size of API has been previously reported, and thus similar parameters were selected to reduce the particle size of TAF [54,55]. As reported, homogenization speed is an indicator of the amount of energy that is applied to the system, as determined by the velocity of the rotating mixing heads. Further, the mechanical impingement of the particles against the wall, due to the high acceleration of the fluid and shear stress in the gap between the rotor and stator, leads to the reduction in the particle size of the drug substance [54].

The average particle size of the pure TAF was observed to be 53.39 $\pm$ 16.15 μm under the optical microscope (Figure 3A). After homogenization at 30,000 rpm for 20 min, the particle size of TAF was reduced to 11.51 $\pm$ 2.89 μm ($p < 0.05$) (Figure 3B). Further increasing the speed to 32,000 rpm and the homogenization time to 30 min. significantly reduced the particle size of TAF to 6.0 $\pm$ 1.8 μm as compared to that of homogenization at 30,000 rpm for 20 min. ($p < 0.05$) (Figure 3C). The observation of the reduction in the particle size of API after the increase in the speed and time of homogenization was in concordance with those that were reported previously [54]. As shown in Figure 3C, some of the particles were smaller and were not visible under the microscope, and thus could not be measured using the software. However, it was found that homogenization at 32,000 rpm over 5 min. resulted in a similar reduction in drug particle size (5.28 ± 2.71 µm, Figure 3D) to that of homogenization at 32,000 rpm for 30 min. Collectively, the speed of homogenization had relatively more impact on the reduction in particle size as compared to the duration of the homogenization. Thus, 32,000 rpm and 5 min. were the homogenization parameters that were employed for the formulation of SP7-SP13.

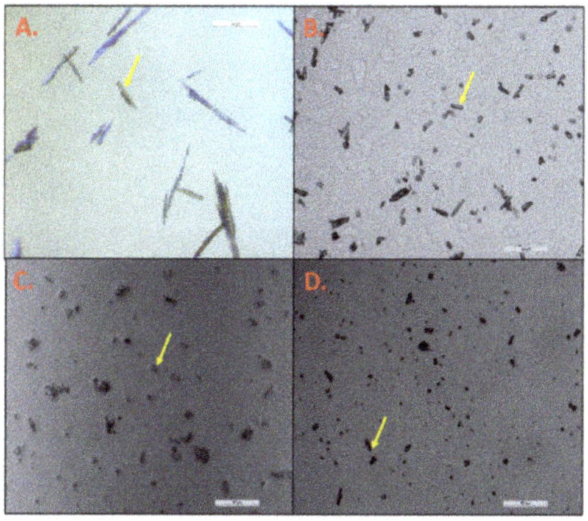

Figure 3. (**A**) Microscopic images of pure TAF particles suspended in heptane; TAF particles in suspension formulation, before addition of adhesive (at 20×) after homogenization at: (**B**) 30,000 rpm for 20 min; (**C**) 32,000 rpm for 30 min; (**D**) 32,000 rpm for 5 min.

3.4. Coat Weight and Drug Content

Coat weight and drug content of the patches evaluated for the in vitro permeation studies were measured and are reported in Table 3. These measured properties are influenced by various factors, such as percentage of non-volatile components in the adhesive blend, scale set-up, surface level of the casting knife, textural properties, of release liner and backing membrane, and they are reflective of the coating efficiency of the patches. The results showed that all of the patches (acrylate, silicone, PIB) had uniformity in both coat weight and drug assay. Additionally, the values of theoretical and experimental drug content were close for all of the patches that confirmed the stability of the drug after exposure to the temperature conditions used for the processing of respective patches.

3.5. In Vitro Permeation Studies

3.5.1. Epidermal Integrity and Thickness Assessment

The integrity of human epidermis samples must be evaluated prior to permeation studies, as the procedures that are employed for procurement of human skin, such as surgical removal, dermatoming, and storage, as well as technical procedure for the separation of epidermis from dermatomed skin, can damage the skin, and thus ultimately influence drug permeation. The physical integrity of stratum corneum has been reported to be indicated by its electrical properties. Therefore, the measurement of electrical conductivity is used as a means of assessing the barrier integrity for full thickness skin as well as epidermal membranes [27,56]. Hence, for the selection of epidermal pieces with optimum barrier integrity, electrical resistance of skin was measured and considered to be the main selection criterion.

Epidermal pieces with the resistance above 10 kΩ, were selected for the study. Further, the thickness of human epidermis used for the permeation studies ranged from 50–150 μm.

3.5.2. Selection of Receptor Solution

Table 6 shows the percentage degradation of TAF in different receptor solutions at 37 °C. It was evident that TAF degrades in PBS, pH 7.4, with or without PEG 400. Lowering the pH to 6.0 and using non-aqueous media, such as pure PEG 400, reduced the degradation of TAF. However, due to limitations, such as high viscosity as well as low drug sensitivity in HPLC, PEG 400 was not selected as the receptor media. Phosphate buffer, pH 6.0 was thus selected as the receptor solution, and it was completely replaced every 24 h during the seven-day skin permeation studies. The degradation of TAF at both acidic and basic pH has been previously reported in the literature. Additionally, TAF has been found to be stable at pH of around 5–6.8 [57]. In addition, the solubility of TAF in PBS was found to be 4.95 ± 0.07 mg/mL. Experimental conditions ensured the maintenance of sink conditions for our in vitro permeation studies.

Table 6. Stability of TAF in different receptor solutions.

TAF Concentration (μg/mL)	% Degradation of TAF in 24 h at 37 °C			
	PBS, pH 7.4	Phosphate Buffer, pH 6.0	PEG 400: PBS, pH 7.4 (1:1)	PEG 400
0.5	22.28	3.65	41.85	Not detected due to low LOD
5	28.29	1.61	32.40	0.00
50	30.54	2.27	33.84	0.85

3.5.3. Determination of Permeation Flux of TAF Transdermal Patches

Table 3 describes the different patches that were evaluated in seven day in vitro permeation studies.

Permeation of TAF from Acrylate-Based Clear Patches: Effect of Drug Concentration

The amount of TAF that permeated from patch AP1 (simplest matrix comprising of 2% *w*/*w* TAF dissolved in mixture of OA in acrylate adhesive) after seven days was found to be 19.20 ± 2.92 μg/cm^2. The average permeation flux over duration of seven days was calculated to be 0.12 ± 0.01 μg/cm^2/h. To evaluate the effect of drug concentration in the patch on its skin permeation, patch AP5 with 10% *w*/*w* TAF, which was found to be stable for about two weeks at RT, was selected for an in vitro permeation study. As shown in Figure 4, increasing the concentration of TAF to 10% *w*/*w* (AP5) significantly enhanced the drug permeation to 137.76 ± 10.42 μg/cm^2 after seven days and flux rate was calculated to be 0.88 ± 0.07 μg/cm^2/h as compared to that of AP1 ($p < 0.05$). Thus, increasing the drug concentration significantly enhanced its permeation across human epidermis. However, since AP5, as well as other acrylate adhesive-based patches with a TAF concentration of more than 2% (*w*/*w*), crystallized at RT over a period of time, transdermal patches with 2% *w*/*w* TAF, and consisting of combination of chemical penetration enhancers as well as higher coat weight, were prepared and evaluated for skin permeation testing (AP12 to AP15).

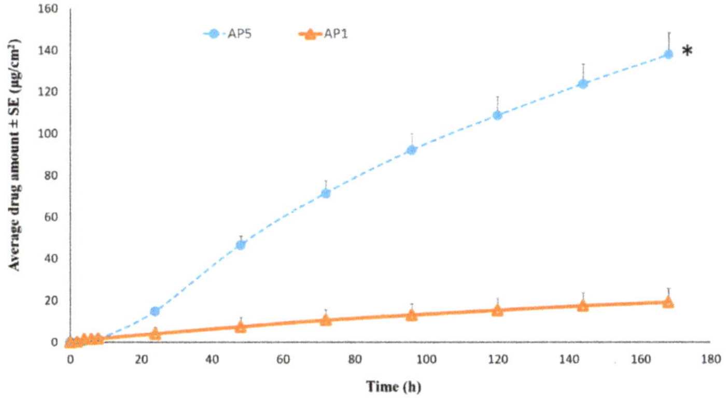

Figure 4. Permeation profiles of TAF from acrylate based patches across human epidermis. * Represents statistical significant difference, Student's t test ($p < 0.05$). AP1: Acrylate patch containing 2% TAF, AP5: Acrylate patch containing 10% TAF.

Permeation of TAF from Acrylate-Based Clear Patches: Effect of Coat Weight and Chemical Penetration Enhancers

As shown in Figure 5A, the cumulative amount of TAF permeation after seven days in the patches AP12 (96.40 ± 14.83 μg/cm^2), AP13 (87.15 ± 8.60 μg/cm^2), and AP14 (89.56 ± 17.28 μg/cm^2) was found to be significantly higher than that of AP1 (19.20 ± 2.92 μg/cm^2, $p < 0.05$). Additionally, the permeation flux that was observed from the three aforementioned groups was 0.60 ± 0.10 μg/cm^2/h, 0.54 ± 0.05 μg/cm^2/h, and 0.56 ± 0.12 μg/cm^2/h, respectively, which was about five-fold greater than that observed from AP1 (0.12 ± 0.01 μg/cm^2/h). Thus, increasing the patch coat weight to 400 gsm from 200 gsm along with the incorporation of a combination of chemical enhancers (OA+ PG+ oleyl alcohol, OA+ PG+ triacetin, or OA+ PG+ octisalate) resulted in the significant enhancement in drug permeation. However, no significant difference in TAF permeation between AP12, AP13, and AP14 was observed. As shown in Figure 5B, the amount of drug permeation over 6 h was significantly higher in patch AP12 (6.15 ± 0.41 μg/cm^2) than AP13 (3.38 ± 1.40 μg/cm^2), as well as AP14 (3.17 ± 1.30 μg/cm^2, $p < 0.05$). Patch AP15 was formulated by replacing 5% *w*/*w* OA in AP12 with oleyl alcohol, thus having total of 10% *w*/*w* oleyl alcohol and a coat weight of 400 gsm. However, as shown in Figure 6, there was no significant difference between the amount of TAF permeation after seven days between the two groups (AP12: 96.40 ± 14.83 μg/cm^2 and AP15: 122.69±28.34 μg/cm^2, $p > 0.05$).

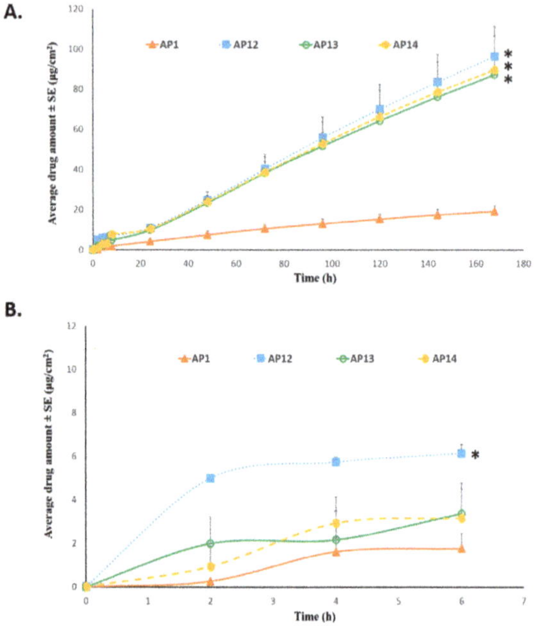

Figure 5. Effect of combination of enhancers and higher patch coat weight on permeation of TAF across human epidermis from acrylate patches: (**A**) Seven day permeation profile (**B**) Six hours permeation profile. * Represents statistically significant difference, ANOVA one way test ($p < 0.05$). AP1: Acrylate based patch containing 2% TAF and 5% oleic acid. AP12: Acrylate based patch containing 2% TAF, 5% oleic acid, 5% PG, and 5% oleyl alcohol. AP13: Acrylate based patch containing 2% TAF, 5% oleic acid, 5% PG, and 5% triacetin. AP14: Acrylate based patch containing 2% TAF, 5% oleic acid, 5% PG, and 8.5% octisalate.

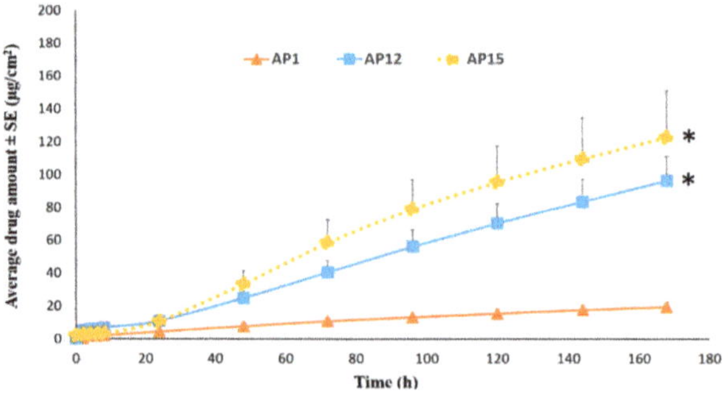

Figure 6. Permeation profile of acrylate based TAF patches comprising of oleyl alcohol and comparison with the control patch through human epidermis. * Represents statistically significant difference, ANOVA one way test ($p < 0.05$). AP1: Acrylate based patch containing 2% TAF and 5% oleic acid. AP12: Acrylate based patch containing 2% TAF, 5% oleic acid, 5% PG, and 5% oleyl alcohol. AP15: Acrylate based patch containing 2% TAF, 5% PG, and 10% oleyl alcohol.

The effects of various chemical enhancers that were used in the patches, such as OA [36–38], oleyl alcohol [39,40], triacetin [41–43], and octisalate [44,45] on skin permeation enhancement are well-known. Synergistic permeation enhancing effect of the combination of PG and OA has also been reported earlier [25,39,58]. In our study, PG is expected to act as a cosolvent, and thus enhance the concentrations of the drug substance as well as the enhancer in the stratum corneum. OA, on the other hand, has the potential to delipidize the stratum corneum, and therefore facilitate the partitioning of the drug molecules and PG into skin [38,59]. Additionally, the synergistic effect of PG and oleyl alcohol was found to markedly increase the skin permeation of tenoxicam across hairless mouse skin [60]. Furthermore, a combination use of PG and oleyl alcohol, under occlusive conditions, has been reported to result in considerable enhancement in the permeation of testosterone across full thickness neonatal porcine skin. The mechanism of penetration enhancement by octisalate is not fully understood. However, due to the structural similarity, as well as similar observations in differential scanning calorimetry and attenuated total reflectance spectroscopy, in terms of the effect of delipidization on stratum corneum, octisalate is assumed to mechanistically work in the similar way to azone [44]. Additionally, octisalate has been incorporated as a penetration enhancer in sunscreens, cosmetic products, as well as in metered spray transdermal systems [45]. Further, triacetin has been incorporated as a penetration enhancer in the commercially available Oxytrol® patch for overactive bladder [43,61]. Patel et al. also reported an enhancement in diclofenac permeation through human cadaver skin while using 10% triacetin [41].

Permeation of TAF from Suspension-Type Patches: PIB vs Silicone-Based

The PIB based TAF suspension patch (SP14) showed a permeation of 90.40 ± 10.18 $\mu g/cm^2$ after seven days through human epidermis. A TAF permeation flux rate of 0.60 ± 0.07 $\mu g/cm^2/h$ and a lag time of about 19 h was observed with this patch. The results showed that the flux rate from SP14 was significantly higher than AP1 ($p < 0.05$), but was similar to that of AP12, AP13, AP14, and AP15 ($p > 0.05$). Further, silicone patch (SP1) with the same composition as PIB (SP14) was prepared. The cumulative amount of TAF that permeated from SP1 across human epidermis in seven days was observed to be 432.21 ± 41.25 $\mu g/cm^2$ and it was significantly higher than all of the other patches evaluated previously ($p < 0.05$). Figure 7 shows the skin permeation profiles of PIB (SP14) and silicone (SP1) based patch. The resultant permeation flux from SP1 for seven days was found to be 3.39 ± 0.03 $\mu g/cm^2/h$ and the lag time was observed to be only 1.2 h. Further, with the addition of 10% *w*/*w* oleyl alcohol, as well as increased coat weight of ~250 gsm in patch SP5, the permeation flux of TAF was observed to significantly increase to 5.97 ± 0.47 $\mu g/cm^2$ /h ($p < 0.05$). The total amount of TAF observed in the receptor after seven days was 958.07 ± 82.81 $\mu g/cm^2$ and the lag time was about 1 h.

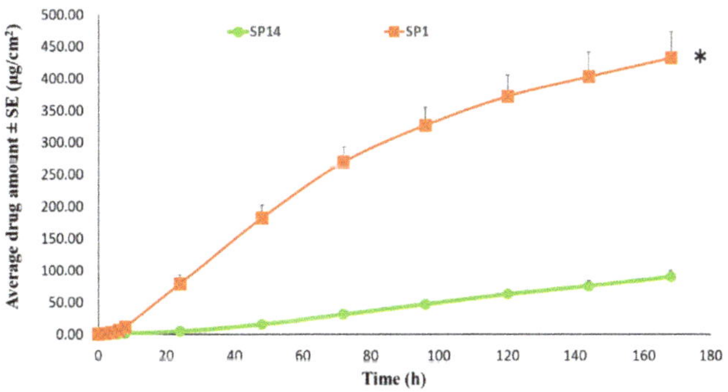

Figure 7. Comparison of permeation profile of silicone and PIB based TAF patches through human epidermis. * Represents statistical significant difference, Student's t test ($p < 0.05$). SP1: Silicone based 15% TAF patch. SP14: Poly isobutylene based 15% TAF patch.

Permeation of TAF from Silicone-Based Suspension-Type Patches: Effect of Homogenization

Briefly, the suspension blend for patch SP5 was homogenized at 30,000 rpm for 20 min, while homogenization at 32,000 rpm for 30 min. was performed for the preparation of SP6, to investigate the effect of homogenization parameters on the skin permeation. However, the amount of TAF permeation after seven days was not found to be significantly different between the two patch formulations (SP5: 958.07 ± 82.81 μg/cm^2 and SP6: 1,054.38 ± 214.40 μg/cm^2). It was found that a permeation flux rate of TAF from patch SP6 was 6.60 ± 1.48 μg/cm^2/h and lag time of about 1 h was observed. Furthermore, the patch SP7 blend was prepared by homogenization at 32,000 rpm for only 5 min., and it did not show any significant difference in the amount of TAF permeation as well as the lag time when compared to those of SP5 and SP6 ($p > 0.05$), as shown in Figure 8. The permeation flux of TAF observed from patch SP7 was 7.24 ± 0.47 μg/cm^2/h. Additionally, it can be inferred that the use of different backing materials in patches SP5/6 and SP7 did not impact the drug permeation across human epidermis.

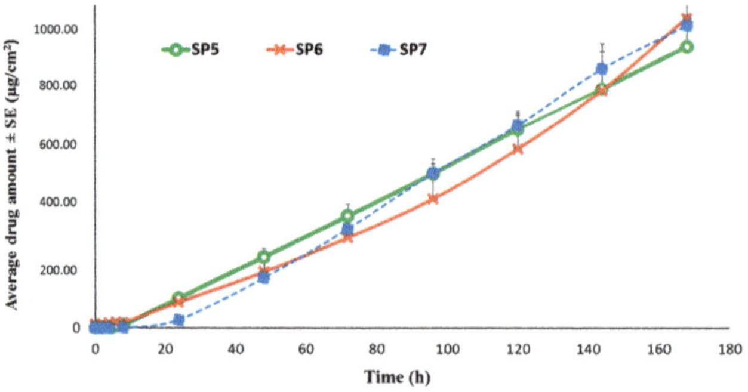

Figure 8. Comparison of permeation profile of silicone based TAF patches through human epidermis. SP5: Silicone-based 15% TAF patch prepared from formulation blend homogenized at 30,000 rpm for 20 min. SP6: Silicone-based 15% TAF patch prepared from formulation blend homogenized at 32,000 rpm for 30 min. SP7: Silicone-based 15% TAF patch prepared from formulation blend homogenized at 32,000 rpm for 5 min.

Permeation of TAF from Silicone-Based Suspension-Type Patches: Effect of Drug Concentration

Silicone based suspension patches, SP10-13, were prepared to investigate the effect of concentration of TAF on its permeation profile across human epidermis. As shown in Figure 9, the cumulative amount of TAF permeated after seven days was found to be 191.51 ± 23.26, 695.09 ± 68.85, 1,028.69 ± 78.63, 1,158.86 ± 57.96, and 1,266.19 ± 162.04 μg/cm^2 from SP10, SP11, SP7, SP12, and SP13, respectively. The permeation flux values for each of these patches were: 1.24 ± 0.13, 4.32 ± 0.47, 7.23 ± 0.58, 7.82 ± 0.48, and 8.44 ± 1.16 μg/cm^2/h. A significant increase in permeation was observed with the increase in TAF concentration from 5 (SP10) to 10% *w/w* (SP11), as well as from 10 (SP11) to 15% *w/w* (SP7), ($p < 0.05$). However, there was no statistically significant difference between the amount of TAF that permeated from patches containing 15 (SP7), 20 (SP12), and 25% *w/w* (SP13) drug ($p > 0.05$). Thus, increasing the TAF concentration from 5 to 15% *w/w* resulted in a higher amount of undissolved drug as a reservoir in the latter that probably provided an additional and constant driving force, eventually aiding in considerably enhancing the permeation of TAF. However, increasing the drug concentration from 15 to 25% *w/w* did not result in any further enhancement in permeation. Therefore, 15% drug concentration (SP7) was found to be optimum in order to achieve the targeted permeation flux.

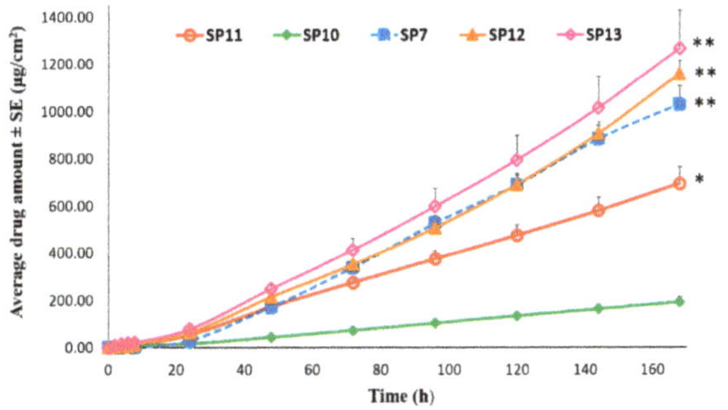

Figure 9. Effect of concentration of TAF in the silicone-based suspension patches on its permeation through human epidermis. * represents significant difference between 5 and 10% patch. ** represents significant difference as compared to 5 and 10% patch, ANOVA one-way test ($p < 0.05$). SP7: Silicone-based 15% TAF patch. SP10: Silicone-based 5% TAF patch. SP11: Silicone-based 10% TAF patch. SP12: Silicone-based 20% TAF patch. SP13: Silicone-based 25% TAF patch.

Permeation of TAF from Acrylate-Based Clear vs Silicone-Based Suspension-Type Patches

The aim of this study was to demonstrate a proof of concept for feasibility of developing a transdermal patch of TAF and achieve a clinically relevant flux of about 7 $\mu g/cm^2/h$ across human epidermis. From an in vivo perspective, as blood vessels are present in the dermis layer, once the drug crosses the stratum corneum and the viable epidermis, it reaches the systemic circulation. Hence, human epidermis was used in this study for the in vitro permeation testing. Owing to the desirable physicochemical properties (molecular weight= 476.47 g/mol, log P= 1.8), TAF showed permeation from the simplest transdermal patch formulation (AP1), across the human epidermis, depicting its potential to passively permeate through human skin. However, for the acrylate-based TAF patch formulations, it was only possible to achieve a flux rate of 0.6 $\mu g/cm^2/h$, with a combination of different chemical penetration enhancers at a low drug concentration (2% *w*/*w*), with no issues regarding drug crystallization. However, the silicone based TAF suspension patch formulations containing 15–25% *w*/*w* drug showed higher permeation flux of about 7 $\mu g/cm^2/h$, as well as a lower lag time (about 1h), which may be attributed to the release characteristics of TAF from the silicone adhesive matrix, and a higher drug loading in a suspension form, which can provide constant concentration gradient/driving force for the drug release and continuous permeation through epidermis. In addition, all of the excipients that were included in the TAF suspension patch—mineral oil, OA, and oleyl alcohol—have been previously reported for transdermal permeation enhancing effects, which may contribute to higher flux.

In our study, it was possible to achieve a relatively high flux rate of 7 $\mu g/cm^2/h$, from silicone based TAF suspensions, and the optimized patch "SP7" was characterized for drug release, coat weight and thickness, and drug content after storage at RT and 40 °C for three months. This prototype patch was further characterized for physical properties, such as tack, shear, and peel adhesion.

3.6. In Vitro Drug Release Studies

Figure 10 presents the seven-day in vitro release profiles of TAF from the optimized patch "SP7". The average percentage of TAF released after 7 days was observed to be 64.27 ± 5.47, in the study performed during week 1 after the preparation of the patch. The release profile of TAF from the same batch of suspension patch was also studied after 1.5 and 3 months of storage at RT and 40 °C. Table 7 shows the results. No significant difference in the total percentage of TAF released from the patches that are exposed to different test conditions was observed ($p>0.05$). This depicted uniformity in the release profile and percentage of drug released after exposure to higher temperature for three months, further indicating the stability of TAF in the suspension-based patches.

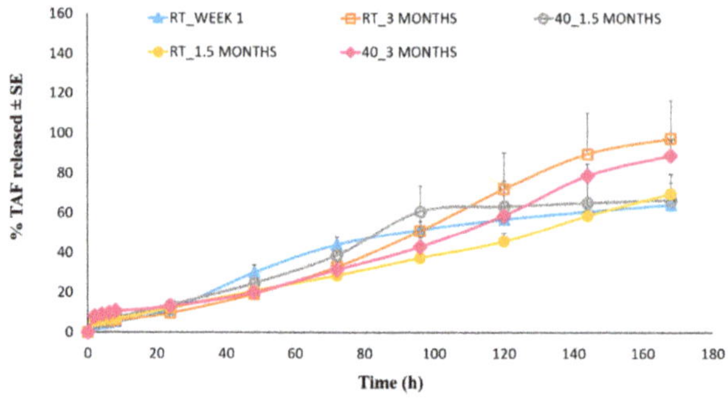

Figure 10. In vitro release profiles of TAF from the optimized suspension patch.

Table 7. Characterization parameters of optimized patch "SP7" up to three months at ambient and accelerated storage conditions.

Parameters	Week 1 (RT)	After 1.5 months (RT)	After 1.5 months (40 °C)	After 3 months (RT)	After 3 months (40 °C)
% TAF released after 7 days ± SE	64.27 ± 5.47	69.71 ± 5.63	66.82 ± 12.83	97.43 ± 19.03	88.86 ± 7.88
Coat weight (mg/cm2) ± SD	29.60 ± 1.94	30.38 ± 0.95	32.73 ± 3.06	29.39 ± 4.57	33.55 ± 3.14
Coat thickness (μm) ± SD	182.66 ± 1.15	186.67 ± 3.06	188.00 ± 3.46	186.00 ± 4.00	189.33 ± 3.06
Drug content (mg/cm2) ± SD	4.60 ± 1.01	4.95 ± 0.31	4.92 ± 0.39	5.25 ± 0.32	5.36 ± 0.43

3.7. Coat Thickness, Coat Weight, and Drug Content of Optimized Patch

Table 7 presents the average patch thickness, coat weight, and drug content of patch SP7, as measured after exposure to different temperature conditions for different durations. No significant difference in the measured parameters was observed after exposure of the patches to different test conditions ($p > 0.05$), signifying the stability and integrity of the suspension-based patches over time. Furthermore, as depicted by the results of drug content analysis, no degradation of TAF was observed after 1.5 and 3 months of patch storage at RT as well as 40 °C, further indicating the stability of TAF in the silicone-based suspension patches.

3.8. Physical Characterization of the Optimized Patch

3.8.1. Peel Adhesion

An ideal transdermal patch should peel off after application on skin, without causing delamination. Peel adhesion is not only affected by the intrinsic adhesiveness of the PSA, but it also involves the stretching and the bending of the patch matrix, and also the backing layer prior to the separation. The force that is required to peel the patch should be consistent for different batches and the value of peel adhesion obtained in the test varies the width of the test material [62]. No significant difference ($p > 0.05$) in the average force that is required to peel the patch from human dermatomed skin was observed when tested during week 1 (202.78 ± 43.07 g) and three months after patch storage at RT (149.35 ± 30.76 g) and 40 °C (171.17 ± 13.56 g). No delamination was observed for the tested patches.

3.8.2. Tack Properties

The adhesion efficiency of a transdermal patch can be tested by evaluating its tack, which is a measure of the force of debonding upon the application of a light pressure for a short time. A probe tack test was employed in this study, where the force that is required to separate a probe from the adhesive surface of a transdermal patch was measured. Tack was then expressed as the maximum value of the force that is required to break the bond between the probe and transdermal patch after a brief period of contact [62]. The average absolute positive force (adhesive or sticky property), average positive area (work of adhesion), and average separation distance (degree of legging), recorded for three replicates during the first week after patch preparation, was found to be 1,116.88 ± 167.79 g/cm^2, 34.30 ± 10.98 g.s, and 1.27 ± 0.50 mm, respectively. Storage at RT (1,012.03 ± 220.59 g/cm^2, 24.77 ± 8.50 g.s, and 0.8 ± 0.1 mm, respectively) and 40 °C (983.72 ± 201.56 g/cm^2, 23.33 ± 6.20 g.s, and 1.0 ± 0.17 mm, respectively) for three months did not impact the tack properties of the optimized patch formulation ($p > 0.05$). Additionally, the average absolute positive force that was observed in our study was comparable to that observed for the 9.5% fentanyl containing silicone-based suspension patch laminate (1,499 g/cm^2) reported in literature [63].

3.9. Evaluation of Skin Irritation Potential of Optimized Patch

The principle of irritation assay using the in vitro skin model is based on the premise that irritant chemicals can penetrate the stratum corneum by diffusion and they are cytotoxic to the cells in the underlying layers. In a transdermal patch, the drug by itself or in combination with other additives can be irritant to skin [30]. Hence, the entire patch was tested for its irritation potential, and not just the drug by itself. The tested transdermal patch (SP7) resulted in a mean relative cell viability of 104.64 ± 7.42%, which was comparable to the negative control (100.00 ± 6.09) and significantly higher than the positive control (14.27 ± 8.09). Hence, it can be concluded that the final optimized silicone suspension patch (SP7) is non-irritant to human skin.

4. Conclusions

Several solution and suspension-type patch formulations were developed for the transdermal delivery of TAF. Ultimately, the optimized silicone-based suspension patch design successfully achieved the target release and duration profile for TAF and it was found to be suitable for weekly drug dosing. Drug permeation flux of 7 μg/cm^2/h, as observed from the optimized patch, when extrapolated to 50 cm^2 patch size, indicates the delivery of 8.4 mg TAF/day, surpassing the target dose of 8 mg/day. Further studies are underway to characterize its safety and pharmacokinetic profiles in vivo.

Author Contributions: Conceptualization, A.P., W.Z. and A.K.B.; Methodology, A.P., S.A.B., W.Z. and A.K.B.; Formal Analysis, A.P. and S.A.B.; Investigation, A.P. and S.A.B.; Writing-Original Draft Preparation, A.P. and S.A.B.; Writing-Review & Editing, G.F.D., M.C. and O.N.S.; Visualization, A.P.; Supervision, A.K.B., O.N.S. and W.Z.; Project Administration, M.C., O.N.S. and W.Z.; Funding Acquisition, G.F.D. and M.C.

Funding: This work was supported by a subagreement (MAPS2-15-061) between Eastern Virginia Medical School, on behalf of CONRAD, and the Corporation of Mercer University, with funds provided by the USAID/PEPFAR through a cooperative agreement (AID-OAA-A-14-00010) with CONRAD.

Conflicts of Interest: The authors declare no conflict of interest.

Abbreviations

ANOVA	Analysis of variance
EPI-200-SIT	EpiDerm^{TM} skin irritation test
ELISA	Enzyme linked immunosorbent assay
EVAC	Ethylene vinyl acetate copolymer
HPLC	High performance liquid chromatography
PBS	Phosphate buffered saline
PDMS	Polydimethyl-siloxane
PEG	Polyethylene glycol
PET	poly-ethylene terephthalate
PIB	Polyisobutylene
PSA	Pressure sensitive adhesives
RhE	Reconstructed human epidermis
RT	Room temperature
STRs	Single tablet regimens
TFV	Tenofovir
TAF	Tenofovir alafenamide
THF	Tetrahydrofuran

References

1. De Clercq, E. Role of tenofovir alafenamide (TAF) in the treatment and prophylaxis of HIV and HBV infections. *Biochem. Pharmacol.* **2018**, *153*, 2–11. [CrossRef]
2. DeJesus, E.; Haas, B.; Segal-Maurer, S.; Ramgopal, M.N.; Mills, A.; Margot, N.; Liu, Y.-P.; Makadzange, T.; McCallister, S. Superior Efficacy and Improved Renal and Bone Safety After Switching from a Tenofovir Disoproxil Fumarate- to a Tenofovir Alafenamide-Based Regimen Through 96 Weeks of Treatment. *AIDS Res. Hum. Retrovir.* **2018**, *34*, 337–342. [CrossRef] [PubMed]
3. Kahn, P. AIDS Vaccine Handbook: Global Perspectives. Available online: http://www.avac.org/ (accessed on 2 November 2019).
4. Nelson, K.; Varadarajan, P.; Rajappan, M.; Chikkanna, N. Advances in Drug Delivery of Anti-HIV Drugs—An Overview. *Am. J. PharmTech Res.* **2012**, *2*, 231–244.
5. Vedha Hari, B.N.; Devendharan, K.; Narayanan, N. Approaches of novel drug delivery systems for Anti-HIV agents. *Int. J. Drug Dev. Res.* **2013**, *5*, 16–24.
6. Dou, H.; Destache, C.J.; Morehead, J.R.; Mosley, R.L.; Boska, M.D.; Kingsley, J.; Gorantla, S.; Poluektova, L.; Nelson, J.A.; Chaubal, M.; et al. Development of a macrophage-based nanoparticle platform for antiretroviral drug delivery. *Blood* **2006**, *108*, 2827–2835. [CrossRef] [PubMed]
7. Gunawardana, M.; Remedios-Chan, M.; Miller, C.S.; Fanter, R.; Yang, F.; Marzinke, M.A.; Hendrix, C.W.; Beliveau, M.; Moss, J.A.; Smith, T.J.; et al. Pharmacokinetics of long-acting tenofovir alafenamide (GS-7340) subdermal implant for HIV prophylaxis. *Antimicrob. Agents Chemother.* **2015**, *59*, 3913–3919. [CrossRef]
8. Ball, C.; Krogstad, E.; Chaowanachan, T.; Woodrow, K.A. Drug-Eluting Fibers for HIV-1 Inhibition and Contraception. *PLoS ONE* **2012**, *7*, e49792. [CrossRef] [PubMed]
9. Rohan, L.C.; Devlin, B.; Yang, H. Microbicide Dosage Forms. In *Microbicides for Prevention of HIV Infection*; Springer: Berlin/Heidelberg, Germany, 2013; pp. 27–54. ISBN 978-3-662-44596-9.
10. Nelson, A.G.; Zhang, X.; Ganapathi, U.; Szekely, Z.; Flexner, C.W.; Owen, A.; Sinko, P.J. Drug delivery strategies and systems for HIV/AIDS pre-exposure prophylaxis and treatment. *J. Control. Release* **2015**, *219*, 669–680. [CrossRef]
11. Karim, Q.A.; Karim, S.S.A.; Frohlich, J.A.; Grobler, A.C.; Baxter, C.; Mansoor, L.E.; Kharsany, A.B.M.; Sibeko, S.; Mlisana, K.P.; Omar, Z.; et al. Effectiveness and safety of tenofovir gel, an antiretroviral microbicide, for the prevention of HIV infection in women. *Science* **2010**, *329*, 1168–1174. [CrossRef]
12. Puri, A.; Sivaraman, A.; Zhang, W.; Clark, M.R.; Banga, A.K. Expanding the domain of drug delivery for HIV prevention: Exploration of the transdermal route. *Crit. Rev. Ther. Drug Carrier Syst.* **2017**, *34*, 551–587. [CrossRef] [PubMed]
13. Karim, S.S.A.; Kashuba, A.D.; Werner, L.; Karim, Q.A. Drug concentrations after topical and oral antiretroviral pre-exposure prophylaxis: Implications for HIV prevention in women. *Lancet* **2011**, *378*, 279–281. [CrossRef]

14. Banga, A.K. *Transdermal and Intradermal Delivery of Therapeutic Agents: Application of Physical Technologies*; CRC Press: Boca Raton, FL, USA, 2011; ISBN 9781439805107.
15. Ham, A.S.; Buckheit, R.W. Current and emerging formulation strategies for the effective transdermal delivery of HIV inhibitors. *Ther. Deliv.* **2015**, *6*, 217–229. [CrossRef]
16. Alexander, A.; Dwivedi, S.; Giri, T.K.; Saraf, S.; Saraf, S.; Tripathi, D.K. Approaches for breaking the barriers of drug permeation through transdermal drug delivery. *J. Control. Release* **2012**, *164*, 26–40. [CrossRef]
17. Pastore, M.; Kalia, Y.; Hortsmann, M.; Roberts, M. Transdermal patches: history, development and pharmacology. *Br. J. Phamacol.* **2015**, *172*, 2179–2209. [CrossRef]
18. De Clercq, E. Tenofovir alafenamide (TAF) as the successor of tenofovir disoproxil fumarate (TDF). *Biochem. Pharmacol.* **2016**, *119*, 1–7. [CrossRef]
19. Murphy, R.A.; Valentovic, M.A. Factors Contributing to the Antiviral Effectiveness of Tenofovir. *J. Pharmacol. Exp. Ther.* **2017**, *363*, 156–163. [CrossRef]
20. Babusis, D.; Phan, T.K.; Lee, W.A.; Watkins, W.J.; Ray, A.S. Mechanism for effective lymphoid cell and tissue loading following oral administration of nucleotide prodrug GS-7340. *Mol. Pharm.* **2013**, *10*, 459–466. [CrossRef]
21. Foreman, P.; Hansen, A.; Silverberg, E.; Manegold, T.; Yang, H.; Li, J.; Jacobson, S. Predicting Drug Solubility in Transdermal Adhesives. American Association of Pharmaceutical Scientists. In Proceedings of the 37th AAPS Annual Pharmaceutical Technologies Conference, New York, NY, USA, 13–18 January 2002.
22. Sachdeva, V.; Bai, Y.; Kydonieus, A.; Banga, A.K. Formulation and optimization of desogestrel transdermal contraceptive patch using crystallization studies. *Int. J. Pharm.* **2013**, *441*, 9–18. [CrossRef]
23. Banerjee, S.; Chattopadhyay, P.; Ghosh, A.; Bhattacharya, S.; Kundu, A.; Veer, V. Accelerated stability testing of a transdermal patch composed of eserine and pralidoxime chloride for prophylaxis against (±)-anatoxin A poisoning. *J. Food Drug Anal.* **2014**, *22*, 264–270. [CrossRef]
24. Kassis, V.; Søndergaard, J. Heat-separation of normal human skin for epidermal and dermal prostaglandin analysis. *Arch. Dermatol. Res.* **1982**, *273*, 301–306. [CrossRef]
25. Puri, A.; Murnane, K.S.; Blough, B.E.; Banga, A.K. Effects of chemical and physical enhancement techniques on transdermal delivery of 3-fluoroamphetamine hydrochloride. *Int. J. Pharm.* **2017**, *528*, 452–462. [CrossRef]
26. Lakhani, P.; Bahl, R.; Bafna, P. Transdermal Patches: Physiochemical and in-vitro Evaluation Methods. *Int. J. Pharm. Sci. Res.* **2015**, *6*, 1826–1836.
27. Cai, B.; Söderkvist, K.; Engqvist, H.; Bredenberg, S. A new drug release method in early development of transdermal drug delivery systems. *Pain Res. Treat.* **2012**, *2012*, 953140. [CrossRef]
28. Wokovich, A.M.; Prodduturi, S.; Doub, W.H.; Hussain, A.S.; Buhse, L.F. Transdermal drug delivery system (TDDS) adhesion as a critical safety, efficacy and quality attribute. *Eur. J. Pharm. Biopharm.* **2006**, *64*, 1–8.
29. Lu, Z.; Fassihi, R. Influence of colloidal silicon dioxide on gel strength, robustness, and adhesive properties of diclofenac gel formulation for topical application. *AAPS PharmSciTech* **2015**, *16*, 636–644. [CrossRef]
30. Kandárová, H.; Hayden, P.; Klausner, M.; Kubilus, J.; Sheasgreen, J. An in vitro skin irritation test (SIT) using the EpiDerm reconstructed human epidermal (RHE) model. *J. Vis. Exp. JoVE* **2009**, *29*, e1336.
31. Banerjee, S.; Chattopadhyay, P.; Ghosh, A.; Datta, P.; Veer, V. Aspect of adhesives in transdermal drug delivery systems. *Int. J. Adhes. Adhes.* **2014**, *50*, 70–84.
32. Pocius, A. Adhesives. In *Kirk-Othmer Encyclopedia of Chemical Technology*; Howe-Grants, M., Ed.; Wiley-Interscience: New York, NY, USA, 1991; pp. 445–466.
33. Lobo, S.; Sachdeva, S.; Goswami, T. Role of pressure-sensitive adhesives in transdermal drug delivery systems. *Ther. Deliv.* **2016**, *7*, 33–48. [CrossRef]
34. Weng, W.; Quan, P.; Liu, C.; Zhao, H.; Fang, L. Design of a Drug-in-Adhesive Transdermal Patch for Risperidone: Effect of Drug-Additive Interactions on the Crystallization Inhibition and In Vitro/In Vivo Correlation Study. *J. Pharm. Sci.* **2016**, *10*, 3153–3161. [CrossRef]
35. Jain, P.; Banga, A.K. Induction and inhibition of crystallization in drug-in-adhesive-type transdermal patches. *Pharm. Res.* **2013**, *30*, 562–571. [CrossRef]
36. Saini, S.; Shikha, B.; Chauhan, S.S. Recent development in penetration enhancers and techniques in transdermal drug delivery system. *J. Adv. Pharm. Educ. Res.* **2014**, *4*, 31–40.
37. Pathan, I.B.; Setty, C.M. Chemical penetration enhancers for transdermal drug delivery systems. *Trop. J. Pharm. Res.* **2009**, *8*, 173–179. [CrossRef]

38. Mitragotri, S. Synergistic effect of enhancers for transdermal drug delivery. *Pharm. Res.* **2000**, *17*, 1354–1359. [CrossRef]
39. Santoyo, S.; Arellano, A.; Ygartua, P.; Martin, C. Penetration enhancer effects on the in vitro percutaneous absorption of piroxicam through rat skin. *Int. J. Pharm.* **1995**, *117*, 219–224. [CrossRef]
40. Andega, S.; Kanikkannan, N.; Singh, M. Comparison of the effect of fatty alcohols on the permeation of melatonin between porcine and human skin. *J. Control. Release* **2001**, *77*, 17–25. [CrossRef]
41. Patel, K.N.; Patel, H.K.; Patel, V.A. Formulation and characterization of drug in adhesive transdermal patches of diclofenac acid. *Int. J. Pharm. Pharm. Sci.* **2012**, *4*, 296–299.
42. Moghimipour, E.; Rezaee, S.; Omidi, A. The effect of formulation factors on the release of oxybutynin hydrochloride from transdermal polymeric patches. *J. Appl. Pharm. Sci.* **2011**, *1*, 73–76.
43. Nicoli, S.; Penna, E.; Padula, C.; Colombo, P.; Santi, P. New transdermal bioadhesive film containing oxybutynin: In vitro permeation across rabbit ear skin. *Int. J. Pharm.* **2006**, *325*, 2–7. [CrossRef]
44. Nicolazzo, J.A.; Morgan, T.M.; Reed, B.L.; Finnin, B.C. Synergistic enhancement of testosterone transdermal delivery. *J. Control. Release* **2005**, *103*, 577–585. [CrossRef]
45. Fraser, I.S.; Weisberg, E.; Kumar, N.; Kumar, S.; Humberstone, A.J.; McCrossin, L.; Shaw, D.; Tsong, Y.Y.; Sitruk-Ware, R. An initial pharmacokinetic study with a Metered Dose Transdermal System® for delivery of the progestogen Nestorone® as a possible future contraceptive. *Contraception* **2007**, *76*, 432–438. [CrossRef]
46. European Meedicines Agency. Guideline on quality of transdermal patches Guideline on quality of transdermal patches Table of contents. *Eur. Med. Agency* **2014**, *44*, 1–28.
47. Sekikawa, H.; Nakano, M.; Arita, T. Inhibitory effect of polyvinylpyrrolidone on the crystallization of drugs. *Chem. Pharm. Bull.* **1978**, *26*, 118–126. [CrossRef]
48. Gong, K.; Viboonkiat, R.; Rehman, I.U.; Buckton, G.; Darr, J.A. Formation and characterization of porous indomethacin-PVP coprecipitates prepared using solvent-free supercritical fluid processing. *J. Pharm. Sci.* **2005**, *94*, 2583–2590. [CrossRef]
49. Ohm, A. Interaction of Bay t 3839 coprecipitates with insoluble excipients. *Eur. J. Pharm. Biopharm.* **2000**, *49*, 183–189. [CrossRef]
50. Ma, X.; Taw, J.; Chiang, C.M. Control of drug crystallization in transdermal matrix system. *Int. J. Pharm.* **1996**, *142*, 115–119. [CrossRef]
51. Karande, P.; Mitragotri, S. Enhancement of transdermal drug delivery via synergistic action of chemicals. *Biochim. Biophys. Acta-Biomembr.* **2009**, *1788*, 2362–2373. [CrossRef]
52. Dhankhar, P. Homogenization Fundamentals. *IOSR J. Eng.* **2014**, *4*, 1–8. [CrossRef]
53. Chu, K.R.; Lee, E.; Jeong, S.H.; Park, E.-S. Effect of particle size on the dissolution behaviors of poorly water-soluble drugs. *Arch. Pharm. Res.* **2012**, *35*, 1187–1195. [CrossRef]
54. Anarjan, N.; Jafarizadeh-Malmiri, H.; Nehdi, I.A.; Sbihi, H.M.; Al-Resayes, S.I.; Tan, C.P. Effects of homogenization process parameters on physicochemical properties of astaxanthin nanodispersions prepared using a solvent-diffusion technique. *Int. J. Nanomed.* **2015**, *10*, 1109–1118.
55. Sharma, N.; Madan, P.; Lin, S. Effect of process and formulation variables on the preparation of parenteral paclitaxel-loaded biodegradable polymeric nanoparticles: A co-surfactant study. *Asian J. Pharm. Sci.* **2016**, *11*, 404–416. [CrossRef]
56. Lawrence, J.N. Electrical resistance and tritiated water permeability as indicators of barrier integrity of in vitro human skin. *Toxicol. Vitr.* **1997**, *11*, 241–249. [CrossRef]
57. Golla, V.M.; Kurmi, M.; Shaik, K.; Singh, S. Stability behaviour of antiretroviral drugs and their combinations. 4: Characterization of degradation products of tenofovir alafenamide fumarate and comparison of its degradation and stability behaviour with tenofovir disoproxil fumarate. *J. Pharm. Biomed. Anal.* **2016**, *131*, 146–155. [CrossRef]
58. Larrucea, E.; Arellano, A.; Santoyo, S.; Ygartua, P. Combined effect of oleic acid and propylene glycol on the percutaneous penetration of tenoxicam and its retention in the skin. *Eur. J. Pharm. Biopharm.* **2001**, *2*, 113–119. [CrossRef]
59. Benson, H.A.E. Transdermal drug delivery: Penetration enhancement techniques. *Curr. Drug Deliv.* **2005**, *2*, 23–33. [CrossRef]
60. Chatterjee, D.J.; Li, W.Y.; Koda, R.T. Effect of vehicles and penetration enhancers on the in vitro and in vivo percutaneous absorption of methotrexate and edatrexate through hairless mouse skin. *Pharm. Res.* **1997**, *14*, 1058–1065. [CrossRef]

61. Quan, D.; Deshpanday, N.; Venkateshwaran, S.; Ebert, C. Triacetin as a Penetration Enhancer for Transdermal Delivery of a Basic Drug. U.S. Patent No 5,834,010, 10 November 1998.
62. Cilurzo, F.; Gennari, C.G.M.; Minghetti, P. Adhesive properties: A critical issue in transdermal patch development. *Expert Opin. Drug Deliv.* **2012**, *9*, 33–45. [CrossRef]
63. Miller, K.; Govil, S.; Bhatia, K. Fentanyl Suspension-Based Silicone Adhesive Formulations and Devices for Transdermal Delivery of Fentanyl. U.S. Patent 7556823B2, 28 October 2008.

pharmaceutics

MDPI

Article

Characterization of a Reservoir-Style Implant for Sustained Release of Tenofovir Alafenamide (TAF) for HIV Pre-Exposure Prophylaxis (PrEP)

Leah M. Johnson [1,*], Sai Archana Krovi [1], Linying Li [1], Natalie Girouard [1], Zach R. Demkovich [2], Daniel Myers [3], Ben Creelman [3] and Ariane van der Straten [2]

1 Engineered Systems, RTI International, 3040 E. Cornwallis Road, Research Triangle Park, NC 27709, USA
2 Women's Global Health Imperative, RTI International, 351 California Street, Suite 500, San Francisco, CA 94104, USA
3 PATH, 2201 Westlake Ave, Suite 200, Seattle, WA 98121, USA
* Correspondence: leahjohnson@rti.org; Tel.: +1-919-541-7233

Received: 31 May 2019; Accepted: 29 June 2019; Published: 4 July 2019

Abstract: Long-acting (LA) HIV pre-exposure prophylaxis (PrEP) offers the potential to improve adherence by lowering the burden of daily or on-demand regimens of antiretroviral (ARV) drugs. This paper details the fabrication and in vitro performance of a subcutaneous and trocar-compatible implant for the LA delivery of tenofovir alafenamide (TAF). The reservoir-style implant comprises an extruded tube of a biodegradable polymer, poly(ε-caprolactone) (PCL), filled with a formulation of TAF and castor oil excipient. Parameters that affect the daily release rates of TAF are described, including the surface area of the implant, the thickness of the PCL tube walls (between 45 and 200 μm), and the properties of the PCL (e.g., crystallinity). In vitro studies show a linear relationship between daily release rates and surface area, demonstrating a membrane-controlled release mechanism from extruded PCL tubes. Release rates of TAF from the implant are inversely proportional to the wall thickness, with release rates between approximately 0.91 and 0.15 mg/day for 45 and 200 μm, respectively. The sustained release of TAF at 0.28 ± 0.06 mg/day over the course of 180 days in vitro was achieved. Progress in the development of this implant platform addresses the need for new biomedical approaches to the LA delivery of ARV drugs.

Keywords: poly(ε-caprolactone) (PCL); tenofovir alafenamide (TAF); pre-exposure prophylaxis (PrEP); long-acting drug delivery systems; implant

1. Introduction

HIV pre-exposure prophylaxis (PrEP) with antiretroviral (ARV) drugs is promising among the biomedical strategies to address the global HIV epidemic. Tenofovir-based PrEP has demonstrated landmark successes with daily [1–4] and on-demand dosing [3,5] in men who have sex with men (MSM) and transgender women (TGW). Despite these advancements, adherence to time- or event-driven regimens for PrEP remains an incessant struggle [6–11]. The long-acting (LA) delivery of ARV drugs simplifies traditional dosing regimens for PrEP by alleviating the emotional and logistical burden of user-dependent methods. For example, a LA-injectable formulation of the integrase inhibitor, cabotegravir (CAB), is currently under investigation in a pair of phase two and three HIV PrEP trials [12,13]. Although injectable methods are acceptable to many users [14,15], and offer key advantages such as a bi-monthly dosing regimen and discretion, drawbacks do exist. Injectable formulations cannot be removed in the event of an adverse drug-related event and the potential exists for a long plasma "tail" of sub-therapeutic drug levels [16,17]. An alternative injectable formulation in preclinical development involves an in situ forming polymer depot for LA delivery of dolutegravir for HIV PrEP, which shows the capacity for removal within the period of drug delivery, if needed [18].

A promising biomedical approach for LA-PrEP involves implants that reside under the skin to continuously release the drug, which supports adherence over longer time periods, enables discretion of use, lowers the burden of the regimen, and remains reversible during the therapeutic duration. Polymeric implants can comprise different architectures that each have advantages for drug delivery [19–21]. Matrix-style implants contain a drug dispersed within a polymer that controls the rate of the drug exiting the implant. For example, an FDA-approved matrix-style implant for the six-month maintenance treatment of opioid addiction (Probuphine®) contains buprenorphine distributed through four individual poly(ethylene-vinyl acetate) (EVA) rods [22]. Reservoir-style implants involve a formulated drug core encapsulated by a polymeric barrier that control drug release rates. Notable examples of implants with a core-sheath configuration include the collection of subdermal contraceptive implants: Norplant and Jadelle [23,24] for the delivery of levonorgestrel (LNG) using a rod of silicone-based polymer, as well as Implanon [25] and Nexplanon [26] for the delivery of etonogestrel (ENG) using a rod of EVA-based polymer. The low dosages required for the subcutaneous delivery of hormonal contraceptives enable these implants to last multiple years. Implants also show utility for indications in ophthalmology, including intraocular implants for the delivery of ganciclovir for the treatment of cytomegalovirus retinitis (Vitrasert) [27], dexamethasone for the treatment of macular edema (Ozurdex) [28,29] and fluocinolone acetonide for the treatment of noninfectious posterior uveitis (Retisert) [30].

Several implants are currently under development for HIV PrEP, with each implant system holding unique configurations and features. A subdermal, silicone implant that delivers tenofovir alafenamide (TAF) from orthogonal channels coated with polyvinyl alcohol (PVA) showed 40-days of drug delivery in beagle dogs without observed adverse events [31]. A non-polymeric, refillable implant designed to deliver TAF and emtricitabine (FTC) from separate devices showed sustained levels of tenofovir diphosphate (TFV-DP) in peripheral blood mononuclear cells (PBMCs) over 83 days in rhesus macaques, but only 28 days for FTC-triphosphate (FTC-TP) due to the large dosing required and short plasma half-life [32]. Intarcia is developing a titanium osmotic pump system, called the Medici Drug Delivery System™, for PrEP and for type-2 diabetes [33]. A matrix-style PrEP implant for delivery of 4′-ethylnyl-2-fluoro-2′-dexoyadenosine (EFdA) has shown promising efficacy for HIV treatment and prevention, as demonstrated in animal models [34].

Concurrently with these other innovative technologies, RTI is developing a subcutaneous biodegradable implant for HIV PrEP as a single indication and as a multipurpose prevention technology (MPT) for HIV and pregnancy prevention [35,36]. The implant uses a semi-crystalline aliphatic polyester, poly(ε-caprolactone) (PCL), pioneered by Pitt et al. at RTI International in the 1980s [37] and largely neglected for nearly 20 years [38]. Renewed appeal for PCL has surfaced in light of biomedical applications, including tissue engineering [39,40] and drug delivery [41,42], that demand materials with long-term functionality, mechanical integrity, biocompatibility, and capacity for biodegradation and bioresorption. PCL is currently used in FDA-approved products for root canal fillings (Resilon) [43] and sutures (Monocryl) [44] and was previously explored for use as a 1-year contraceptive implant (Capronor) [45]. In terms of HIV PrEP, PCL implants can advantageously offer LA delivery of ARVs, while also enabling bioresorption at the end of the implant drug delivery period. An implant that is biodegradable could benefit health care systems by eliminating the need for a clinic visit to remove a depleted implant, whereby a minor surgical procedure would be required to remove the implant when discontinuing PrEP. In the case of the implant described in this paper, reversibility and retrievability can be maintained, potentially throughout the duration of treatment.

Herein, we report advancements to the fabrication and performance of a reservoir-style subcutaneous PCL implant for sustained release of TAF. We detail parameters that control release rates of TAF from the implant, including wall thickness and surface area, and further describe the effects of crystallinity on the performance of the implant. We demonstrate the fabrication and processing steps that align the implant with future manufacturing requirements, while keeping preferences of the end user in mind.

2. Materials and Methods

2.1. Implant Fabrication

PCL pellets were purchased in research-grade from Sigma Aldrich, referred to as "Sigma-PCL" throughout this paper (number average molecular weight (M_n) = 103 kDa, Cat# 440744, St. Louis, MO, USA) and in medical-grade from Corbion, referred to as "PC-12" throughout this paper (M_n = 51 kDa, PURASORB PC 12, Amsterdam, The Netherlands). PCL tubes were fabricated via a hot-melt, single screw extrusion process using solid PCL pellets at GenX Medical (Chattanooga, TN, USA). All tubes were 2.5 mm in outer diameter (OD) and had wall thicknesses of 45, 70, 100, 150 or 200 μm, as measured with a 3-axis laser measurement system and light microscopy at GenX Medical.

PCL tubes were first sealed at one end using two different approaches: impulse heat sealing and injection sealing. Implants fabricated with the impulse heat sealing were used for certain in vitro studies, such as the implants used in the surface area studies that comprised different lengths (e.g., 70 mm length), because they were too long to fit the injection sealing apparatus. Importantly, no significant differences in release rates were observed for TAF implants sealed with either approach. For the first approach, an impulse heat sealer (AIE-110T, American International Electric Sealer Supply, South El Monte, CA, USA) was used to clamp the tube flat and then apply a pulse of heat for a few seconds. The tubing was then allowed to cool for about 10 s. Thicker tubes were sealed with longer heat pulses. The sealing step fused the PCL tube wall together through melting and created a flat-shaped seal. The seal was trimmed with scissors to remove excess PCL. For the injection sealing, the PCL tube was marked and trimmed to the correct length to achieve an implant with a 40-mm paste length with 3 mm of headspace at both ends for sealing. The initial seal was then created on one end of the implant by placing the tube over a stainless steel rod that filled all the tube except for a 3 mm headspace at one end, placing a Teflon collar around the headspace to support the tube wall and injecting molten PCL into the cavity of the headspace. After the injected PCL was solidified, excess PCL was trimmed, and the collar was removed to form a cylindrical seal approximately 2 mm long that is compatible with commercial contraceptive trocars.

TAF was graciously provided by Gilead Sciences (Foster City, CA, USA). TAF was mixed with pharmaceutical grade, Super Refined™ Castor Oil (Croda, Cat# SR40890, Snaith, UK) at 2:1 mass ratio immediately prior to loading into the implant. The mixture was first ground with a mortar and pestle to create a smooth paste, and then back loaded into a 1 mL syringe fitted with a 14-gauge blunt tip needle. The TAF and castor oil paste was then extruded through the needle into the empty tube. Otherwise, the TAF formulation was loaded into the PCL tube using a modified spatula. After the filled formulation reached the 40-mm mark, the interior tube wall was cleaned with a rod and sealed in a similar manner to the first seal. After fabrication, all devices were weighed to determine the total payload and photographed with a ruler to record the final dimensions. Paste area was measured with ImageJ (Version 1.50e, NIH, Bethesda, MD, USA) and release rates were normalized to the surface area of a full-sized implant (2.5 mm OD, 40 mm in length), 314 mm^2. The end of the implants (i.e., end-seals) were not included in calculations of the implant surface area.

2.2. Device Sterilization

All implants were fabricated and handled under aseptic conditions using a biosafety cabinet. Certain devices were exposed to gamma irradiation, as indicated in the text. Devices exposed to gamma irradiation were first packed in amber glass vials and then irradiated with a dose range of 18–24 kGy at room temperature, using a Cobalt-60 gamma-ray source (Nordion Inc., Ottawa, Canada) at Steris (Mentor, OH, USA). Samples were exposed to the source on a continuous path for a period of 8 h.

2.3. In Vitro Release Studies

In vitro release characterization involved incubation of the implants in 40 mL 1X phosphate buffered saline (PBS) (pH 7.4) at 37 °C and placed on an orbital shaker. TAF species in the release media was measured by ultraviolet-visible (UV) spectroscopy at 260 nm using the Synergy MX multi-mode plate reader (BioTek Instruments, Inc, Winooski, VT, USA). The release buffer was sampled three times per week during which the devices were transferred to 40 mL of fresh buffer to maintain sink conditions. TAF quantity released in each PBS buffer during the time interval was calculated and cumulative mass of drug release as a function of time was determined.

2.4. Stability Analysis of TAF Formulation

The purity of TAF formulations inside the device reservoir was evaluated by opening a device, extracting the entire reservoir contents into an organic solution, and measuring TAF chromatographic purity using ultra performance liquid chromatography coupled with UV spectroscopy (UPLC/UV). The analysis was performed using a Waters BEH C18 column (2.1 mm × 50 mm, 1.7 μm) under gradient, reversed phase conditions with detection at 260 nm. For each device, one single aliquot was prepared and quantitated by linear regression analysis against a five-point calibration curve. TAF purity was calculated as % peak area associated with TAF relative to total peak area of TAF related degradation products (detected above the limit of detection (LOD) ≥ 0.05%). The TAF formulations within the implant were analyzed after exposure of the implant to a simulated physiological condition (i.e., 1X PBS, pH 7.4 at 37 °C) for up to 180 days.

2.5. Characterization of PCL Extruded Tubes

2.5.1. Differential Scanning Calorimetry (DSC)

The melting behavior of PCL samples was assessed with modulated differential scanning calorimetry (MDSC) (TA Instruments Q200, RCS90 cooling system, New Castle, DE, USA). Approximately 8 mg of extruded polymer tubing was placed in a Tzero^{TM} Pan and sealed with a Tzero^{TM} Lid and a dome-shaped die, resulting in a crimped seal. Samples were then placed in a nitrogen-purged DSC cell, cooled to 0 °C, then heated to 120 °C at a rate of 1 °C/min with an underlying heat-only modulation temperature scan of ± 0.13 °C every 60 s. The melting temperature (T_m) of the polymer was determined by the peak temperature of the melting endotherm, and the enthalpy associated with melting was determined by integrating linearly the area of the melt peak (between 25 and 65 °C) using the TA Universal Analysis software (version 4.5A, TA Instruments, New Castle, DE, USA). PCL samples did not exhibit exothermic peaks in the non-reversing heat flow signal indicating that PCL did not experience cold-crystallization during the melting process; therefore, the total heat flow curve was used to assess the mass % crystallinity. The mass % crystallinity was calculated using Equation (1), where X_c represents the mass fraction of crystalline domains in PCL, ΔH_m represents the enthalpy of melting measured by the DSC, and ΔH_{fus} represents the theoretical enthalpy of melting for 100% crystalline PCL, reported as 139.5 J/g [46,47].

$$X_C = \frac{\Delta H_m}{\Delta H_{fus}} \times 100 \tag{1}$$

The peak melting temperatures of polymers were used calculate crystallite sizes within the sample using the Thompson–Gibbs equation (Equation (2)) [48,49]:

$$L = \frac{2\sigma_e T_m^o}{\Delta H_m^o (T_m^o - T_m)} \tag{2}$$

where L is the crystallite size in nm, σ_e is the free energy of chain folds in mJ/m^2, T_m^o is the equilibrium melting temperature in K, T_m is the melting temperature measured by DSC in K, and ΔH_m^o is the enthalpy of fusion for 100% crystalline polymer in J/g. T_m^o and ΔH_m^o were taken from the ATHAS data bank as 342.2 K and 139.5 J/g, respectively. The free energy associated with chain folding was taken as 60 mJ/m^2 [50].

2.5.2. X-ray Diffraction (XRD)

The extruded PCL tubes at wall thickness of 100 μm were cryo-grinded in a freezer mill using liquid nitrogen. The material was ground for 1.5 min after cooling for three minutes before initiating the grinding cycle. The X-ray diffraction (XRD) patterns were acquired using a Bruker AXS, Inc D8 Advance model utilizing standard Bragg-Brentano geometry and a LynxEye XE-T high resolution detector (Bruker, Billerica, MA, USA). Samples were packed into a zero background sample holder and scanned at 40 kV and 40 mA power settings (1600 Watts) for a scan covering 5° to 70°, with a step size of 0.02° and a dwell time of 2 s per step. The MDI Jade version 9.6 software (MDI, Livermore, CA, USA) was used to analyze results and the 2019 International Center for Diffraction Data (ICDD) PDF 4+ database was used to search match crystalline phases present in the materials. The crystallite size was determined via the Scherrer equation (Equation (3)):

$$L = \frac{K\lambda}{\beta cos\theta} \tag{3}$$

where L = crystallite size, K = Scherrer constant (0.94 from literature [51,52]), λ = X-ray wavelength, β = full-width at half maximum of a crystallographic peak, and θ = Bragg angle.

2.5.3. Gel Permeation Chromatography (GPC)

The molecular weight of PCL was analyzed via GPC by first dissolving samples in tetrahydrofuran (THF) to 10 mg/mL injecting 40 μL of sample using an Agilent 1100/1200 HPLC-UV instrument (Santa Clara, CA, USA, flow rate of 1.0 mL/min). Polystyrene polymer standards (MWs of 2460 to 0.545 kDa) were used to calibrate the MW of samples.

2.5.4. Statistical Analysis

Where indicated, significance testing was performed with GraphPad Prism 7.00 (GraphPad Software, San Diego, CA, USA) using an unpaired, parametric, two tailed, *t*-test with a confidence level of 95%. Probability (p)-values $\leq$ 0.05 were considered statistically significant.

3. Results and Discussion

3.1. *Tuning TAF Release Rates: Surface Area and Wall Thickness*

These studies involve a reservoir-style PCL implant (Figure 1), that can deliver TAF at sustained, zero-order release kinetics. Once inserted subcutaneously, biological fluid from the surrounding environment transports through the PCL membrane into the reservoir and can solubilize TAF. TAF can partition into the PCL and transport passively through the PCL membrane to exit the implant. Transport of a drug through the PCL material is dictated by many parameters, such as the diffusion coefficient and partition coefficient, as described elsewhere [53]. As an aliphatic polyester, PCL undergoes bulk hydrolysis through random chain scission as water permeates through the polymer [37,54]. However, biodegradation of PCL is slow and can require years (e.g., 1–2 years) for complete bioresorption [37], depending on the starting MW. Because bulk erosion of PCL is slow, the faster process of drug delivery is decoupled from biodegradation, enabling zero-order release profiles of drug from the implant. At this zero-order release profile, the daily drug delivery rates are controlled by various parameters: surface area of the device, thickness of the device wall, polymer properties, and drug formulation. A digital camera image of RTI's trocar-compatible reservoir-style implant is shown in Figure 1.

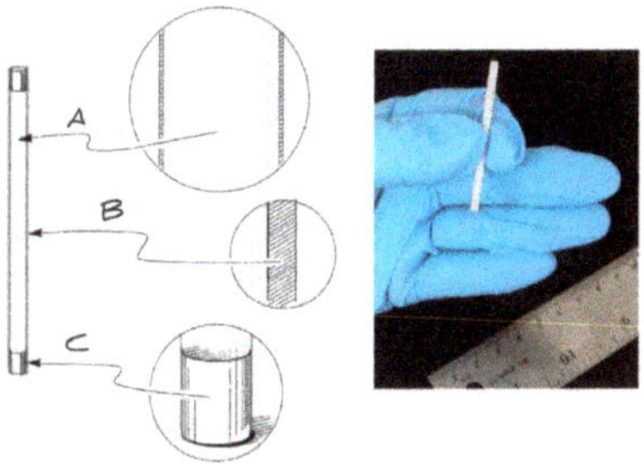

Figure 1. (Left) A schematic of a PCL reservoir-style device for delivery of TAF, which comprises a formulated drug core (A) encapsulated by a rate-controlling PCL membrane (B). The device is end-sealed using PCL material (C) for trocar compatibility. (Right) A digital camera image of the biodegradable implant.

To evaluate the relationship between release rates and the surface area of the extruded PCL tubes, implants were fabricated with three different surface areas, as generated by varying the implant length: 82 ± 1 mm^2, 311 ± 4 mm^2, and 543 ± 5 mm^2, with an average of 31, 124, and 216 mg of TAF loaded into the implant, respectively. All devices comprised Sigma-PCL with a wall thickness of 100 μm, an OD of 2.5 mm, and a formulation of 2:1 TAF:castor oil. The cumulative release of TAF from the implants were monitored for approximately 30 days, as shown in Figure 2a. As expected for a membrane-controlled system, the higher surface area results in a higher release rate of TAF from the implant. Furthermore, the linear relationship between daily release rates and surface area supports the mechanism of membrane-controlled release from these implants (Figure 2b). These results align with a previous report using reservoir-style devices with thinner walls of PCL (8.5 μm) fabricated via solvent film-casting [55]. In the current study, devices were fabricated using PCL tubes prepared via melt extrusion, which produced thicker walled tubes (between 45–200 μm). Despite the thicker PCL wall and different fabrication approach, these devices also maintained membrane-controlled release in this range of wall thickness, demonstrating the robustness of the PCL-based drug delivery platform. For the remainder of this paper, the cylindrical geometry was fixed at 2.5 mm OD and 40 mm length to accommodate commercially available trocars utilized for contraceptive implants [56,57], and the release rates were normalized to the surface area of 314 mm^2.

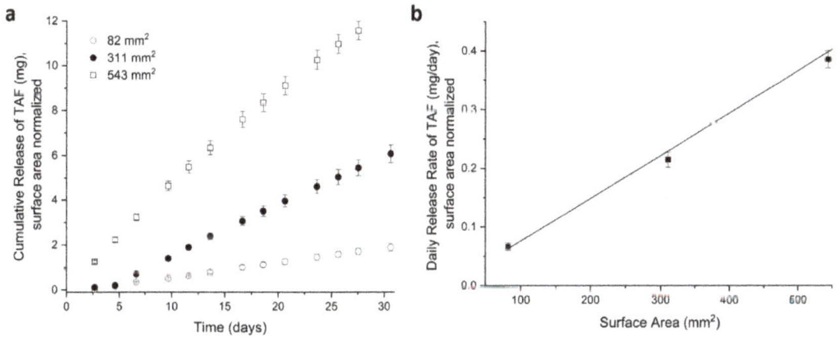

Figure 2. In vitro release studies showing (**a**) cumulative release of TAF from implants of differing surface areas and (**b**) daily release rates of TAF at day = 24 for implants with different surface areas. All implants were fabricated with Sigma-PCL, 100 μm wall thickness and a formulation of 2:1, TAF:castor oil. Surface areas were normalized according to the theoretical surface area for implant of 10, 40, 70 mm in length. Three implants were tested per condition.

The thickness of the implant walls was another attribute that affected release rates of drug. Figure 3 shows the daily release rates of TAF from implants comprising PCL of different wall thicknesses and containing a formulation of 2:1 TAF:castor oil excipient. The release rates of TAF inversely correlated with thickness of the PCL walls: 0.91 ± 0.23 mg/day (45 μm), 0.61 ± 0.09 mg/day (70 μm), 0.29 ± 0.05 mg/day (100 μm), 0.19 ± 0.04 mg/day (150 μm), and 0.15 ± 0.03 mg/day (200 μm). As the wall thickness increased from 45 to 200 μm, the release rates approach a plateau wherein the release rates of TAF show minimal change. Importantly, the daily release rates were calculated over the first 35 days of TAF release from the implants, which included a burst release that is more pronounced in thinner walled implants that results in a higher standard deviation (e.g., 45 μm walled implant). We speculate the burst release may arise from transport of drug into the PCL. This inverse relationship between the thickness of the PCL walls and the release rates of TAF was also demonstrated for thin-walled PCL implants fabricated by a solvent casting approach despite considerable differences in the device processing technique [55]. To reserve adequate volume in the reservoir for drug load, this study only investigated wall thickness up to 200 μm. Overall, these experiments demonstrate the ability to employ two parameters, surface area or wall thickness, to tailor the release rates of TAF from a reservoir-style implant fabricated with extruded PCL tubes.

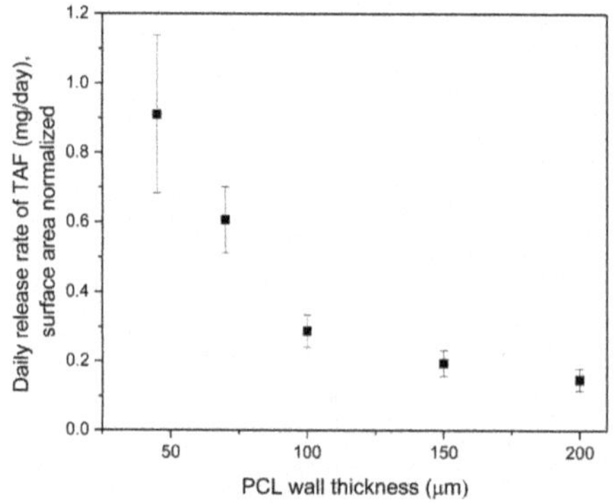

Figure 3. Daily release of TAF (mg/day) from implants with different wall thicknesses as calculated over 35 consecutive days within an in vitro assay. Implants comprised a 2:1 TAF-castor oil formulation (dimensions: 2.5 mm outer diameter (OD) by 40 mm length) fabricated with Sigma-Grade PCL. Three implants were tested per condition.

3.2. *Effects of PCL Properties on Implant Performance*

PCL is a semi-crystalline, hydrophobic polymer with biodegradation kinetics that depend on the initial MW, typically occurring in the order of 1–2 years [38], which supports a LA PrEP implant. In these studies, PCL starting material with two different MWs were selected to potentially support an implant with target duration of 6–12 months: Sigma-PCL (M_n of 103 kDa) and PC-12 PCL (M_n of 51 kDa). PCL tubes of different wall thicknesses (70, 100, 200 μm) were extruded with either Sigma-PCL or PC-12 and subsequently filled with a formulation of 2:1 TAF:castor oil. Evaluation of these implants using in-vitro release assays revealed two important concepts (Figure 4). First, the release rates of drug from the implant depended on the selection of PCL; TAF releases at a higher rate from implants comprising Sigma-PCL as compared to implants comprising PC-12. Interestingly, the influence of PCL type on TAF release rates is minimal in tubes with thicker walls (e.g., 200 μm) versus thinner walls (e.g., 70 μm). Second, Figure 4 also shows that irrespective of the PCL type used to fabricate the implant, the release rates of TAF still scales inversely with wall thickness, as also shown in Figure 3.

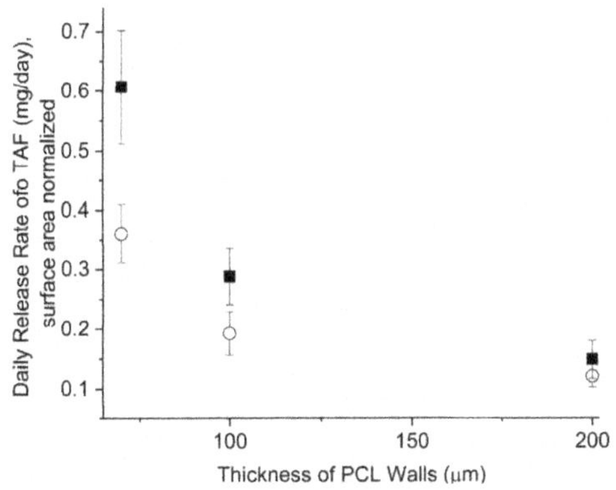

Figure 4. Effect of PCL type on daily release rates of TAF (mg/day) from implants with different wall thicknesses and fabricated with (■) Sigma -PCL or (○) PC-12. Implants contained a formulation of 2:1 TAF-castor Oil. Daily release rates were calculated from release over at 35 days and three implants were tested per condition.

It is possible that differences in crystallinity between the two types of PCL starting materials could affect the release rates of TAF from the implant. Therefore, to further understand the effect of polymer properties on release rates of drug, extruded tubes comprising PC-12 or Sigma-PCL were evaluated with DSC and XRD. Analysis by DSC showed that all PCL tubes exhibit a melting endotherm with a peak near 60 °C (Figure 5A, Figure S1), the characteristic melting temperature (T_m) of PCL [58,59]. However, notable differences in the melting endotherms were also evident, such as a narrower melt transition of PC-12 compared to Sigma-PCL and the presence of a small shoulder peak around 50 °C in Sigma-PCL which was absent in PC-12. Quantitatively, the specific T_m values also differed; Sigma-PCL showed a slightly higher T_m compared to PC-12 for all thicknesses of the tube walls (Table 1 and Figure S1). For each sample, Equation (1) was used to calculate the mass % crystallinity and Equation (2) (Thompson–Gibbs equation) was used to calculate the crystallite sizes. Results in Table 1 show that irrespective of the wall thickness, the crystallite size of PC-12 was slightly lower than the crystallite size of Sigma-PCL. Moreover, the crystallite size of Sigma-PCL slightly varied with different tube thicknesses, whereas PC-12 remained consistent. The % crystallinity was slightly higher in certain cases for PC-12 compared to Sigma-PCL, showing statistically significant differences for extruded tubes with 70 and 200 μm wall thicknesses.

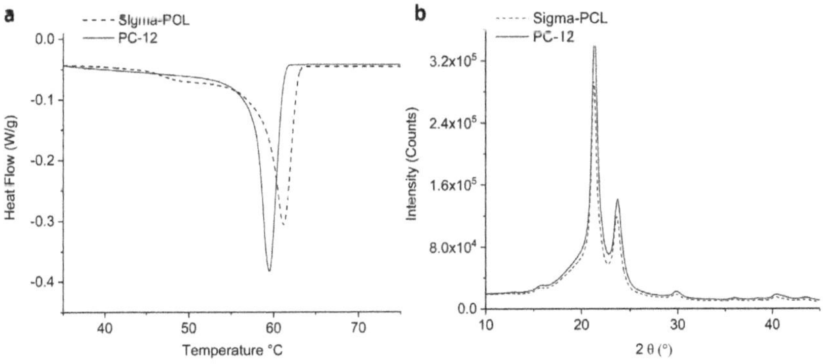

Figure 5. Exemplary graphs of (**a**) DSC heat flow curves and (**b**) XRD profiles of PCL tubes with 100 μm wall thickness.

Table 1. Thermal properties of PCL extruded tubes from DSC analysis.

PCL Type	Wall Thickness (μm)	T_m (°C)	% Crystallinity	Crystallite Size (nm)
	70	59.4 ± 0.1	56 ± 1.0	27 ± 0.2
PC-12	100	59.4 ± 0.1	53 ± 2.0	27 ± 0.4
	200	59.7 ± 0.4	56 ± 1.0	27 ± 1.2
	70	60.7 ± 0.1	53 ± 0.3	31 ± 0.2
Sigma-PCL	100	61.1 ± 0.2	52 ± 1.2	32 ± 0.6
	200	61.3 ± 0.1	53 ± 0.1	33 ± 0.3

XRD analysis was also performed to further examine the crystallite size of PCL extruded tubes using the Scherrer equation (Equation (3)). Extruded tubes (100 μm wall thickness) fabricated from Sigma-PCL and PC-12 showed similar diffraction patterns that include intense Bragg peaks at 2θ near 21.3° and 23.7°, correlating to diffraction of the (110) and (200) planes of the PCL crystallite, respectively (Figure 5B) [60,61]. Results from XRD analysis (Table 2) show that the crystallite sizes of PC-12 were slightly smaller than Sigma-PCL, where Sigma-PCL total crystallite sizes was 25 nm (14.2 + 10.8) and PCL-12 was 23.4 nm (13.2 + 10.2), which also agrees with DSC data. Both techniques used to measure crystallite size indicate a similar order of magnitude from the two PCL types, therefore it is unlikely that crystal size alone was responsible for the differences in drug diffusion kinetics from the materials considered in this study, however the observation that crystallite size increased with tube thickness for Sigma-PCL (as measured by DSC) may play a role in release kinetics.

Table 2. Thermal properties of PCL tubes * from XRD analysis.

PCL Type	Crystallite Size (nm)	
	L_{110}	L_{200}
PC-12	13.2	10.2
Sigma-PCL	14.2	10.8

* Extruded tubes comprised 100 μm wall thickness.

Taken together, these data highlight the importance in considering properties of drug transport in products comprising semi-crystalline polymers, which contain both amorphous regions amenable to drug transport when above the polymer glass transition temperature (T_g), and crystalline regions which pose a diffusive mass transport barrier. These data indicate that PCL is an ideal polymer suited for membrane-controlled drug diffusion applications given its material properties and semi-crystalline nature. For example, PCL has a T_g of −60 °C which allows for drug transport at physiological conditions (37 °C) where the amorphous regions exhibit adequate free volume for passive diffusion of small molecules and fluid driven by concentration gradients. Concurrently, PCL crystals impart structural integrity to the implant and act as a transport barrier which modulate drug diffusion and allow for sustained release of TAF. The DSC and XRD results presented here suggest that crystallite size, quantity of crystallinity, and ultimately polymer free volume within PCL will impact transport properties of TAF through the polymer, as also supported by studies with other systems [62]. Our results show that extruded tubes with lower MW (PC-12) contain smaller sizes of crystals and slightly higher % crystallinity (statistically significant for 70 and 200 μm tubes, $p = 0.008$ and $p = 0.007$, respectively) as compared to PCL with higher MW (Sigma-PCL). This suggests that higher degree of crystallinity and smaller crystallites could create a more tortuous path for diffusion of the drug, leading to a lower release rate from the implant. At 37 °C, TAF likely diffuses through the amorphous regions of PCL, where the polymer exhibits greater segmental mobility to facilitate passage of small molecules. The size and quantity of the crystal regions would affect the spatial arrangement and quantity of these amorphous regions, ultimately affecting transport kinetics. These findings are supported by the mathematical relationship between membrane flux through a given area which is inversely proportional to distance traveled (wall thickness) by the constant of mass diffusivity, i.e., Fick's first law of diffusion. The diffusion constant is a function of temperature, molecular size, and viscosity. For polymers, the viscosity term describes polymer free volume, which is impacted by crystallinity, hence the differences in material properties and resultant release rates were observed here.

In addition to the quantitative differences in polymer physicochemical properties observed in this study, three important qualitative findings are also of note. First, Sigma-PCL exhibited irregularities in the melting endotherm as evidenced by a small but apparent shoulder peak prior to the melt whereas PC-12 did not. It was likely a result of thermal history incurred in Sigma-PCL processing from the manufacturer. Given this consideration, the shoulder peak likely did not represent the PCL crystalline phase and may have contributed to an over-estimation of the crystalline content of the Sigma-PCL. A second difference was noted when comparing the width of the melt transition where PC-12 exhibited a narrower melt endotherm compared to Sigma-PCL, suggesting a tighter distribution of polymer molecular weight comprising the crystalline phase. Finally, another difference between these two grades of material was the crystallite size as a function of tube thicknesses. While PC-12 demonstrated more consistency in crystallite size, Sigma-PCL crystallite size increased with tube thickness, indicating a lack of control on the final properties during processing and perhaps explaining the large variability in release rates at low thicknesses and more consistent release rates at higher thicknesses. Sigma grade PCL also exhibited a steeper decline in release rate with respect to wall thickness, while PC-12 demonstrated a more gradual decline in release rate as a function of thickness. It was hypothesized that the concomitant increase in wall thickness and crystallite size observed with Sigma PCL was responsible for the attenuation in release rate values for the two grades of PCL at the higher wall thicknesses studied here. Taken together, these observations highlight the importance of material choices in the design of drug delivery devices from an engineering and quality control perspective.

3.3. *Performance and Fabrication of a LA PCL Implant for Delivery of TAF*

The duration of this reservoir-style implant for TAF is dictated by two parameters: the drug quantity within the reservoir and the rate of drug release from the implant. Using selected implant dimensions (2.5 mm OD, 40 mm length), TAF payload within the reservoir for the 2:1 TAF:castor oil formulation was approximately 115 mg for an implant with wall thicknesses of 100 μm. Within these constraints of drug payload, the duration of a single TAF implant for PrEP ultimately depends on the daily drug release required for protection as administered via the subcutaneous route, which is currently unknown. In this manuscript, in-vitro release rates from prototype implants were tailored to the range between approximately 0.2 and 0.8 mg/day from a single device, as suggested from previous animal studies and in-silico modelling set with TFV-DP target concentrations of >48 fmol/10^6 cells among 500 virtual healthly women [31,63].

Using these dimensional parameters, a batchwise process was developed to fabricate TAF implants from extruded PCL tubes, which entails loading the drug formulation into the cavity of a PCL tube and sealing the ends. Given the low melting temperature of PCL (60 °C), the implant was readily end-sealed by controllably heating PCL into the desired geometry using an in-house customized polymer extruder. Care was given to avoid contamination of the interior walls of the tube near the sealing points, which could hinder the formation of end seals during the melt sealing. This fabrication process was used to generate implants for a six-month in-vitro study to assess the release of TAF. As shown in Figure 6, implants released TAF at a rate of 0.28 ± 0.06 mg/day over the course of 180 days. After 180 days, approximately 68 mg of TAF remained within the implant, with a chromatographic purity of 89.2 ± 0.8% (Table S1). The trend of decreased TAF stability over time results from ingress of water into the implant as drug depletes, which, in turn, facilitates hydrolytic degradation of TAF [64]. We are currently evaluating other formulations of TAF that help optimize stability. The implant maintained structural integrity throughout the 180-day release period in simulated physiological conditions.

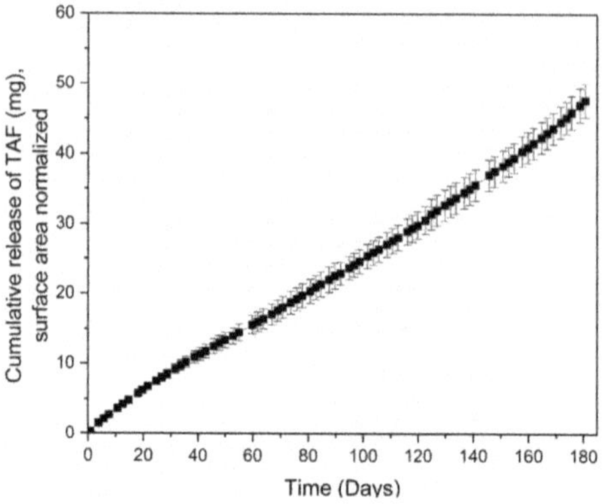

Figure 6. Cumulative release of TAF (mg) from an implant comprising Sigma-PCL of 100 μm wall thickness at 2.5 mm outer diameter and 40 mm length. Implants contained a 2:1 TAF-castor oil formulation. Three implants were tested per condition.

To support in vivo use of the implant, gamma irradiation was used to sterilize the implant after fabrication. Since gamma irradiation has been shown to affect the chemical and physical properties of PCL [65,66], studies were performed to evaluate its potential effects on the implant performance. Table S2 shows the GPC analysis of PCL, including samples of PCL raw material used for the extrusion process and extruded PCL tubes before and after gamma irradiation. Both PCL types (Sigma-PCL and PC-12) showed a slight decrease in M_n after gamma irradiation, as expected, but the extrusion process minimally affected the M_n of PCL. To further ascertain release rates of TAF from the implant after gamma irradiation, in vitro release assays were performed on implants with and without gamma irradiation at dosages between 18–24 kGy. As shown in Table 3, the release rates were comparable irrespective of treatment with gamma irradiation and the difference in release rates were not statistically significant when comparing non-irradiated and gamma irradiated release rates (p = 0.27 and p = 0.42 for Sigma-PCL at 70 and 100 μm, respectively; p = 0.11 and p > 0.99 for PC-12 at 70 and 100 μm, respectively).

Table 3. Daily TAF release rates from implants pre- and post-gamma irradiation.

PCL Type	Wall Thickness (μm)	Release Rates of TAF (mg/Day)	
		Non-Irradiated	Gamma Irradiated
Sigma-PCL	70	0.62 ± 0.09	0.54 ± 0.06
	100	0.29 ± 0.05	0.32 ± 0.03
PC-12	70	0.37 ± 0.05	0.30 ± 0.03
	100	0.20 ± 0.03	0.20 ± 0.02

Formulation of 2:1, TAF:castor oil; Daily release rates calculated from 30 days of consecutive release.

In summary, the in vitro studies presented in this manuscript demonstrate a new fabrication method to produce reservoir-style PCL implants using a batchwise process of polymer tube extrusion, formulation filling, and implant sealing via melt procedures. Although this fabrication process is not currently conducive to high-throughput manufacture, incremental steps detailed here support future production and implementation efforts. For instance, use of melt extrusion to produce PCL tubes presents an improvement over previously reported thin-film casting techniques, in terms of throughput of tube manufacturing and ultimate robustness of the implant. By increasing the implant wall thickness (e.g., from 70 to 200 μm) using an extrusion method, the implants were more aesthetically pleasing, exhibited greater mechanical sturdiness during handling and therapeutic use, while simultaneously retaining sustained release kinetics of drug. Moreover, the injection end-sealing method presented here produces an implant suitable for commercially available trocars, and is loaded with enough of the drug to last longer than six months; all considerations that were deemed preferable by end-users [67,68]. With an eye towards product development needs for clinical implementation, compatibility with applicators already clinically employed offers options to adopt existing procedures to support LA-PrEP rollout.

4. Conclusions

As the field of HIV PrEP rapidly progresses, new LA drug delivery systems must respond to evolving clinical and socio-behavioral insights in terms of effective ARV dosing, desired duration, and product acceptability. This paper describes a reservoir-style implant with the flexibility to adapt to needs of this advancing field by readily tailoring properties of the implant, including surface area, wall thickness, and properties of PCL. The first section of this manuscript describes the ability to control release rates of TAF via dimensions of extruded PCL tubes (i.e., surface area, wall thickness). We report an inverse correlation between wall thickness and drug release rates and this feature was further used to tune TAF dosing between 0.91 ± 0.23 mg/day (45 μm wall thickness) and 0.15 ± 0.03 mg/day (200 μm wall thickness). Our studies further show that the release rates of TAF scale proportionally with the surface area of the implant, demonstrating the membrane-controlled release from extruded PCL tubes. These results build from previous work using implants made of thin-films of PCL, which reported similar trends with the ability to tune release rates from reservoir-style devices between 8 and 15 μm thick [55]. However, the replacement of solvent-cast thin films with extruded tubes, as described here, supports the development of future manufacturing processes and ultimately provides robust implants with thicker walls for greater ease of handling. Although outside the scope of this manuscript, additional strategies to further increase drug stability, dosing and duration from a single implant, while simultaneously maintaining compatibility with market-available trocars, include modifications to the formulations such as selection and quantity of excipient. Although in vivo studies were not reported in this paper, the in vitro assay conditions used here have shown good in vitro/in vivo correlations previously [57]. Future in vivo studies will further advance this implant technology.

We also described a batchwise fabrication of implants using two types of PCL starting materials, either PC-12 or Sigma-PCL. In these studies, the selection of PCL starting material affects the release rates of TAF from the implant; TAF releases faster from implants fabricated with Sigma-PCL as compared to PC-12. Analysis via DSC and XRD were used to probe the reason for these differences. Analysis by DSC revealed slight differences in the mass percent of crystallinity from both PCL types (70 and 200 μm wall thickness) and showed slight differences in the crystalline sizes. Although additional studies are needed to further characterize these differences, our results do suggest that a smaller crystalline size and higher mass % crystallinity within the polymeric material retard the diffusion of TAF through PCL. Interestingly, when using these thicker walled tubes (e.g., 200 μm walls), the effect of PCL type on release rate is less pronounced. Using down-selected parameters, we demonstrate delivery of TAF in vitro for 180 days at a dose of 0.28 ± 0.06 mg/day and the TAF within the implant maintains stability at 89.2% ± 0.8% at the end of this duration. This is the first report of a six-month in vitro study showing sustained release of TAF from a biodegradable trocar-compatible implant. This manuscript also shows that the use of gamma irradiation (dosages 18–24 kGy) as a sterilization technique minimally affects the release rates of TAF, which benefits future in vivo studies. Overall, this implantable drug delivery system holds various parameters that can be tuned to achieve a targeted dose of TAF. Although the final design of this drug delivery system for TAF awaits feedback from dosing requirements, these promising results help to highlight the path towards the goal of developing LA delivery of TAF for HIV PrEP.

Supplementary Materials: The following are available online at http://www.mdpi.com/1999-4923/11/7/315/s1. Figure S1: Heat flow versus temperature for extruded tubes comprising (**A**) Sigma-PCL and (**B**) PC-12 at different tube thicknesses, Table S1: Stability of TAF in implants at different timepoints within in vitro conditions, Table S2: GPC Analysis of PCL.

Author Contributions: Conceptualization and Design, L.M.J., S.A.K., L.L., N.G., A.v.d.S.; Methodology, L.M.J., S.A.K., L.L., N.G., D.M., B.C.; Formal Analysis: L.M.J., S.A.K., L.L., N.G.; Writing: L.M.J., L.L., N.G., Z.R.D., A.v.d.S.; Review & Editing, L.M.J., S.A.K., L.L., Z.R.D., D.M., B.C., A.v.d.S.; Project Administration, Z.R.D.; Funding Acquisition, A.v.d.S.

Funding: This research was funded by the Bill & Melinda Gates Foundation, grant number OPP1149227. This research was also funded by the United States Agency for International Development (USAID) through the U.S. President's Emergency Plan for AIDS Relief (PEPFAR) (AID-OAA-A-14-00012). The APC was funded by the Bill & Melinda Gates Foundation.

Acknowledgments: This research was made possible by the generous support of the American people through the U.S. President's Emergency Plan for AIDS Relief. The contents are the responsibility of the authors and do not necessarily reflect the views of USAID, PEPFAR, or the United States Government. We thank The Bill and Melinda Gates Foundation for support of this work (OPP1149227). We thank Gilead Sciences, Inc. for graciously providing TAF. We thank RTI team members: Christine Areson, Marza Hill, Pafio Johnson, Todd Ennis, and Teresa Jester for data collection, Ellen Luecke for program management, and Dayle Johnson for Figure 1 drawing and the TOC graphic.

Conflicts of Interest: The authors declare no conflict of interest. At the time of publication, the co-author N.G. is employee of Aerie Pharmaceuticals, Inc. The company had no role in the design of the study; in the collection, analyses, or interpretation of data; in the writing of the manuscript, and in the decision to publish the results. The sponsors had no role in the design, execution, interpretation, or writing of the study.

References

1. Grant, R.M.; Lama, J.R.; Anderson, P.L.; McMahan, V.; Liu, A.Y.; Vargas, L.; Goicochea, P.; Casapía, M.; Guanira-Carranza, J.V.; Ramirez-Cardich, M.E.; et al. Preexposure Chemoprophylaxis for HIV prevention in men who have sex with men. *N. Engl. J. Med.* **2010**, *363*, 2587–2599. [CrossRef] [PubMed]
2. Baeten, J.M.; Donnell, D.; Ndase, P.; Mugo, N.R.; Campbell, J.D.; Wangisi, J.; Tappero, J.W.; Bukusi, E.A.; Cohen, C.R.; Katabira, E.; et al. Antiretroviral prophylaxis for HIV prevention in heterosexual men and women. *N. Engl. J. Med.* **2012**, *367*, 399–410. [CrossRef] [PubMed]
3. Molina, J.-M.; Capitant, C.; Spire, B.; Pialoux, G.; Cotte, L.; Charreau, I.; Tremblay, C.; Le Gall, J.-M.; Cua, E.; Pasquet, A.; et al. On-Demand preexposure prophylaxis in men at high risk for HIV-1 infection. *N. Engl. J. Med.* **2015**, *373*, 2237–2246. [CrossRef] [PubMed]
4. McCormack, S.; Dunn, D.T.; Desai, M.; Dolling, D.I.; Gafos, M.; Gilson, R.; Sullivan, A.K.; Clarke, A.; Reeves, I.; Schembri, G.; et al. Pre-exposure prophylaxis to prevent the acquisition of HIV-1 infection (PROUD): Effectiveness results from the pilot phase of a pragmatic open-label randomised trial. *Lancet* **2016**, *387*, 53–60. [CrossRef]
5. Cranston, R.D.; Lama, J.R.; Richardson, B.A.; Carballo-Diéguez, A.; Kunjara Na Ayudhya, R.P.; Liu, K.; Patterson, K.B.; Leu, C.-S.; Galaska, B.; Jacobson, C.E.; et al. MTN-017: A Rectal Phase 2 extended safety and acceptability study of tenofovir reduced-glycerin 1% gel. *Clin. Infect. Dis.* **2016**, *64*, 614–620. [CrossRef]
6. Marrazzo, J.M.; Ramjee, G.; Richardson, B.A.; Gomez, K.; Mgodi, N.; Nair, G.; Palanee, T.; Nakabiito, C.; van der Straten, A.; Noguchi, L.; et al. Tenofovir-based preexposure prophylaxis for HIV infection among African women. *N. Engl. J. Med.* **2015**, *372*, 509–518. [CrossRef] [PubMed]
7. Hosek, S.G.; Rudy, B.; Landovitz, R.; Kapogiannis, B.; Siberry, G.; Rutledge, B.; Liu, N.; Brothers, J.; Mulligan, K.; Zimet, G.; et al. An HIV preexposure prophylaxis demonstration project and safety study for young MSM. *J. Acquir. Immune Defic. Syndr.* **2017**, *74*, 21–29. [CrossRef]
8. Hosek, S.G.; Landovitz, R.J.; Kapogiannis, B.; Siberry, G.K.; Rudy, B.; Rutledge, B.; Liu, N.; Harris, D.R.; Mulligan, K.; Zimet, G.; et al. Safety and feasibility of antiretroviral preexposure prophylaxis for adolescent men who have sex with men aged 15 to 17 years in the United States. *JAMA Pediatrics* **2017**, *171*, 1063–1071. [CrossRef]
9. Morgan, E.; Ryan, D.T.; Newcomb, M.E.; Mustanski, B. High rate of discontinuation may diminish PrEP coverage among young men who have sex with men. *AIDS Behav.* **2018**, *22*, 3645–3648. [CrossRef]
10. Deutsch, M.B.; Glidden, D.V.; Sevelius, J.; Keatley, J.; McMahan, V.; Guanira, J.; Kallas, E.G.; Chariyalertsak, S.; Grant, R.M.; For the iPrEx Study Team. HIV pre-exposure prophylaxis in transgender women: A subgroup analysis of the iPrEx trial. *Lancet* **2015**, *2*, e512–e519. [CrossRef]
11. Van Damme, L.; Corneli, A.; Ahmed, K.; Agot, K.; Lombaard, J.; Kapiga, S.; Malahleha, M.; Owino, F.; Manongi, R.; Onyango, J.; et al. Preexposure prophylaxis for HIV infection among African women. *N. Engl. J. Med.* **2012**, *367*, 411–422. [CrossRef] [PubMed]

12. HPTN. HPTN 083. A Phase 2b/3 Double Blind Safety and Efficacy Study of Injectable Cabotegravir Compared to Daily Oral Tenofovir Disoproxil Fumarate/Emtricitabine (TDF/FTC), for Pre-Exposure Prophylaxis in HIV-Uninfected Cisgender Men and Transgender Women who have Sex with Men. Available online: https://www.hptn.org/research/studies/hptn083 (accessed on 1 July 2019).
13. HPTN. HPTN 084. A Phase 3 Double Blind Safety and Efficacy Study of Long-Acting Injectable Cabotegravir Compared to Daily Oral TDF/FTC for Pre-Exposure Prophylaxis in HIV-Uninfected Women. Available online: https://www.hptn.org/research/studies/hptn084 (accessed on 1 July 2019).
14. van der Straten, A.; Agot, K.; Ahmed, K.; Weinrib, R.; Browne, E.N.; Manenzhe, K.; Owino, F.; Schwartz, J.; Minnis, A.; on behalf of the TRIO Study Team. The Tablets, Ring, Injections as Options (TRIO) study: What young African women chose and used for future HIV and pregnancy prevention. *J. Int. AIDS Soc.* **2018**, *21*, e25094. [CrossRef] [PubMed]
15. Landovitz, R.J.; Li, S.; Grinsztejn, B.; Dawood, H.; Liu, A.Y.; Magnus, M.; Hosseinipour, M.C.; Panchia, R.; Cottle, L.; Chau, G.; et al. Safety, tolerability, and pharmacokinetics of long-acting injectable cabotegravir in low-risk HIV-uninfected individuals: HPTN 077, a phase 2a randomized controlled trial. *PLoS Med.* **2018**, *15*, e1002690. [CrossRef] [PubMed]
16. Markowitz, M.; Frank, I.; Grant, R.M.; Mayer, K.H.; Elion, R.; Goldstein, D.; Fisher, C.; Sobieszczyk, M.E.; Gallant, J.E.; Van Tieu, H.; et al. Safety and tolerability of long-acting cabotegravir injections in HIV-uninfected men (ECLAIR): A multicentre, double-blind, randomised, placebo-controlled, phase 2a trial. *Lancet HIV* **2017**, *4*, e331–e340. [CrossRef]
17. Penrose, K.J.; Parikh, U.M.; Hamanishi, K.A.; Else, L.; Back, D.; Boffito, M.; Jackson, A.; Mellors, J.W. Selection of Rilpivirine-Resistant HIV-1 in a Seroconverter From the SSAT 040 Trial Who Received the 300-mg Dose of Long-Acting Rilpivirine (TMC278LA). *J. Infect. Dis.* **2016**, *213*, 1013–1017. [CrossRef] [PubMed]
18. Kovarova, M.; Benhabbour, S.R.; Massud, I.; Spagnuolo, R.A.; Skinner, B.; Baker, C.E.; Sykes, C.; Mollan, K.R.; Kashuba, A.D.M.; García-Lerma, J.G.; et al. Ultra-long-acting removable drug delivery system for HIV treatment and prevention. *Nat. Commun.* **2018**, *9*, 4156. [CrossRef] [PubMed]
19. Solorio, L.; Carlson, A.; Zhou, H.; Exner, A.A. Implantable drug delivery systems. In *Engineering Polymer Systems for Improved Drug Delivery*, 1st ed.; Bader, R.A., Putnam, D.A., Eds.; John Wiley & Sons, Inc.: Hoboken, NJ, USA, 2014. [CrossRef]
20. Yang, W.-W.; Pierstorff, E. Reservoir-based polymer drug delivery systems. *J. Lab. Autom.* **2012**, *17*, 50–58. [CrossRef] [PubMed]
21. Langer, R. Implantable controlled release systems. *Pharmacol. Ther.* **1983**, *21*, 35–51. [CrossRef]
22. Highlights of Prescribing Information, Probuphine (Buprenorphine) Implant for Subdermal Administration. Available online: https://probuphine.com/prescribing-information/ (accessed on 1 July 2019).
23. Highlights of Prescribing Information, Jadelle (Levonogestrel Implant). Available online: https://www.accessdata.fda.gov/drugsatfda_docs/label/2016/020544s010lbl.pdf (accessed on 1 July 2019).
24. "Norplant® II" Levonorgestrel Implants (Jadelle®), Supplement 003. Council, P. (Ed.) Vol. NDA 20-544. Available online: https://www.accessdata.fda.gov/drugsatfda_docs/label/2002/20544se2-003_jadelle_lbl.pdf (accessed on 1 July 2019).
25. Highlights of Prescribing Information, Implanon (Etonogestrel Implant). Available online: https://www.merck.com/product/usa/pi_circulars/i/implanon/implanon_pi.pdf (accessed on 1 July 2019).
26. Highlights of Prescribing Information, Nexplanon (etonogestrel implants). Available online: https://www.merck.com/product/usa/pi_circulars/n/nexplanon/nexplanon_pi.pdf (accessed on 1 July 2019).
27. Musch, D.C.; Martin, D.F.; Gordon, J.F.; Davis, M.D.; Kuppermann, B.D. Treatment of cytomegalovirus retinitis with a sustained-release ganciclovir implant. *N. Engl. J. Med.* **1997**, *337*, 83–90. [CrossRef]
28. The Highlights of Prescribing Information Ozurdex (Dexamethasone Intravitreal Implant). Available online: https://www.accessdata.fda.gov/drugsatfda_docs/label/2014/022315s009lbl.pdf (accessed on 1 July 2019).
29. Haller, J.A.; Bandello, F.; Belfort, R.; Blumenkranz, M.S.; Gillies, M.; Heier, J.; Loewenstein, A.; Yoon, Y.-H.; Jacques, M.-L.; Jiao, J.; et al. Randomized, sham-controlled trial of dexamethasone intravitreal implant in patients with macular edema due to retinal vein occlusion. *Ophthalmology* **2010**, *117*, 1134–1146. [CrossRef]
30. Jaffe, G.J.; Martin, D.; Callanan, D.; Pearson, P.A.; Levy, B.; Comstock, T. Fluocinolone Acetonide Implant (Retisert) for Noninfectious Posterior Uveitis: Thirty-Four–Week results of a multicenter randomized clinical study. *Ophthalmology* **2006**, *113*, 1020–1027. [CrossRef] [PubMed]

31. Gunawardana, M.; Remedios-Chan, M.; Miller, C.S.; Fanter, R.; Yang, F.; Marzinke, M.A.; Hendrix, C.W.; Beliveau, M.; Moss, J.A.; Smith, T.J.; et al. Pharmacokinetics of long-acting tenofovir alafenamide (GS-7340) subdermal implant for HIV prophylaxis. *Antimicrob. Agents Chemother.* **2015**, *59*, 3913–3919. [CrossRef] [PubMed]
32. Chua, C.Y.X.; Jain, P.; Ballerini, A.; Bruno, G.; Hood, R.L.; Gupte, M.; Gao, S.; Di Trani, N.; Susnjar, A.; Shelton, K.; et al. Transcutaneously refillable nanofluidic implant achieves sustained level of tenofovir diphosphate for HIV pre-exposure prophylaxis. *J. Control. Release* **2018**, *286*, 315–325. [CrossRef] [PubMed]
33. Pipeline and Partners. Available online: https://www.intarcia.com/pipeline-technology/itca-650.html (accessed on 1 July 2019).
34. Barrett, S.E.; Teller, R.S.; Forster, S.P.; Li, L.; Mackey, M.A.; Skomski, D.; Yang, Z.; Fillgrove, K.L.; Doto, G.J.; Wood, S.L.; et al. Extended-Duration MK-8591-Eluting Implant as a Candidate for HIV Treatment and Prevention. *Antimicrob. Agents Chemother.* **2018**, *62*, e01058-18. [CrossRef] [PubMed]
35. Gatto, G.; Girouard, N.; Brand, R.M.; Johnson, L.; Marzinke, M.A.; Rowshan, S.; Engstrom, J.C.; McGowan, I.; Demkovich, Z.; Luecke, E.; et al. Pharmacokinetics of Tenofovir Alafenamide by Subcutaneous Implatn for HIV PrEP. In Proceedings of the Conference on Retrovirusus and Opportunistic Infections (CROI) 2018, Boston, MA, USA, 7 March 2018.
36. New Multipurpose Device to Help Prevent HIV and Pregnancy—RTI International Awarded Project to Develop a Device to Help Women in Africa. Available online: https://www.rti.org/news/new-multipurpose-device-help-prevent-hiv-and-pregnancy (accessed on 1 July 2019).
37. Pitt, C.G.; Chasalow, F.I.; Hibionada, Y.M.; Klimas, D.M.; Schindler, A. Aliphatic Polyesters. I. The Degradation of Poly(ε-caprolactone) In Vivo. *J. Appl. Polym. Sci.* **1981**, *26*, 3779–3787. [CrossRef]
38. Woodruff, M.A.; Hutmacher, D.W. The return of a forgotten polymer—Polycaprolactone in the 21st century. *Prog. Polym.Sci.* **2010**, *35*, 1217–1256. [CrossRef]
39. Dong, L.; Wang, S.-J.; Zhao, X.-R.; Zhu, Y.-F.; Yu, J.-K. 3D-Printed Poly(ε-caprolactone) Scaffold Integrated with Cell-laden Chitosan Hydrogels for Bone Tissue Engineering. *Sci. Rep.* **2017**, *7*, 13412. [CrossRef]
40. Patrício, T.; Domingos, M.; Gloria, A.; Bártolo, P. Characterisation of PCL and PCL/PLA Scaffolds for Tissue Engineering. *Procedia CIRP* **2013**, *5*, 110–114. [CrossRef]
41. Grossen, P.; Witzigmann, D.; Sieber, S.; Huwyler, J. PEG-PCL-based nanomedicines: A biodegradable drug delivery system and its application. *J. Control. Release* **2017**, *260*, 46–60. [CrossRef]
42. Dash, T.K.; Konkimalla, V.B. Polymeric modification and its implication in drug delivery: Poly-epsilon-caprolactone (PCL) as a model polymer. *Mol. Pharm.* **2012**, *9*, 2365–2379. [CrossRef]
43. Shrestha, D.; Wei, X.; Wu, W.-C.; Ling, J.-Q. Resilon: A methacrylate resin-based obturation system. *J. Dental Sci.* **2010**, *5*, 47–52. [CrossRef]
44. Trott, A.T. Chapter 8—Instruments, Suture Materials, and Closure Choices. In *Wounds and Lacerations*, 4th ed.; Trott, A.T., Ed.; W.B. Saunders: Philadelphia, PA, USA, 2012; pp. 82–94. [CrossRef]
45. Ory, S.J.; Hammond, C.B.; Yancy, S.G.; Wayne Hendren, R.; Pitt, C.G. The effect of a biodegradable contraceptive capsule (Capronor) containing levonorgestrel on gonadotropin, estrogen, and progesterone levels. *Am. J. Obstet. Gynecol.* **1983**, *145*, 600–605. [CrossRef]
46. Gupta, B.; Geeta; Ray, A.R. Preparation of poly(ε-caprolactone)/poly(ε-caprolactone-*co*-lactide) (PCL/PLCL) blend filament by melt spinning. *J. Appl. Polym. Sci.* **2012**, *123*, 1944–1950. [CrossRef]
47. Obregon, N.; Agubra, V.; Pokhrel, M.; Campos, H.; Flores, D.; De la Garza, D.; Mao, Y.; Macossay, J.; Alcoutlabi, M. Effect of Polymer Concentration, Rotational Speed, and Solvent Mixture on Fiber Formation Using Forcespinning®. *Fibers* **2016**, *4*, 20. [CrossRef]
48. Su, H.-H.; Chen, H.-L.; Díaz, A.; Casas, M.T.; Puiggalí, J.; Hoskins, J.N.; Grayson, S.M.; Pérez, R.A.; Müller, A.J. New insights on the crystallization and melting of cyclic PCL chains on the basis of a modified Thomson–Gibbs equation. *Polymer* **2013**, *54*, 846–859. [CrossRef]
49. Menczel, J.D.; Judovits, L.; Prime, R.B.; Bair, H.E.; Reading, M.; Swier, S. Chapter 2-Differential Scanning Calorimetry (DSC). In *Thermal Analysis of Polymers: Fundamentals and Applications*; Menczel, J.D., Prime, R.B., Eds.; Wiley: Hoboken, NJ, USA, 2009.
50. Núñez, E. Crystallization in Constrained Polymer Structures: Approaching the Unsolved Problems in Polymer Crystallization. Ph.D. Thesis, KTH, Stockholm, Sweden, 2006.
51. Tuba, F. Towards the Understanding of the molecular weight dependence of essential work of fracture in semi-crystalline polymers: A study on poly(ε-caprolactone). *Express Polym. Lett.* **2014**, *8*, 869–879. [CrossRef]

52. Kosobrodova, E.; Kondyurin, A.; Chrzanowski, W.; Theodoropoulos, C.; Morganti, E.; Hutmacher, D.; Bilek, M.M.M. Effect of plasma immersion ion implantation on polycaprolactone with various molecular weights and crystallinity. *J. Mater. Sci. Mater. Med.* **2017**, *29*, 5. [CrossRef] [PubMed]
53. Schlesinger, E.; Ciaccio, N.; Desai, T.A. Polycaprolactone thin-film drug delivery systems: Empirical and predictive models for device design. *Mater. Sci. Eng. C Mater. Biol. Appl.* **2015**, *57*, 232–239. [CrossRef]
54. Lam, C.X.; Savalani, M.M.; Teoh, S.H.; Hutmacher, D.W. Dynamics of in vitro polymer degradation of polycaprolactone-based scaffolds: Accelerated versus simulated physiological conditions. *Biomed. Mater.* **2008**, *3*, 034108. [CrossRef]
55. Schlesinger, E.; Johengen, D.; Luecke, E.; Rothrock, G.; McGowan, I.; van der Straten, A.; Desai, T. A Tunable, Biodegradable, Thin-Film Polymer Device as a Long-Acting Implant Delivering Tenofovir Alafenamide Fumarate for HIV Pre-exposure Prophylaxis. *Pharm. Res.* **2016**, *33*, 1649–1656. [CrossRef]
56. Steiner, M.J.; Boler, T.; Obhai, G.; Hubacher, D. Assessment of a disposable trocar for insertion of contraceptive implants. *Contraception* **2010**, *81*, 140–142. [CrossRef] [PubMed]
57. Gatto, G.J.; Brand, R.M.; Girouard, N.; Li, L.A.; Johnson, L.; Marzinke, M.A.; Krogstad, E.; Siegel, A.; Helms, E.; Demkovich, Z.; et al. Development of an End-user Informed Tenofovir Alafenamide (TAF) Implant for Long-acting (LA)-HIV Pre-exposure Prophylaxis (PrEP). In Proceedings of the HIV Research for Prevention (HIVR4P) 2018, Madrid, Spain, 25 October 2018.
58. Wang, Y.; Rodriguez-Perez, M.A.; Reis, R.L.; Mano, J.F. Thermal and Thermomechanical Behaviour of Polycaprolactone and Starch/Polycaprolactone Blends for Biomedical Applications. *Macromol. Mater. Eng.* **2005**, *290*, 792–801. [CrossRef]
59. Speranza, V.; Sorrentino, A.; De Santis, F.; Pantani, R. Characterization of the Polycaprolactone Melt Crystallization: Complementary Optical Microscopy, DSC, and AFM Studies. *Sci. World J.* **2014**, *2014*, 9. [CrossRef] [PubMed]
60. Hu, H.; Dorset, D.L. Crystal structure of poly(iε-caprolactone). *Macromolecules* **1990**, *23*, 4604–4607. [CrossRef]
61. Lv, Q.; Wu, D.; Xie, H.; Peng, S.; Chen, Y.; Xu, C. Crystallization of poly(ε-caprolactone) in its immiscible blend with polylactide: Insight into the role of annealing histories. *RSC Adv.* **2016**, *6*, 37721–37730. [CrossRef]
62. Scheler, S. The polymer free volume as a controlling factor for drug release from poly(lactide-*co*-glycolide) microspheres. *J. Appl. Polym. Sci.* **2014**, *131*. [CrossRef]
63. Rajoli, R.; Demkovish, Z.; van der Straten, A.; Flexner, C.; Owen, A.; Siccardi, M. In Silico simulation of long-acting tenofovir alafenamide subcutaneous implant. In Proceedings of the CROI 2019, Seatle, WA, USA, 4–7 March 2019.
64. Golla, V.M.; Kurmi, M.; Shaik, K.; Singh, S. Stability behaviour of antiretroviral drugs and their combinations. 4: Characterization of degradation products of tenofovir alafenamide fumarate and comparison of its degradation and stability behaviour with tenofovir disoproxil fumarate. *J. Pharm. Biomed. Anal.* **2016**, *131*, 146–155. [CrossRef]
65. Navarro, R.; Burillo, G.; Adem, E.; Marcos-Fernández, A. Effect of Ionizing Radiation on the Chemical Structure and the Physical Properties of Polycaprolactones of Different Molecular Weight. *Polymers* **2018**, *10*, 397. [CrossRef]
66. Cooke, S.L.; Whittington, A.R. Influence of therapeutic radiation on polycaprolactone and polyurethane biomaterials. *Mater. Sci. Eng. C* **2016**, *60*, 78–83. [CrossRef]
67. Krogstad, E.A.; Montgomery, E.T.; Atujuna, M.; Minnis, A.M.; O'Rourke, S.; Ahmed, K.; Bekker, L.-G.; van der Straten, A. Design of an Implant for Long-Acting HIV Pre-Exposure Prophylaxis: Input from South African Health Care Providers. *AIDS Patient Care STDs* **2019**, *33*, 157–166. [CrossRef]
68. Krogstad, E.A.; Atujuna, M.; Montgomery, E.T.; Minnis, A.; Ndwayana, S.; Malapane, T.; Shapley-Quinn, M.K.; Manenzhe, K.; Bekker, L.-G.; van der Straten, A. Perspectives of South African youth in the development of an implant for HIV prevention. *J. Int. AIDS Soc.* **2018**, *21*, e25170. [CrossRef] [PubMed]

pharmaceutics

MDPI

Article

Design of Poly(lactic-*co*-glycolic Acid) (PLGA) Nanoparticles for Vaginal Co-Delivery of Griffithsin and Dapivirine and Their Synergistic Effect for HIV Prophylaxis

Haitao Yang [1,2], Jing Li [1,2], Sravan Kumar Patel [1,2], Kenneth E. Palmer [3], Brid Devlin [4] and Lisa C. Rohan [1,2,5,*]

1 Department of Pharmaceutical Sciences, School of Pharmacy, University of Pittsburgh, Pittsburgh, PA 15261, USA; hyang96@its.jnj.com (H.Y.); jil132@pitt.edu (J.L.); patels10@mwri.magee.edu (S.K.P.)
2 Magee-Womens Research Institute, Pittsburgh, PA 15213, USA
3 Center for Predictive Medicine and Department of Pharmacology and Toxicology, University of Louisville School of Medicine, Louisville, KY 40202, USA; kenneth.palmer@louisville.edu
4 International Partnership for Microbicides, Silver Spring, MD 20910, USA; bdevlin@ipmglobal.org
5 Department of Obstetrics, Gynecology and Reproductive Sciences, University of Pittsburgh, Pittsburgh, PA 15261, USA
* Correspondence: rohanlc@upmc.com; Tel.: +1-412-641-6108

Received: 21 February 2019; Accepted: 11 April 2019; Published: 16 April 2019

Abstract: Long-acting topical products for pre-exposure prophylaxis (PrEP) that combine antiretrovirals (ARVs) inhibiting initial stages of infection are highly promising for prevention of HIV sexual transmission. We fabricated core-shell poly(lactide-*co*-glycolide) (PLGA) nanoparticles, loaded with two potent ARVs, griffithsin (GRFT) and dapivirine (DPV), having different physicochemical properties and specifically targeting the fusion and reverse transcription steps of HIV replication, as a potential long-acting microbicide product. The nanoparticles were evaluated for particle size and zeta potential, drug release, cytotoxicity, cellular uptake and in vitro bioactivity. PLGA nanoparticles, with diameter around 180–200 nm, successfully encapsulated GRFT (45% of initially added) and DPV (70%). Both drugs showed a biphasic release with initial burst phase followed by a sustained release phase. GRFT and DPV nanoparticles were non-toxic and maintained bioactivity (IC_{50} values of 0.5 nM and 4.7 nM, respectively) in a cell-based assay. The combination of drugs in both unformulated and encapsulated in nanoparticles showed strong synergistic drug activity at 1:1 ratio of IC_{50} values. This is the first study to co-deliver a protein (GRFT) and a hydrophobic small molecule (DPV) in PLGA nanoparticles as microbicides. Our findings demonstrate that the combination of GRFT and DPV in nanoparticles is highly potent and possess properties critical to the design of a sustained release microbicide.

Keywords: microbicides; HIV; combination ARVs; sustained release; synergism; PLGA nanoparticles; co-delivery

---

1. Introduction

Human immunodeficiency virus (HIV) infection is one of the world's most serious health challenges, and the development of means to prevent its spread is urgently needed. By the end of 2017, approximately 36.9 million people were living with HIV, with 1.8 million more becoming newly infected globally [1]. Women, especially young (15–24 years) and adolescent (10–19 years), are disproportionately affected by HIV due to social, cultural, economic, and biological vulnerabilities [2,3]. Sexual transmission is the principal cause of new HIV-1 infections in women in the U.S. and worldwide [2].

To protect women from this global epidemic, it is critical to develop effective biomedical interventions for HIV/AIDS prevention and treatment. However, there is no effective vaccine available. In addition, the low instance of condom use by men and other social factors could significantly increase HIV-1 infection in women [4,5]. With these considerations in mind, oral or topical pre-exposure prophylaxis (PrEP) with antiretroviral (ARV) drugs could be the most promising avenue for preventing sexually transmitted HIV infection. Topical PrEP products could be applied by women in the vaginal or rectal tract before sexual intercourse to inactivate pathogens, including HIV, thus providing a female-controlled strategy against HIV-1 infection [6–8]. Rectal-specific topical PrEP products are equally applicable to men who have sex with men (MSM).

HIV-specific ARV drugs that can target specific steps in HIV's life cycle and then effectively inhibit its replication have been selected in the development of current microbicides [9]. To reach maximum protection effect of microbicides, ARVs inhibiting early stage of HIV replication can be beneficial. A combination strategy that includes multiple ARVs targeting different steps of HIV infection lifecycle could potentially improve the antiretroviral activity of each product with synergistic effect as a viable PrEP product for preventing acquisition of HIV-1. It has been proven that combination ARV (cARV) therapy can increase the efficacy of each drug in clinic, reduce resistance, decrease the risk of both HIV disease progression and death, reduce doses and cost, and improve patient adherence and compliance [10].

Dapivirine (DPV) is a diarylaminopyrimidine compound with high activity against HIV-1, and it belongs to the family of non-nucleoside reverse transcriptase inhibitors (NNRTIs). It is extremely potent in inhibiting both wild-type and mutant HIV-1 in vitro [11]. No vaginal or systemic toxicity of DPV was observed in sheep, indicating a desirable safety profile of DPV as a microbicide candidate [12]. In an intravaginal ring dosage form, DPV has shown great promise in Phase III clinical trials, ASPIRE and The Ring study [13,14]. Another promising molecule is Griffithsin (GRFT), a 121–amino acid and13-kDa-molecular-mass lectin that was originally isolated from the red alga *Griffithsia* sp. [15]. Binding to the mannose-rich glycans gp120 and gp41, GRFT inhibits HIV-1 gp120 and exposes the CD4 binding site [16,17]. GRFT is shown to exhibit high potency in inhibition of both X4- and R5-tropic HIV-1 virus at subnanomolar concentration, with high stability in cervical/vaginal lavage fluid [18]. It has been shown to be safe and tolerable in subcutaneous treatment in both guinea pigs and mice at 10 mg/kg and in rabbit vaginal irritation studies [19,20]. Furthermore, our previous investigation of the effect of GRFT rectal gel formulations on proteome and microbiome in non-human primates was supportive of their safety [21]. Importantly, GRFT has shown strong synergistic activity when combined with nucleoside reverse transcriptase inhibitor (NRTI), NNRTI and fusion inhibitor compounds, such as tenofovir, maraviroc, and enfuvirtide in vitro [22]. Furthermore, the broad-spectrum anti-HIV activity and unique safety profile of GRFT and DPV support their development as combined topical microbicides. Therefore, we hypothesized that GRFT will display strong synergistic activity with DPV when combined in application as a microbicide product.

The extreme opposite properties of these two molecules greatly hinder the development of a formulation for the simultaneous delivery of both drugs. The proteinaceous nature of GRFT and quick tissue elimination and low water solubility of DPV, pose limitations to their topical microbicide potential [23]. Additionally, developing a long-acting drug delivery system, which is more favorable for patient adherence of the PrEP product [24,25], will meet with additional challenges. Therefore, our goal was to develop a novel drug delivery system that overcomes formulation limitations and co-delivers GRFT and DPV with a sustained release profile to solve the clinical need in microbicides.

Nanoparticle drug delivery systems not only provide the sustained or controlled delivery of APIs, but also improve drug solubility, protect drug payloads, and enhance mucosal drug permeability [26]. With these advantages, nanoparticle delivery systems have been explored in the design of vaginal microbicides [27–31]. Polymeric nanoparticles can provide controlled or sustained release profiles for the payloads. Poly(ethylene oxide) (PEO) modified poly(caprolactone) (PCL) nanoparticles have been developed for the delivery of DPV as an alternative vaginal microbicide [31] and showed enhanced mucosal penetration and improved local pharmacokinetic profiles of DPV [29]. Nanoparticles made from PEO modified poly(lactic-*co*-glycolic acid) (PLGA) were applied to deliver rilpivirine and demonstrated improved vaginal tissue concentration [32]. Previous studies have utilized nanoparticles for delivery of DPV [33]. Here, we used a PLGA-based nanoparticle delivery system to deliver GRFT/DPV simultaneously in a sustained release pattern. Furthermore, encapsulation of GRFT and DPV in PLGA nanoparticles overcomes the limitations of poor tissue permeability and low aqueous solubility respectively. PLGA is one of the most widely used biodegradable polymers for drug delivery applications due to its superior biocompatibility and biodegradability [34]. It is an FDA-approved polymer that has been widely used for the delivery of proteins/peptides and small molecules [35–38]. Importantly, PLGA nanoparticles were found to have no immunogenicity in vivo, which is critical for the safety of an anti-HIV drug delivery system [39–42]. Additionally, PLGA nanoparticles have been studied in the delivery of several ARV drugs for HIV/AIDS treatment and prevention [38,43–46]. However, to our knowledge, no studies have yet been reported to deliver both a macromolecule and a small hydrophobic molecule simultaneously in one nanoparticle system for PrEP. This attempt on the co-delivery of a macromolecule and a small hydrophobic molecule with PLGA nanoparticles will have implications for the development of future microbicides.

In this study, we investigated the ability of PLGA nanoparticles to deliver both GRFT and DPV simultaneously and their anti-HIV efficacy in combined application as ARVs. We show that GRFT and DPV can be individually and simultaneously fabricated into biodegradable PLGA nanoparticles with a high encapsulation efficiency. The ARV-containing nanoparticles were nontoxic in cell culture. We also observed synergistic effects in the combined application of GRFT and DPV. Collectively, our data show that core-shell type PLGA nanoparticles are a promising strategy for delivering multiple ARV drugs with extreme physicochemical diversity. Our results support a viable nanoparticle platform for the delivery of multi-ARV combinations for sustained HIV-1 prevention.

2. Materials and Methods

2.1. Materials

PLGA with lactic acid to glycolic acid ratios of 50:50 was purchased from Sigma-Aldrich (Resomer® RG 502 H, MW, 30 kDa; St. Louis, MO, USA). Cell culture reagents were obtained from GIBCO, Invitrogen by Life Sciences, Inc. (Lenexa, KS, USA). DPV was obtained from International Partnership for Microbicides (Silver Spring, MD, USA). Recombinant GRFT (~13 kDa) was kindly provided by Kentucky Bioprocessing, LLC (Owensboro, KY, USA). The protein was supplied in a solution of phosphate buffered saline. Phosphate buffered saline 10× molecular biology grade was purchased from Mediatech, Inc. (Manassas, VA, USA). All other reagents used for nanoparticle formulation were purchased from Thermo Fisher Scientific (Pittsburgh, PA, USA).

2.2. Methods

2.2.1. Fabrication of ARV-Loaded Nanoparticles

The ARV loaded nanoparticles were prepared by using double emulsion-solvent evaporation technique as shown in Figure 1. Blank nanoparticles (vehicle control) were prepared using this technique at ambient temperature, as previously described, with some modifications [38]. Briefly, 100 μL Milli-Q water as inner water phase was added into a solution of PLGA (20 mg) in ethylacetate (EA) (1 mL). Then, the primary water-in-oil (W/O) emulsion was obtained by homogenization using a 6-mm diameter Vibra-Cell probe sonicator (Sonics and Materials, Newton, CT, USA) for 40 s at 50 W. The primary W/O emulsion was then added into 2 mL of 2% *w*/*v* aqueous solution of polyvinyl alcohol (PVA) with sonication for 50 s at 50 W to form the secondary water-in-oil-in-water (W/O/W) emulsion.

The W/O/W nanoparticle solution was then diluted with 10 mL of Milli-Q water under magnetic stirring for 4 h in an ice water bath to allow EA to evaporate. The hardened nanoparticles were washed with deionized water three times by centrifugation for 15 min at 15,000× *g* (Sorvall Ultra 80, Waltham, MA, USA). Nanoparticles were then re-suspended in 1 mL of Milli-Q water after removing the PVA supernatant and were then lyophilized overnight (approximately 12 h) under vacuum at 0.120 mbar and at −50 °C using a FreeZone 6 lyophilizer (Labconco, Kansas City, MO, USA). Nanoparticles loaded with GRFT, GRFT/DPV, and DPV were fabricated similar to the blank nanoparticles. GRFT (50 μL, 10 mg/mL) was dissolved in the inner water phase for GRFT nanoparticle preparation. DPV was dissolved in EA (0.033 mg/mL) with PLGA polymer in DPV nanoparticles. GRFT/DPV nanoparticles were prepared by dissolving GRFT in inner water phase and DPV in EA. GRFT and DPV loading levels were kept the same in each preparation unless noted otherwise. The lyophilized nanoparticles were stored in aliquots in glass vials at 4 °C until use. To prepare fluorescent nanoparticles, fluorescein isothiocyanate–dextran 70 (FITC-dextran, MW 70,000; Sigma LLC, St. Louis, MO, USA) and Nile red (Sigma LLC) were added to inner phase and EA respectively during the manufacture of blank NPs.

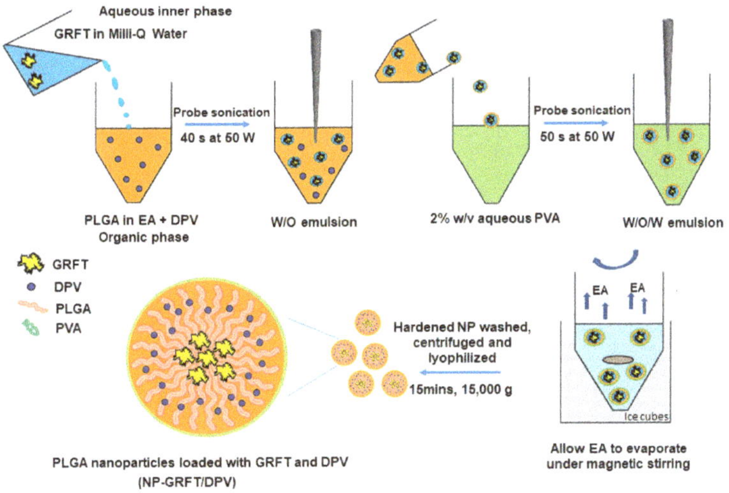

Figure 1. Schematic representation of DPV and GRFT-loaded PLGA nanoparticle preparation by double emulsion-solvent evaporation method.

2.2.2. Characterization of Nanoparticles

Size and zeta potential of the fabricated nanoparticles were determined using a Zetasizer Nano ZS90 (Malvern Instruments, Malvern, UK). Size and morphology of the nanoparticles were also confirmed by Transmission Electron Microscope (TEM), visualized with a JEM 1011 (JEOL, Sheboygan, WI, USA) scanning electron microscope. Samples of nanoparticles were negatively stained with ammonium phosphomolybdate and imaged using an 80 kV electron beam at the Center for Biologic Imaging of the University of Pittsburgh. The zeta potential of the PLGA nanoparticles, both drug-free and drug-loaded, in Milli-Q water, was measured using the zeta potential analysis mode in the Zetasizer.

2.2.3. Drug Loading

The amount of GRFT and DPV encapsulated in the nanoparticles was determined by analyzing the GRFT and DPV content in the supernatant after centrifugation by high-performance liquid chromatography (HPLC) to quantify encapsulation efficiency (Equation (1)). For GRFT analysis, an HPLC system (Waters Corporation, Milford, MA, USA) equipped with an auto injector (model 717), a quaternary pump (model 600), an ultraviolet detector (model 2487) at 280 nm, and a multi λ fluorescence detector (model 2475) at 370 nm was used. Separation of GRFT was achieved by using a Jupiter 5 μ 300 Å (250 × 4.6 mm) column (Phenomenex, Torrance, CA, USA) protected by a guard cartridge Jupiter C18 (4.0 × 3 mm). A gradient consisting of mobile phase A (0.1% trifluoroacetic acid [TFA] in Milli-Q water), and mobile Phase B (0.1% TFA in acetonitrile), at a flow rate of 1.0 mL/min, was used. Retention time of GRFT was approximately 16 min, and the total run time was 30 min. Empower Pro 3 software (Waters Corp., Milford, MA) was used to control the HPLC system.

DPV was assayed by using an Acquity ultra-performance liquid chromatography (UPLC) system H-class (Waters Corp., Milford, MA). Separation of DPV was achieved by using an Acquity UPLC BEH C18 (1.7 μm, 2.1 × 50 mm) column (Waters Corp., Milford, MA). A gradient consisting of mobile phase A (0.1% TFA in Milli-Q water) and mobile Phase B (0.1% TFA in acetonitrile), at a flow rate of 0.25 mL/min, was used. Retention time of DPV was approximately 8.5 min, and the total run time was 15 min. Empower Pro 3 software was used to control the UPLC system. Drug encapsulation efficiency is calculated by following equation as described by Papadimitriou et al. [47]:

$$\begin{aligned} \text{Drug encapsulation efficiency (\%)} &= \frac{\text{weight of drug in nanoparticles}}{\text{weight of drug fed initially}} \\ &= \frac{\text{weight of drug fed initially weight of drug in supernantant}}{\text{weight of drug fed initially}} \end{aligned} \quad (1)$$

2.2.4. In Vitro Release of ARVs from PLGA Nanoparticles

To determine the extent and rate of GRFT and DPV released from the nanoparticles, an in vitro release study was conducted over a 7-day period. In this study, vaginal fluid simulant (VFS, pH 4.5) was selected as dissolution medium to evaluate drug release from the fabricated nanoparticles because of its physiological relevance to the cervicovaginal fluid [48]. ARV-loaded nanoparticles were dispersed in 3 mL of VFS, with continuous shaking, at a temperature of 37 °C. At regular intervals, the nanoparticles were isolated via centrifugation (12,000× *g*, 4 °C, 30 min), and the entire VFS solution was decanted for analysis using chromatographic methods describe above. The nanoparticles were then re-suspended in fresh VFS and returned to the in vitro release set-up.

To determine the impact of medium and pH on drug release, in vitro release studies were also performed in 1 M phosphate-buffered saline (PBS) at pH 7.4 and 1 M PBS at pH 4.5 (1 M PBS adjusted to pH 4.5 with 10% hydrochloric acid). The amount of GRFT and DPV released into the release media was determined by HPLC or UPLC methodology as previously detailed.

2.2.5. Anti-HIV-1 Activity and Cellular Viability Assay of ARVs

The cytotoxicity and anti-HIV-1 activity of free and nanoparticle-formulated ARVs against HIV-1_{BaL} was determined in TZM-bl cells by luciferase quantification of cell lysates [49–51]. The HIV-1_{BaL} and TZM-bl indicator cell line were kindly provided by Dr. Charlene Dezzutti of Magee-Womens Research Institute. TZM-bl is a HeLa cell line derivation that stably expresses high levels (% positive cells) of CD4 (32%), CCR5 (82%) and CXCR4 (85%) [52]. The cells contain HIV-1 Tat-regulated reporter genes for firefly luciferase and β-galactosidase for quantitative analysis of HIV-1, simian immunodeficiency virus (SIV), and simian/human immunodeficiency virus (SHIV) infection [53]. The evaluation of cytotoxicity and anti-HIV-1 activity were conducted following methods previously described [54]. The cells were regularly cultured in Dulbecco's Modified Eagle Medium (DMEM) with 10% fetal bovine serum (FBS), 100 U/mL penicillin, 100 μg/mL streptomycin, and 1% 200 mM L-glutamine at 37 °C in 5% CO_2 atmosphere.

The TZM-bl cells were seeded into a 96-well clear-view plate, at a concentration of 5 × 10^4 cells per well, in 100 μL of DMEM medium (10% FBS) and allowed to adhere for 24 h at 37 °C [54]. Identical, but separate, plates were set up to measure efficacy and cellular toxicity. 100 μL of medium was removed and replaced with 100 μL of DMEM dilutions of test articles i.e., each drug and ARV-loaded nanoparticles (NP-ARVs). Cells exposed to media without test articles was used as a control. For toxicity testing, additional 100 μL of medium (without test articles) was added to each well to bring the total volume to 200 μL. The next day, 100 μL of medium was removed and replaced with 100 μL of CellTiter-Glo® (Promega, Madison, WI, USA) and the luminescence measured using SpectraMax M3 plate reader (Molecular Devices, Sunnyvale, CA, USA). Viability was determined on the basis of deviations from the cell-only control and presented as the percentage viability ± SD (standard deviation).

For efficacy testing, following addition of media with and without test articles, 100 μL of medium containing HIV-1 was added to each well to reach a total volume in each well of 200 μL. HIV-1_{BaL} was added at an approximate $TCID_{50}$ (50% tissue culture infectious dose) of 3000 per well in the presence of 40 μg/mL of Diethylaminoethyl cellulose- Dextran (DEAE-Dextran, Sigma, St. Louis, MO, USA). After 48 h, 100 μL of medium was removed and replaced with 100 μL of Bright-Glo™ (Promega, Madison, WI, USA) and the luminescence measured as mentioned above. Inhibition of HIV infection was determined on the basis of luminescence deviations from the HIV-1-only control and presented as the percentage inhibition ± SEM. Untreated wells with cells only served as the negative infectivity control (100% inhibition), while wells with cells and HIV-1 served as positive infectivity control (0% inhibition). The percent inhibition was calculated for all test and control cultures to determine the 50% inhibition concentration (IC_{50}) value of each drug. The IC_{50} values of NP-ARVs were calculated via GraphPad Prism (V 5.02) software, using drug concentrations that corresponded to the actual drug loading determined by HPLC.

2.2.6. Combination Effects

The combined anti-HIV-1 activity of the dual-drug combination was evaluated by using the median-effect analysis described by Chou and Talalay [55]. Briefly, median values (IC_{50}) of each drug were obtained using the TZM-bl antiviral activity assay as described in the previous section. The drugs were then mixed according to the ratio of their individual IC_{50} values (1:8, due to the molecular weight difference between DPV and GRFT). For NP-ARVs, amounts of the individual compounds used in combinations were determined by measured drug loading. The drug mixtures were then serially diluted, and IC_{50} values were determined with the TZM-bl assay. Combination effects of ARVs were evaluated by (1) comparing HIV-1 inhibition between the combination and each individual ARV; and (2) identifying combination indices (CI) to quantify drug synergy, using CompuSyn software (ComboSyn, Inc., Paramus, NJ, USA). The CI of drug combination was plotted as a function of the fractional inhibition (Fa), by computer simulation, from Fa = 0.10 to 0.95. In this analysis, the combined effect at the 50% fractional inhibition (CI_{50}) and 90% fractional inhibition (CI_{90}) were reported as synergistic, additive, or antagonistic when CI < 1, = 1, or >1, respectively.

2.2.7. Cellular Uptake Assay

The mouse macrophage cell line RAW264.7 (ATCC, Manassas, VA, USA) was maintained in DMEM culture medium with 10% FBS, 100 U/mL penicillin, 100 μg/mL streptomycin, and 1% 200 mM L-glutamine at 37 °C in 5% CO_2 atmosphere. RAW264.7 cells were cultured on a 1-cm diameter coverslip pre-coated with polylysine (BD BioCoat, BD Biosciences, Bedford, MA, USA) in DMEM culture medium for 48 h before uptake assay. Fluorescent dye-loaded NPs were added to the cells at 2.5% (*w*/*v*) of the total 200 μL of cell culture media for 4 h at 37 °C. NPs were then washed from cells with PBS for 4 times. The coverslips with cells grown on them were first stained with 4′,6-diamidino-2-phenylindole (DAPI, Molecular Probes, and Eugene) in PBS in order to visualize the nuclei and were then mounted on glass slides for fluorescent imaging. The fluorescent images were obtained by ApoTome Confocal Microscope Observer Z1 (Carl Zeiss Microscopy, Zaventem, Belgium). Cell images were taken at 358/461 nm for DAPI, 495/519 nm for FITC-dextran, and 552/636 nm for Nile red.

2.2.8. Statistical Analyses

All differences were evaluated by Student's t-test or ANOVA analysis. $p < 0.05$ was considered statistically significant. Drug release profiles were analyzed using DD Solver, an Excel add-in package [56]. All error bars represent standard deviations unless otherwise noted.

3. Results

3.1. PLGA Nanoparticle Characterization

In this study, PLGA core-shell nanoparticles were successfully developed that simultaneously encapsulate and deliver both a protein drug and a small-molecule hydrophobic drug without compromising the effectiveness of either. The manufactured PLGA nanoparticles are near-spherical in shape, possess high drug-loading capability, and are of a reproducible size. It has been reported that the duration and intensity of sonication can significantly affect the size of the nanoparticles produced using double-emulsion method [57,58]. The targeted particle size and distribution can be obtained by optimizing sonication intensity and time within a given system. With the sonication parameters described in the Methods section, the diameters of the fabricated nanoparticles were confined to a narrow range (Table 1): 182.8 ± 1.7 nm (placebo nanoparticles) to 188.8 ± 1.7 nm (GRFT, DPV, or GRFT/DPV nanoparticles) and had low polydispersity index (PDI < 0.1), indicating that nearly monodispersed nanoparticles were manufactured (Figure 2). Zeta potential measurements showed minimal change between unloaded and loaded nanoparticles, falling within a range of −23.4 ± 0.3 mV to −24.9 ± 1.3 mV, respectively, at a pH of 7.4. The high similarity in colloidal properties of all nanoparticles such as size, PDI, and zeta potential demonstrates that the formed nanoparticles are composed of PLGA-PVA system. It is highly unlikely that the particles are primarily composed of GRFT alone. Table 1 lists properties of placebo nanoparticles (vehicle) and ARV-loaded nanoparticles fabricated using double emulsion evaporation.

A GRFT encapsulation efficiency of 40.7 ± 5.9% and 45.9 ± 13.7% was obtained from the manufactured single (NP-GRFT) and combination nanoparticles (NP-GRFT/DPV), respectively. A DPV encapsulation efficiency of 70.1 ± 4.4% and 69.4 ± 5.1% was obtained from NP-DPV and NP-GRFT/DPV, respectively. Our results indicate that the encapsulation efficiency of DPV and GRFT are independent of each other in the combined NPs. Our findings are similar to those describing the manufacture of PLGA nanoparticles via double-emulsion or solvent displacement methods [30,38,43]. Our results suggest that the double-emulsion technique is suitable for encapsulating both hydrophilic macromolecules and hydrophobic small molecules in PLGA-based core-shell type nanoparticles simultaneously.

Table 1. Physicochemical properties of PLGA nanoparticles loaded with ARVs.

Drugs	Size (d.nm ± SD)	PDI	Zeta Potential (mV ± SD)	Encapsulation Efficiency (% ± SD)
Placebo	182.8 ± 1.7	0.066	−23.7 ± 0.6	-
GRFT	188.8 ± 1.7	0.069	−23.5 ± 0.3	40.7 ± 5.9
DPV	186.6 ± 1.6	0.079	−24.9 ±1.3	70.1 ± 4.4
GRFT/DPV	184.3 ± 1.0	0.063	−23.4 ± 0.3	45.9 ± 13.7 (GRFT) 69.4 ± 5.1 (DPV)

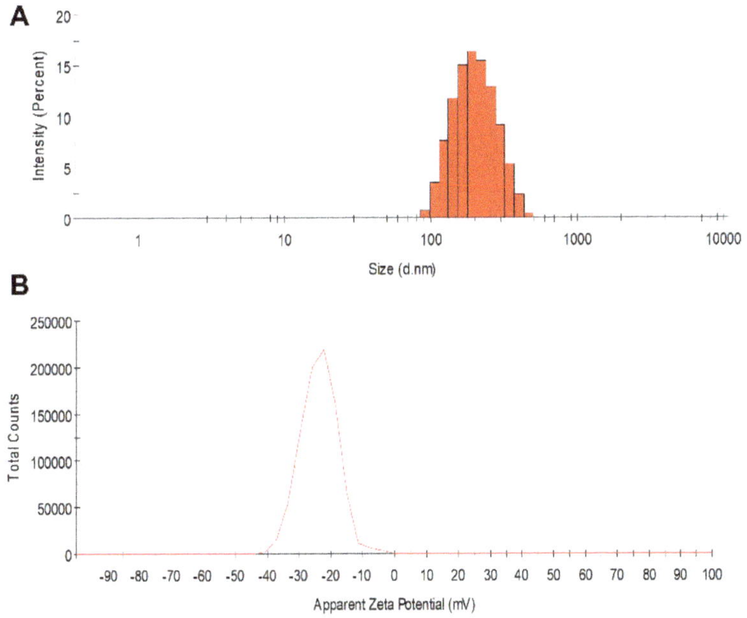

Figure 2. Characterization of PLGA nanoparticles loaded with GRFT and DPV. (**A**) Size and (**B**) zeta potential distribution graphs of nanoparticles encapsulated with anti-HIV drugs GRFT and DPV (NP-DPV/GRFT).

3.2. *In Vitro Release Studies of ARVs from PLGA Nanoparticles Resulting in Sustained Drug Release*

The in vitro drug release profile of GRFT and DPV from NP-ARVs was measured over 7 days in VFS at pH 4.5 that mimics the composition, pH, and viscoelastic properties of vaginal fluid produced by healthy, adult, nonpregnant women [48,59,60], as well as in buffers at pH 7.4 and pH 4.5. Our findings indicate that the combination of ARVs and media pH/composition affect the release profile of individual ARVs from core-shell PLGA nanoparticles as shown in Figure 3.

The in vitro release of GRFT and DPV from both single-entity and combination nanoparticles followed a biphasic release profile (Figure 3A,B). In VFS, the combination nanoparticles showed an initial burst release within first 24 h, where 15.1 ± 2.6% of the GRFT had been released. A sustained release of GRFT was obtained over the remaining 7 days. The total amount of drug released over the 7-day period was 16.7 ± 2.8%. Although not statistically significant, less GRFT was released from NP-GRFT, in both burst phase (11.1 ± 2.2%) and total amount of GRFT released (12.1 ± 2.8%) than that from combination nanoparticles.

The in vitro release of DPV showed a different and more prolonged initial burst release profiles (Figure 3C,D). In VFS, after 1 day, 51.7 ± 7.8% of DPV was released from the combination nanoparticles (NP-DPV/GRFT). DPV was released continuously over the next 7 days. The total amount of drug released over the 7-day period was 76.9 ± 13.1%. Similar to GRFT release, DPV release was lower from single entity nanoparticles in both burst phase (45.9 ± 10.0%), as well as the total amount of DPV released in 7 days (69.0 ± 12.8%) compared to combination nanoparticles in VFS and pH 7.4 media.

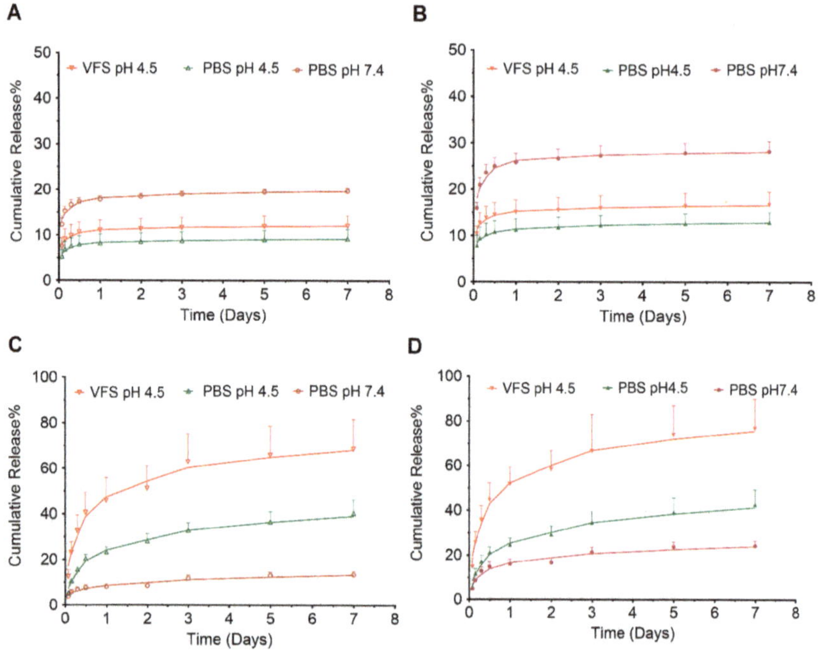

Figure 3. In vitro release of ARVs from PLGA nanoparticles in VFS at pH 4.5, and buffers at pH 4.5 and pH 7.4. (**A**) GRFT release from NP-GRFT and (**B**) NP-DPV/GRFT. (**C**) DPV release from NP-DPV and (**D**) NP-DPV/GRFT. NP-nanoparticles.

We also investigated the effect of pH and release media on the in vitro release of the PLGA nanoparticles. Total release of DPV from NP-DPV/GRFT was significantly higher in pH 4.5 media than that in pH 7.4 media (42.7 ± 6.7% vs. 24.3 ± 2.2%, $p < 0.05$). Even more DPV was released in VFS (pH 4.5) (76.9 ± 13.1%, $p < 0.05$) than in pH 4.5 media. However, pH showed an opposite effect on GRFT's release from nanoparticles. The total release of GRFT from NP-DPV/GRFT was significantly less in pH 4.5 media than that in pH 7.4 (12.9 ± 2.2% vs. 28.2 ± 2.2%, $p < 0.05$), while the total release of GRFT in VFS (16.7 ± 2.8%) was only slightly higher than that in pH 4.5 media. The order of release in the media remained similar for the single entity nanoparticles (NP-DPV and NP-GRFT) compared to combination nanoparticles; however, stronger statistical differences were noted. The order of release rate of NP-ARVs is pH 7.4 > VFS > pH 4.5 for GRFT and VFS > pH 4.5 > pH 7.4 for DPV. These results suggest that co-encapsulation could lead to enhanced release of each ARV from PLGA nanoparticles in all the media investigated.

To reveal the mechanism of drug release from fabricated nanoparticles, we used mathematical modeling to analyze the in vitro release profiles of GRFT and DPV by various kinetic models. Three empirical and semi-empirical models used were the Higuchi model, the Peppas–Sahlin model, and the Weibull model [56]. Higher correlation was observed in the Peppas model and Weibull model (Figures 4 and 5). Therefore, our results indicate that release of ARVs from PLGA nanoparticles is not predominantly driven by a solo mechanism, but a combined mechanism of Fickian (pure diffusion phenomenon) and non-Fickian release (due to the relaxation of the polymer chains between the networks).

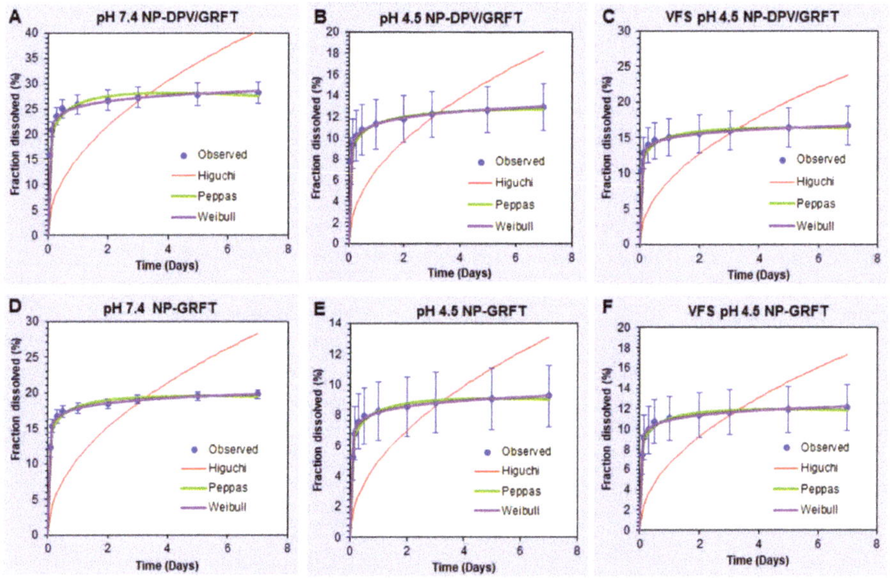

Figure 4. Mathematical model fitting of GRFT release from PLGA nanoparticles. (**A–C**) GRFT release from combination nanoparticles (NP-DPV/GRFT) at pH 7.4, 4.5 and VFS pH 4.5 media. (**D–F**) GRFT release from single entity nanoparticles (NP-GRFT) at pH 7.4, 4.5 and VFS pH 4.5 media. Dissolution profiles were fitted to Higuchi, Peppas, and Weibull models.

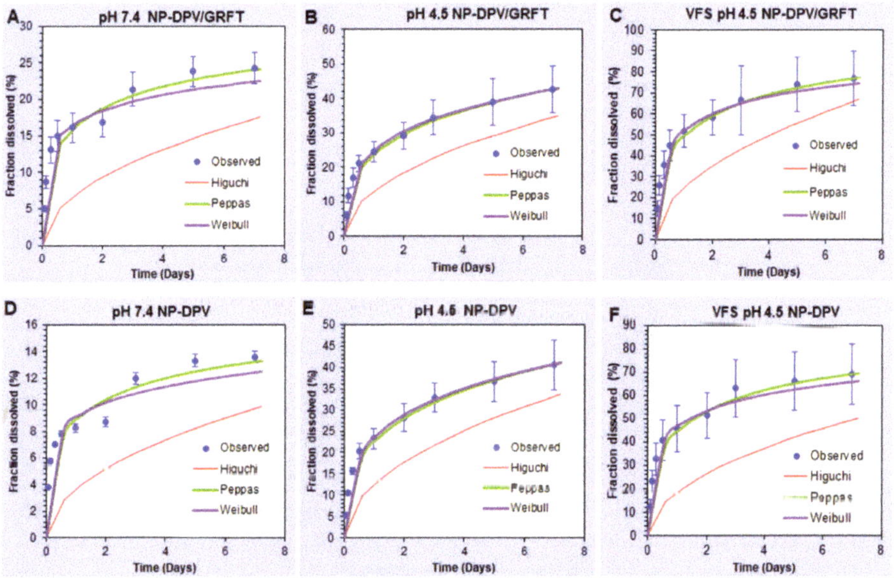

Figure 5. Mathematical model fitting of DPV release from PLGA nanoparticles. (**A–C**) DPV release from combination nanoparticles (NP-DPV/GRFT) at pH 7.4, 4.5 and VFS pH 4.5 media. (**D–F**) DPV release from single entity nanoparticles (NP-DPV) at pH 7.4, 4.5 and VFS pH 4.5 media. Dissolution profiles were fitted to Higuchi, Peppas, and Weibull models.

3.3. *PLGA NP-ARVs Potently Inhibit HIV-1$_{BaL}$ Infection In Vitro*

To ensure that the anti-HIV-1 activity of the ARVs was retained after being loaded into nanoparticles, we tested the protection of NP-ARVs and unformulated ARVs against HIV-1$_{BaL}$ infection, using the TZM-bl assay. Anti-HIV-1 efficacy of NP-ARVs and unformulated ARVs was measured as percent of HIV inhibition, which is calculated using percent relative luminescence units of TZM-bl cells treated with ARVs or NP-ARVs compared to untreated (media only) TZM-bl cells after exposure to HIV-1$_{BaL}$ as noted in the Methods section.

The anti-HIV-1 activity of blank PLGA nanoparticles was not observed in the TZM-bl assay, which was comparable to the negative media control. After exposure of TZM-bl cells to NP-ARVs or free ARVs; however, we observed potent antiviral activity against HIV-1$_{BaL}$, with estimated IC_{50} values in the nanomolar ranges for GRFT and DPV, respectively (Table 2). Compared with free DPV, NP-DPV showed slightly enhanced HIV inhibitory activity, with an IC_{50} of 4.7 nM for the unformulated DPV to 3.6 nM for NP-DPV (Figure 6). The encapsulation of GRFT in PLGA nanoparticles slightly shifted the IC_{50} from 0.5 nM for the unformulated GRFT to 0.8 nM for the NP-GRFT. However, no statistically significant difference was observed between NP-GRFT and unformulated GRFT.

Table 2. IC_{50} of ARVs and NP-ARVs estimated by TZM-bl assay.

Drug	Alone (nM)	
	Unformulated	NP-GRFT/NP-DPV
DPV	4.7 ± 2.9	3.6 ± 2.9
GRFT	0.5 ± 0.3	0.8 ± 0.7

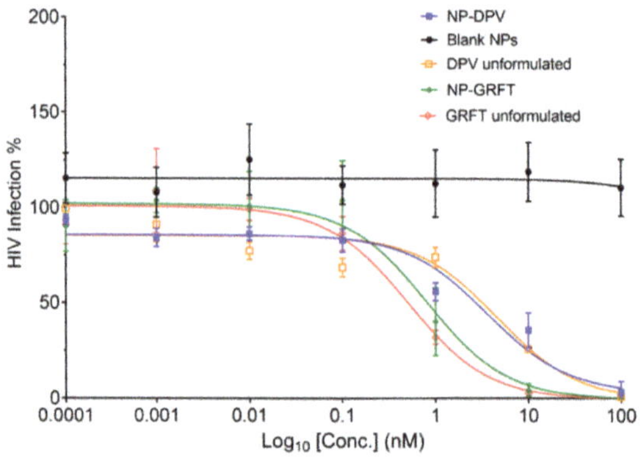

Figure 6. Anti-HIV activity of GRFT or DPV in PLGA nanoparticles. Anti-HIV activity of unformulated ARVs and fabricated NP-ARVs were evaluated in TZM-bl cells exposed to HIV-1. Luciferase luminescent readings of treated TZM-bl cells were compared against untreated cells infected with HIV-1 to obtain % HIV infection. Results are reported on a log scale of ARVs dosing levels versus percent of infection.

We hypothesized that NP-DPV can increase the intracellular concentration of DPV, resulting in greater potency due to enhanced internalization and intracellular uptake of PLGA nanoparticles [46,61]. For GRFT, whose target is on the cell membrane, the internalization and intracellular uptake of nanoparticles may cause the loss of bioactivity as we observed in the experiment, which is consistent with other reports on protein fusion inhibitors, such as RANTES [38]. However, no significant loss of GRFT's bioactivity was observed. Together, our results suggest that both GRFT and DPV can be loaded into nanoparticles without compromising their potent bioactivity.

3.4. NP-ARV Combination Exhibits Strong Synergistic Bioactivity

To identify the interactive effects on our ARVs' bioactivity, we evaluated the activity of their combination in nanoparticle and unformulated drugs, using the TZM-bl assay. The equipotency ratio of GRFT and DPV (i.e., 1:1 ratios of IC_{50} values), were used to assess the effect of the drug combination, when drugs were combined as unformulated or as distinct nanoparticles. This equipotency translated to a 1:8 molar ratio for GRFT:DPV. Notably, this exact ratio cannot be maintained in the manufactured combination nanoparticles (NP-DPV/GRFT), therefore these nanoparticles were tested at the loaded ARV levels.

The efficacy of GRFT and DPV was significantly improved when tested in combination compared with when tested alone, either as unformulated GRFT and DPV or combined as distinct nanoparticles (NP-GRFT and NP-DPV) or when co-encapsulated into combination nanoparticles. Figure 7 shows the dose-response relationships of unformulated GRFT or DPV alone versus unformulated GRFT/DPV combination (Figure 7A), NP-GRFT or NP-DPV alone vs NP-GRFT and NP-DPV combination (Figure 7B). The results demonstrate that GRFT and DPV combined as unformulated or distinct nanoparticles or in a combination nanoparticle system display similar reduction in HIV infection. Interestingly, although the combination system does not include GRFT and DPV at equipotency ratio, a similar reduction in HIV infection was observed. IC_{50} values for combinations were not calculated, because even at the lowest tested concentration, high potency was observed, i.e., <50% HIV infection. Overall, the combination of GRFT and DPV showed significant leftward shift in the dose-response curve compared to single ARVs, indicating a strong synergy. The results also suggest that the efficacy is comparable between combination in unformulated, NP-GRFT and NP-DPV combination, and NP-GRFT/DPV, as shown in Figure 8.

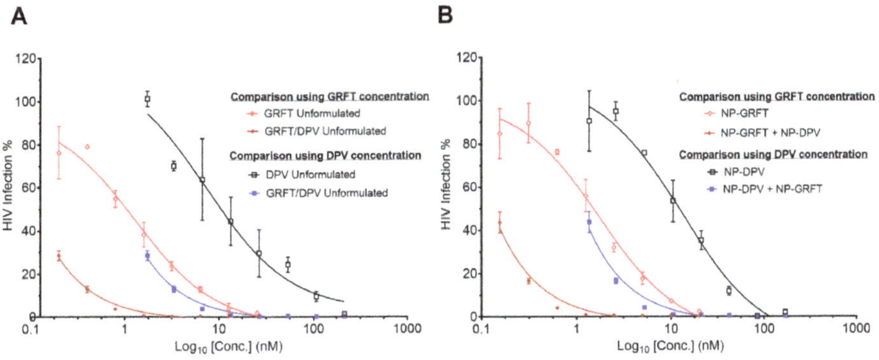

Figure 7. Strong synergistic effect of GRFT and DPV in free drug combination, in NP-GRFT and NP-DPV combination. The dose-response curve shows antiviral activity of (**A**) free GRFT, free DPV or combination of free GRFT and free DPV. (**B**) GRFT-loaded nanoparticles (NP-GRFT), DPV-loaded nanoparticles (NP-DPV) alone or in combination (NP-GRFT + NP-DPV). At a 1:8 molar ratio (GRFT and DPV), the antiviral activity of each drug in combination showed a significant reduction in %HIV infection (IC_{50} cannot be computed) compared to single ARV drug in unformulated or encapsulated forms. No difference in anti-HIV activity of GRFT or DPV were found in unformulated or nanoparticle combinations.

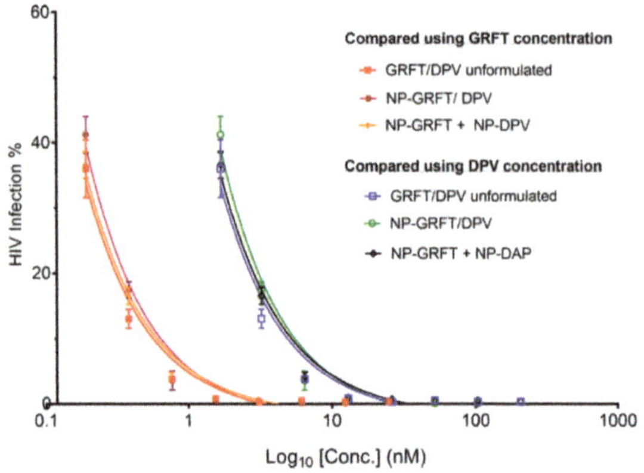

Figure 8. Potent and comparable anti-HIV effect of GRFT and DPV combinations. The dose-response curve shows potent antiviral activity of the combination of free drugs (GRFT/DPV unformulated), combination of single ARV nanoparticles (NP-GRFT + NP-DPV), and nanoparticles encapsulating both GRFT and DPV (NP-GRFT/DPV). Comparable inhibition of HIV infection was achieved with combination nanoparticles although the ARV ratio is not maintained at equipotency values (1:1 IC_{50} or 1:8 molar ratio for GRFT:DPV).

The CI was determined for the ARV drugs combined at molar ratios to achieve anti-HIV equipotency (1:1 ratios of IC_{50} values), leading to a 1:8 (GRFT: DPV) molar ratio in concentration. The CI of each unformulated ARV or NP-ARV combination was plotted as a function of the fractional inhibition (Fa) from 0.10 to 0.95 (Figure 9). The CI values were very low over the range from 0.10 to 0.95 of Fa. We interpreted the combination effects, at the CI_{50} and CI_{90} values, of unformulated GRFT/DPV and NP-GRFT/NP-DPV in combination ($CI_{50/90} < 0.1$) as demonstrating an extremely strong synergistic effect of anti-HIV infection.

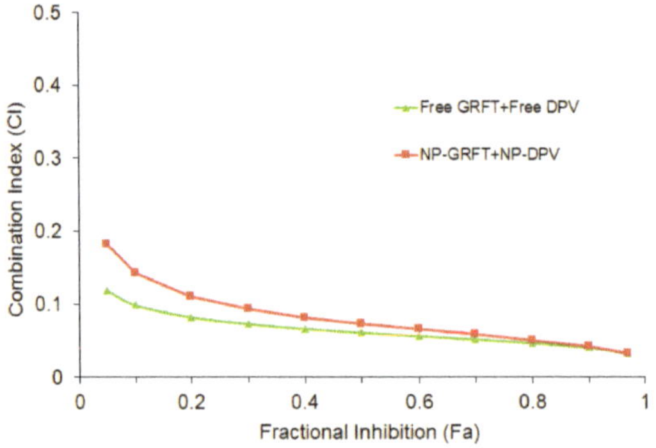

Figure 9. Combination index (CI) of free GRFT with free DPV or GRFT nanoparticles (NP-GRFT) with DPV nanoparticles (NP-DPV) were quantified using the TZM-bl infectivity assay. CI < 1, = 1, and >1 indicate synergistic, additive, and antagonistic effects, respectively. Combination of GRFT and DPV at a 1:8 molar ratio (GRFT:DPV) demonstrated very strong synergism, with CI at 50% inhibition (CI_{50}) of 0.086 and 0.066 for unformulated and nanoparticle combinations respectively. CI 0.90–1.10: nearly additive; CI 0.85–0.95: slight synergism; CI 0.7–0.85: moderate synergism; CI 0.3–0.7: synergism; CI 0.1–0.3 strong synergism; CI < 0.1: Very strong synergism.

3.5. NP-ARVs Are Nontoxic to In Vitro Cell Lines

Neither blank PLGA nanoparticles nor ARV-loaded PLGA nanoparticles were observed to be cytotoxic to cells over the range of concentrations evaluated in the TZM-bl assay. The cytotoxicity of NP-ARVs was evaluated during testing of their bioactivity in order to exclude any effects of the nanoparticles on the viability of TZM-bl cells. Cytotoxicity of our NP-ARVs was measured over a range of ARV concentrations, from 0.0001 to 1000 nM, after 24 h of exposure, leading to polymer concentration less than 0.1 mg/mL. No significant reduction ($p < 0.05$) in viability of the TZM-bl cells was observed in NP-ARV-treated cells as compared to untreated TZM-bl cells (Figure 10). In addition, we tried to evaluate the upper limit of cytotoxicity of nanoparticle vehicles. Compared with the negative control (media only), vehicle-control/blank NPs, at concentrations of 5 mg polymer/mL, showed no reduction of viability (100% ± 8%), suggesting that PLGA nanoparticles alone are not cytotoxic below this concentration. Since anti-HIV bioactivity was measured at doses far below 5 mg/mL polymer concentrations, we did not expect toxicity to confound the outcome of the antiviral activity assays.

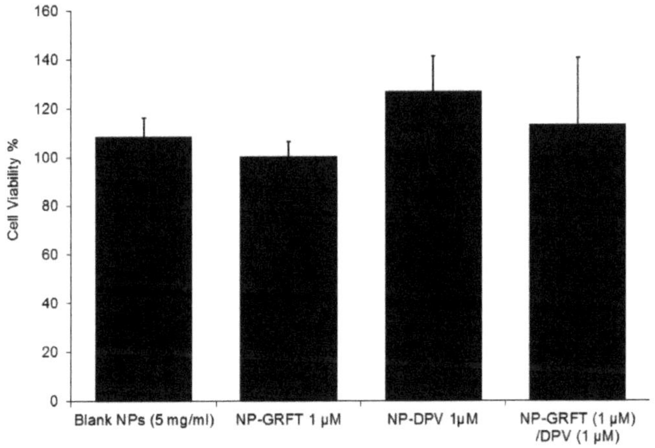

Figure 10. Toxicity of GRFT and DPV loaded PLGA nanoparticles. Viability of TZM-bl cells was measured by the CellTiter-Glo® (Promega, Madison, WI, USA) viability assay kit demonstrating non-toxic nature (≥80% viability) of NP-GRFT, NP-DPV, and NP-GRFT/DPV. Vehicle control (blank nanoparticles) at the concentrations tested showed no reduction of viability (108% ± 7%), indicating non-cytotoxicity of PLGA polymer itself. Negative control = media only. Vehicle control for all nanoparticles was evaluated at 5.0 mg of polymer/mL.

3.6. Cellular Uptake of NPs In Vitro

PLGA nanoparticles loaded with GRFT/DPV maintained the core-shell structure after cellular uptake. FITC-dextran and Nile red were incorporated into NPs as to evaluate the cellular uptake of NP-ARVs using fluorescence imaging as shown in Figure 11. FITC-dextran (Green) was used to mimic GRFT due to its water solubility and its molecular weight of 70 kDa. Nile red (Red) was used to mimic DPV due to its hydrophobicity. Coincident red and green fluorescence signals could be observed in the cytoplasm of RAW264.7 after 4 h incubation. The results demonstrate that the nanoparticles were taken up intact. Therefore, they could be delivered simultaneously in vivo with their synergistic anti-HIV effect.

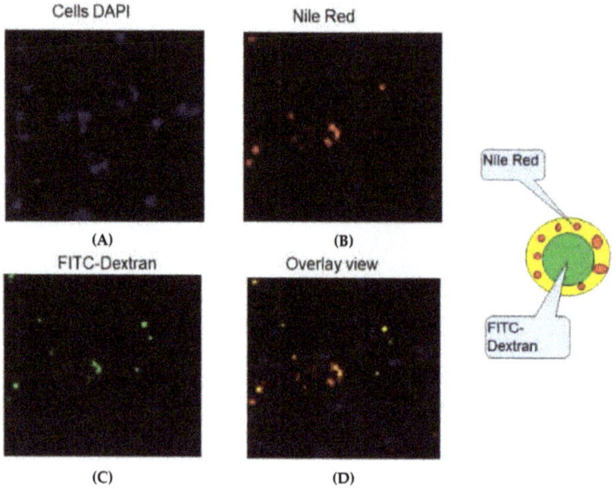

Figure 11. The cellular uptake of nanoparticles loaded with both FITC-Dextran (green, GRFT mimic) and Nile Red (red, DPV mimic) was investigated in RAW264.7 cells. Representative fluorescent microscopy images of RAW264.7 cells post nanoparticle exposure at 40× magnification. (**A**) Fluorescence images show RAW264.7 cells nuclei stained with DAPI (blue) for positioning purpose, (**B**) images of Nile red (red) and (**C**) FITC-dextran (green) of same field were taken through different channels to illustrate the position of FITC-dextran (green)/Nile red (red) encapsulated nanoparticles in cells. (**D**) Overlay image showing coincident green and red signals as yellow color indicating uptake of intact nanoparticles.

4. Discussion

Combination and long-acting drug products are emerging approaches for the prevention of HIV-1 infection, due to the clinical need for effectiveness, safety, and patient adherence. To date, no combination PrEP is available, either as a vaginal or a rectal product, although efforts are underway. The only oral PrEP currently approved by the FDA (in 2004) is Truvada, which is a combination of emtricitabine and tenofovir disoproxil fumarate. Patient adherence plays a key role in the effectiveness of PrEP products [62], a lesson to be learned from several HIV prevention trials, which found that higher adherence resulted in lower HIV incidence [13,14,63]. Long-acting formulations offer a promising paradigm to overcome the adherence challenge in the development of PrEP products [64,65].

Two PrEP candidates have been intensively studied, individually, as potential microbicides in pre-clinical evaluations. GRFT, a fusion inhibitor, has demonstrated high efficacy against HIV-1 infection at nanomolar concentrations, with enhanced anti-HIV-1 activity and extended contact time with HIV-1 [18,20]. However, cervicovaginal secretions tend to inhibit GRFT-gp120 binding, as well as oxidize its methionine at position 78, resulting in compromise of GRFT in vivo [66]. DPV is an NNRTI with extreme potency against HIV-1 infection. Yet, its quick elimination from tissue may lower its anti-HIV-1 activity as a microbicide [23]. We hypothesized that core-shell polymeric nanocarriers could enhance the anti-HIV-1 effect of these two candidate ARVs via simultaneous co-delivery and sustained release for long-acting prevention of HIV-1 infection via vaginal route.

In this study, we demonstrated that a protein drug and a small hydrophobic drug can be encapsulated within polymeric nanoparticles to provide prophylaxis in combination. Anti-HIV-1 drugs were encapsulated into fabricated PLGA nanoparticles without compromising their anti-HIV-1 bioactivity. The nanoparticles were near-monodispersed, mostly spherical in shape. The constant negative zeta potential of the nanoparticles from the unloaded to the loaded state suggests that the hydrophilic microbicide is encapsulated in the core structure of nanoparticles rather than adsorbed onto the surface. High encapsulation efficiency of both GRFT and DPV suggests that the double emulsion method would be a viable approach for the fabrication of PLGA nanoparticles for other microbicide candidates.

We were able to demonstrate the sustained release of both GRFT and DPV, one of the defining characteristics of fabricated PLGA nanoparticles and indicating a potential long-acting nanoparticle delivery system. The in vitro drug release profiles of GRFT and DPV from the nanoparticles were characterized by two stages: the initial burst release followed by a sustained release over the time of the experiment. In vitro release studies showed that a significant amount of drug was released over a 7-day period, which may provide a weekly-based regimen.

The in vitro release of drugs from PLGA nanoparticles can be affected by environmental conditions such as changes in pH. Generally, PLGA polymer degrades faster at lower pH than at neutral pH [67,68]. The normal vaginal environment has an acidic pH, ranging from 3.8 to 4.5, as a protective barrier [69,70]. Therefore, in vitro release of drugs from PLGA nanoparticles was evaluated in buffers and VFS. Our studies showed that pH changes in the environment affected both the release rate and the total amount of GRFT and DPV released from the PLGA nanoparticles. Interestingly, we observed that pH change has opposite effects on the release of DPV and GRFT from the nanoparticles. More DPV was released into dissolution media from PLGA nanoparticles at lower pH (pH 4.5) than that at pH 7.4, which is expected due to the increased hydrolysis or erosion of PLGA polymers at lower pH [71]. Furthermore, DPV is likely to be in the protonated form at low pH, leading to increased solubility and release. Although both media possess similar pH, DPV release in VFS was much higher compared to the pH 4.5 buffer, possibly due to increased solubility of hydrophobic DPV in VFS, which contains bovine serum albumin. On the contrary, less GRFT was released from PLGA nanoparticles at lower pH (pH 4.5) than that at neutral pH (pH 7.4). One possible explanation for this phenomenon is that GRFT is positively charged at pH 4.5 due to its low isoelectric point (pI) of 5.39. Thus, the electrostatic interaction between positively charged GRFT and negatively charged PLGA nanoparticles may suppress the release of GRFT from PLGA nanoparticles. PSC-RANTES, for instance, has a much higher pI (around 9.0), retaining positive charge at both pH 7.4 and pH 4.5 [72]. Therefore, the release of PSC-RANTES is only determined by environmental pH [38]. Further studies are needed to confirm our hypothesis.

This compensatory release profile of our nanoparticles could prove beneficial for HIV-1 prevention. In a regular vaginal environment, more DPV will be released from the nanoparticles at lower pH (pH 4.5); when the presence of semen results in a higher pH environment (pH 7.4), more GRFT will be released. Our nanoparticle system could provide protection against HIV-1 infection in both scenarios. Therefore, it could be used in either a coitally dependent or coitally independent manner and provide protection over a wide range of environmental pH values.

Mathematical models were applied to evaluate the drug release from the nanoparticles. The results suggest that the release may be governed by a mixed mechanism of diffusion and polymer degradation, which is different from the polymer degradation–only mechanism of PSC-RANTES release from PLGA nanoparticles reported by Ham et al. [38]. However, the polymer chain relaxation may be the rate-limiting step for the drug release from the PLGA nanoparticle system. Further detailed studies are needed to elucidate the mechanisms involved, which is beyond the scope of this work.

Since the combination of GRFT and DPV is extremely potent, the HIV infection could not reach 50% in CI study, even at extremely low concentration of GRFT and DPV in combination, which made the estimation of IC_{50} for GRFT and DPV combination not very reliable. However, that also confirmed the potency of our combination strategy using GRFT and DPV. We found that the combination of free GRFT with free DPV at a 1:1 ratio of IC_{50} values demonstrated a strong synergistic effect (Figure 7). Importantly, the same synergistic effect was also demonstrated when the combination of drugs was released from nanoparticles, indicating that our PLGA nanoparticle system is suitable for delivering highly potent anti-HIV-1 drugs in combination. We quantified the synergistic effects of free GRFT and free DPV and NP-GRFT/NP-DPV using the median effect analysis developed by Chou and Talalay [73]. We believe that this synergy suggests a promising combination of ARVs as candidates for HIV-1 PrEP. Importantly, our studies demonstrated that GRFT and DPV possess a strong synergistic effect in vitro and that this synergistic effect can be maintained in PLGA nanoparticle delivery system. It had been reported that GRFT can have synergistic effects when used in combination with other anti-HIV-1 drugs, such as tenofovir [22]. However, it has never been studied in combination with DPV either as free drugs or in a nanoparticle system. Our findings indicate that the antiviral activity against HIV-1_{BaL} of our NP-ARVs was maintained similar to unformulated ARVs. The IC_{50} of DPV and GRFT did not change significantly in nanoparticles compared to unformulated form. Our results, however, show enhanced anti-HIV activity of NP-DPV (but $p > 0.05$), which may be caused by more uptake of nanoparticles [61]. The uptake of nanoparticles was visually confirmed in the RAW267.4 cell model. The combination NP-DPV/GRFT system also showed high anti-HIV-1 activity comparable to combination of unformulated ARVs and distinct NP-ARVs. Even without accomplishing 1:8 molar ratio of ARVs, our developed system displayed high potency and suitable for long-term application.

Interestingly, the potency of a drug may play an important role in the extent of combination effect with other drugs. Chaowanachan et al. reported that only additive effects were identified in a study of tenofovir and efavirenz delivery in combination in PLGA nanoparticles; this may be due to the limited drug release from the nanoparticles within the first hour [43]. Here, both tenofovir and efavirenz have an IC_{50} at a micromolar level, resulting in higher drug concentration needed for efficacy. Therefore, a faster drug release rate is required to achieve effective concentration against HIV infection. On the contrary, in our studies, both GRFT and DPV have an IC_{50} at a nanomolar level, indicating that only a small amount of the released drugs would be sufficient for complete HIV-1 inhibition. The difference between these studies using PLGA nanoparticles shows that the drug release rate needs to be optimized according to the drug potency by altering manufacturing and formulation conditions of the nanoparticles.

5. Conclusions

Although nanoparticles have been intensively investigated in numerous biomedical applications, not many studies have been conducted on nanoparticles as a drug delivery system for PrEP products, especially the combination delivery of a protein drug with a water-insoluble drug. In this study, successful encapsulation of both GRFT and DPV into core-shell nanoparticles resulted in monodispersed, near-spherical particles with low PDI that maintained their anti-HIV-1bioactivity. Furthermore, we have shown that our NP-ARVs act synergistically in preventing HIV-1 infection in vitro, and that the formulated PLGA nanoparticles provide sustained release of the drugs. To our knowledge, this is the first quantitative measure of synergistic effect of GRFT and DPV combination in a nanoparticle system. Additionally, the sustained release property of manufactured NPs provides a delivery system that could potentially reduce the dosing frequency to a weekly based regimen, which could significantly increase the compliance and acceptability in women.

However, additional research is required to further develop PLGA nanoparticles as PrEP products for delivery of combined GRFT and DPV. In particular, the relationship between protein pI and protein release in different media e.g., VFS. The model for drug release study in this paper only provides an empirical explanation on the mechanism of protein drug release from nanoparticles. Further investigation will help to reveal the mechanism of drug release from PLGA-based nanoparticles, particularly for protein drugs. In addition, the distribution of nanoparticles in the reproductive tract may affect the efficacy in vivo and needs to be investigated. These studies are beyond the scope of this paper and will be investigated in our future work.

Author Contributions: Conceptualization, L.C.R. and H.Y.; methodology, L.C.R., H.Y., and J.L.; formal analysis, H.Y., S.K.P., J.L., and L.C.R.; investigation, L.C.R., H.Y., J.L., and S.K.P.; resources, L.C.R., H.Y., K.E.P., and B.D.; data curation, H.Y., L.C.R., J.L., and S.K.P.; writing—H.Y., L.C.R., S.K.P., J.L., D.P., K.E.P.; review and editing, H.Y., L.C.R., J.L., S.K.P.; funding acquisition, L.C.R., H.Y., and K.E.P."

Funding: The work presented was supported through grants from the Eunice Kennedy Shriver National Institute of Child Health and Human Development (NICHD)—"Building Interdisciplinary Research Careers in Women's Health in Pittsburgh" (K12HD043441), and the National Institute of Allergy and Infectious Diseases

(NIAID)—U19AI113182 and U19AI082639. Its contents are solely the responsibility of the authors and do not necessarily represent the official views of the NICHD or NIAID.

Acknowledgments: We would like to thank Charlene Dezzutti (deceased) for her help with anti-HIV testing. We greatly acknowledge Hima Bindu Ruttala for designing the schematic showing nanoparticle preparation and critical review of the manuscript. We greatly acknowledge Won-Bin Young for his help with uptake study. We would like to thank the Center for Biologic Imaging (CBI) of the University of Pittsburgh for access to the electron microscopy instrumentation and for assistance with the execution of this part of the study.

Conflicts of Interest: The authors declare no conflict of interest.

References

1. UNAIDS. Global HIV & AIDS Statistics—2018 Fact Sheet. Available online: http://www.unaids.org/en/resources/fact-sheet (accessed on 19 February 2019).
2. AVERT. Women and Girls, HIV and AIDS. Available online: https://www.avert.org/professionals/hiv-social-issues/key-affected-populations/women (accessed on 19 February 2019).
3. Centers for Disease Control and Prevention. HIV Among Women. Available online: https://www.cdc.gov/hiv/group/gender/women/index.html (accessed on 14 April 2019).
4. Bedimo, A.L.; Bennett, M.; Kissinger, P.; Clark, R.A. Understanding barriers to condom usage among HIV-infected African American women. *J. Assoc. Nurses AIDS Care* **1998**, *9*, 48–58. [CrossRef]
5. Santelli, J.S.; Kouzis, A.C.; Hoover, D.R.; Polacsek, M.; Burwell, L.G.; Celentano, D.D. Stage of behavior change for condom use: the influence of partner type, relationship and pregnancy factors. *Fam. Plan. Perspect.* **1996**, *28*, 101–107. [CrossRef]
6. Moench, T.R.; Chipato, T.; Padian, N.S. Preventing disease by protecting the cervix: the unexplored promise of internal vaginal barrier devices. *AIDS* **2001**, *15*, 1595–1602. [CrossRef]
7. Minnis, A.M.; Padian, N.S. Effectiveness of female controlled barrier methods in preventing sexually transmitted infections and HIV: current evidence and future research directions. *Sex. Transm. Infect.* **2005**, *81*, 193–200. [CrossRef]
8. Matthews, J.; Harrison, T. An update on female-controlled methods for HIV prevention: Female condom, microbicides and cervical barriers. *S. Afr. J. HIV Med.* **2006**, *7*, 7–11.
9. Padian, N.S.; Buve, A.; Balkus, J.; Serwadda, D.; Cates, W., Jr. Biomedical interventions to prevent HIV infection: evidence, challenges, and way forward. *Lancet* **2008**, *372*, 585–599. [CrossRef]
10. Maenza, J.; Flexner, C. Combination antiretroviral therapy for HIV infection. *Am. Fam. Physician* **1998**, *57*, 2789–2798. [PubMed]
11. Ludovici, D.W.; De Corte, B.L.; Kukla, M.J.; Ye, H.; Ho, C.Y.; Lichtenstein, M.A.; Kavash, R.W.; Andries, K.; de Bethune, M.P.; Azijn, H.; et al. Evolution of anti-HIV drug candidates. Part 3: Diarylpyrimidine (DAPY) analogues. *Bioorg. Med. Chem. Lett.* **2001**, *11*, 2235–2239. [CrossRef]
12. Holt, J.D.; Cameron, D.; Dias, N.; Holding, J.; Muntendam, A.; Oostebring, F.; Dreier, P.; Rohan, L.; Nuttall, J. The Sheep as a Model for Preclinical Safety and Pharmacokinetic Evaluations of Candidate Microbicides. *Antimicrob. Agents Chemother.* **2015**, *59*, 3761–3770. [CrossRef] [PubMed]
13. Baeten, J.M.; Palanee-Phillips, T.; Brown, E.R.; Schwartz, K.; Soto-Torres, L.E.; Govender, V.; Mgodi, N.M.; Matovu Kiweewa, F.; Nair, G.; Mhlanga, F.; et al. Use of a Vaginal Ring Containing Dapivirine for HIV-1 Prevention in Women. *N. Engl. J. Med.* **2016**, *375*, 2121–2132. [CrossRef] [PubMed]
14. Nel, A.; van Niekerk, N.; Kapiga, S.; Bekker, L.G.; Gama, C.; Gill, K.; Kamali, A.; Kotze, P.; Louw, C.; Mabude, Z.; et al. Safety and Efficacy of a Dapivirine Vaginal Ring for HIV Prevention in Women. *N. Engl. J. Med.* **2016**, *375*, 2133–2143. [CrossRef] [PubMed]
15. Mori, T.; O'Keefe, B.R.; Sowder, R.C., 2nd; Bringans, S.; Gardella, R.; Berg, S.; Cochran, P.; Turpin, J.A.; Buckheit, R.W., Jr.; McMahon, J.B.; et al. Isolation and characterization of griffithsin, a novel HIV-inactivating protein, from the red alga Griffithsia sp. *J. Biol. Chem.* **2005**, *280*, 9345–9353. [CrossRef]
16. Kagiampakis, I.; Gharibi, A.; Mankowski, M.K.; Snyder, B.A.; Ptak, R.G.; Alatas, K.; LiWang, P.J. Potent strategy to inhibit HIV-1 by binding both gp120 and gp41. *Antimicrob. Agents Chemother.* **2011**, *55*, 264–275. [CrossRef]
17. Alexandre, K.B.; Gray, E.S.; Pantophlet, R.; Moore, P.L.; McMahon, J.B.; Chakauya, E.; O'Keefe, B.R.; Chikwamba, R.; Morris, L. Binding of the mannose-specific lectin, griffithsin, to HIV-1 gp120 exposes the CD4-binding site. *J. Virol.* **2011**, *85*, 9039–9050. [CrossRef] [PubMed]

18. Emau, P.; Tian, B.; O'Keefe, B.R.; Mori, T.; McMahon, J.B.; Palmer, K.E.; Jiang, Y.; Bekele, G.; Tsai, C.C. Griffithsin, a potent HIV entry inhibitor, is an excellent candidate for anti-HIV microbicide. *J. Med. Primatol.* **2007**, *36*, 244–253. [CrossRef]
19. Barton, C.L. Evaluation of the Safety and Pharmacokinetic Profile of the Broad Spectrum Antiviral Lectin Griffithsin. Ph.D. Thesis, University of Louisville, Louisville, KY, USA, 2014.
20. O'Keefe, B.R.; Vojdani, F.; Buffa, V.; Shattock, R.J.; Montefiori, D.C.; Bakke, J.; Mirsalis, J.; d'Andrea, A.L.; Hume, S.D.; Bratcher, B.; et al. Scaleable manufacture of HIV-1 entry inhibitor griffithsin and validation of its safety and efficacy as a topical microbicide component. *Proc. Natl. Acad. Sci. USA* **2009**, *106*, 6099–6104. [CrossRef]
21. Girard, L.; Birse, K.; Holm, J.B.; Gajer, P.; Humphrys, M.S.; Garber, D.; Guenthner, P.; Noel-Romas, L.; Abou, M.; McCorrister, S.; et al. Impact of the griffithsin anti-HIV microbicide and placebo gels on the rectal mucosal proteome and microbiome in non-human primates. *Sci. Rep.* **2018**, *8*, 8059. [CrossRef]
22. Ferir, G.; Palmer, K.E.; Schols, D. Synergistic activity profile of griffithsin in combination with tenofovir, maraviroc and enfuvirtide against HIV-1 clade C. *Virology* **2011**, *417*, 253–258. [CrossRef]
23. Kiser, P.F.; Mesquita, P.M.; Herold, B.C. A perspective on progress and gaps in HIV prevention science. *AIDS Res. Hum. Retroviruses* **2012**, *28*, 1373–1378. [CrossRef] [PubMed]
24. Gupta, S.K.; Nutan. Clinical use of vaginal or rectally applied microbicides in patients suffering from HIV/AIDS. *HIV AIDS (Auckl)* **2013**, *5*, 295–307. [CrossRef]
25. Thurman, A.R.; Clark, M.R.; Hurlburt, J.A.; Doncel, G.F. Intravaginal rings as delivery systems for microbicides and multipurpose prevention technologies. *Int. J. Womens Health* **2013**, *5*, 695–708. [CrossRef]
26. das Neves, J.; Nunes, R.; Rodrigues, F.; Sarmento, B. Nanomedicine in the development of anti-HIV microbicides. *Adv. Drug Deliv. Rev.* **2016**, *103*, 57–75. [CrossRef]
27. Cunha-Reis, C.; Machado, A.; Barreiros, L.; Araujo, F.; Nunes, R.; Seabra, V.; Ferreira, D.; Segundo, M.A.; Sarmento, B.; das Neves, J. Nanoparticles-in-film for the combined vaginal delivery of anti-HIV microbicide drugs. *J. Control. Release* **2016**, *243*, 43–53. [CrossRef]
28. Malik, T.; Chauhan, G.; Rath, G.; Kesarkar, R.N.; Chowdhary, A.S.; Goyal, A.K. Efaverinz and nano-gold-loaded mannosylated niosomes: a host cell-targeted topical HIV-1 prophylaxis via thermogel system. *Artif. Cells Nanomed. Biotechnol.* **2017**. [CrossRef]
29. das Neves, J.; Araujo, F.; Andrade, F.; Amiji, M.; Bahia, M.F.; Sarmento, B. Biodistribution and pharmacokinetics of dapivirine-loaded nanoparticles after vaginal delivery in mice. *Pharm. Res.* **2014**, *31*, 1834–1845. [CrossRef]
30. das Neves, J.; Araujo, F.; Andrade, F.; Michiels, J.; Arien, K.K.; Vanham, G.; Amiji, M.; Bahia, M.F.; Sarmento, B. In vitro and ex vivo evaluation of polymeric nanoparticles for vaginal and rectal delivery of the anti-HIV drug dapivirine. *Mol. Pharm.* **2013**, *10*, 2793–2807. [CrossRef]
31. das Neves, J.; Michiels, J.; Arien, K.K.; Vanham, G.; Amiji, M.; Bahia, M.F.; Sarmento, B. Polymeric nanoparticles affect the intracellular delivery, antiretroviral activity and cytotoxicity of the microbicide drug candidate dapivirine. *Pharm. Res.* **2012**, *29*, 1468–1484. [CrossRef]
32. Kovarova, M.; Council, O.D.; Date, A.A.; Long, J.M.; Nochi, T.; Belshan, M.; Shibata, A.; Vincent, H.; Baker, C.E.; Thayer, W.O.; et al. Correction: Nanoformulations of Rilpivirine for Topical Pericoital and Systemic Coitus-Independent Administration Efficiently Prevent HIV Transmission. *PLoS Pathog.* **2015**, *11*, e1005170. [CrossRef]
33. das Neves, J.; Sarmento, B. Precise engineering of dapivirine-loaded nanoparticles for the development of anti-HIV vaginal microbicides. *Acta Biomater.* **2015**, *18*, 77–87. [CrossRef]
34. Bala, I.; Hariharan, S.; Kumar, M.N. PLGA nanoparticles in drug delivery: The state of the art. *Crit. Rev. Ther. Drug Carrier Syst.* **2004**, *21*, 387–422. [CrossRef]
35. Makadia, H.K.; Siegel, S.J. Poly Lactic-co-Glycolic Acid (PLGA) as Biodegradable Controlled Drug Delivery Carrier. *Polymers* **2011**, *3*, 1377–1397. [CrossRef] [PubMed]
36. Bouissou, C.; Rouse, J.J.; Price, R.; van der Walle, C.F. The influence of surfactant on PLGA microsphere glass transition and water sorption: remodeling the surface morphology to attenuate the burst release. *Pharm. Res.* **2006**, *23*, 1295–1305. [CrossRef] [PubMed]
37. Jain, R.A. The manufacturing techniques of various drug loaded biodegradable poly(lactide-co-glycolide) (PLGA) devices. *Biomaterials* **2000**, *21*, 2475–2490. [CrossRef]

38. Ham, A.S.; Cost, M.R.; Sassi, A.B.; Dezzutti, C.S.; Rohan, L.C. Targeted delivery of PSC-RANTES for HIV-1 prevention using biodegradable nanoparticles. *Pharm. Res.* **2009**, *26*, 502–511. [CrossRef]
39. Menei, P.; Daniel, V.; Montero-Menei, C.; Brouillard, M.; Pouplard-Barthelaix, A.; Benoit, J.P. Biodegradation and brain tissue reaction to poly(D,L-lactide-co-glycolide) microspheres. *Biomaterials* **1993**, *14*, 470–478. [CrossRef]
40. Shive, M.S.; Anderson, J.M. Biodegradation and biocompatibility of PLA and PLGA microspheres. *Adv. Drug Deliv. Rev.* **1997**, *28*, 5–24.
41. McRae, A.; Ling, E.A.; Hjorth, S.; Dahlstrom, A.; Mason, D.; Tice, T. Catecholamine-containing biodegradable microsphere implants as a novel approach in the treatment of CNS neurodegenerative disease. A review of experimental studies in DA-lesioned rats. *Mol. Neurobiol.* **1994**, *9*, 191–205. [CrossRef]
42. Engman, C.; Wen, Y.; Meng, W.S.; Bottino, R.; Trucco, M.; Giannoukakis, N. Generation of antigen-specific Foxp3+ regulatory T-cells in vivo following administration of diabetes-reversing tolerogenic microspheres does not require provision of antigen in the formulation. *Clin. Immunol.* **2015**, *160*, 103–123. [CrossRef]
43. Chaowanachan, T.; Krogstad, E.; Ball, C.; Woodrow, K.A. Drug synergy of tenofovir and nanoparticle-based antiretrovirals for HIV prophylaxis. *PLoS ONE* **2013**, *8*, e61416. [CrossRef]
44. Mallipeddi, R.; Rohan, L.C. Progress in antiretroviral drug delivery using nanotechnology. *Int. J. Nanomed.* **2010**, *5*, 533–547.
45. das Neves, J.; Amiji, M.M.; Bahia, M.F.; Sarmento, B. Nanotechnology-based systems for the treatment and prevention of HIV/AIDS. *Adv. Drug Deliv. Rev.* **2010**, *62*, 458–477. [CrossRef]
46. Destache, C.J.; Belgum, T.; Christensen, K.; Shibata, A.; Sharma, A.; Dash, A. Combination antiretroviral drugs in PLGA nanoparticle for HIV-1. *BMC Infect. Dis.* **2009**, *9*, 198. [CrossRef]
47. Papadimitriou, S.; Bikiaris, D. Novel self-assembled core-shell nanoparticles based on crystalline amorphous moieties of aliphatic copolyesters for efficient controlled drug release. *J. Control. Release* **2009**, *138*, 177–184. [CrossRef]
48. Owen, D.H.; Katz, D.F. A vaginal fluid simulant. *Contraception* **1999**, *59*, 91–95. [CrossRef]
49. Dezzutti, C.S.; Brown, E.R.; Moncla, B.; Russo, J.; Cost, M.; Wang, L.; Uranker, K.; Kunjara Na Ayudhya, R.P.; Pryke, K.; Pickett, J.; et al. Is wetter better? An evaluation of over-the-counter personal lubricants for safety and anti-HIV-1 activity. *PLoS ONE* **2012**, *7*, e48328. [CrossRef]
50. Montefiori, D.C. Measuring HIV neutralization in a luciferase reporter gene assay. *Methods Mol. Biol.* **2009**, *485*, 395–405. [CrossRef]
51. Dezzutti, C.S.; Shetler, C.; Mahalingam, A.; Ugaonkar, S.R.; Gwozdz, G.; Buckheit, K.W.; Buckheit, R.W., Jr. Safety and efficacy of tenofovir/IQP-0528 combination gels - a dual compartment microbicide for HIV-1 prevention. *Antiviral Res.* **2012**, *96*, 221–225. [CrossRef]
52. Kohli, A.; Islam, A.; Moyes, D.L.; Murciano, C.; Shen, C.; Challacombe, S.J.; Naglik, J.R. Oral and vaginal epithelial cell lines bind and transfer cell-free infectious HIV-1 to permissive cells but are not productively infected. *PLoS ONE* **2014**, *9*, e98077. [CrossRef] [PubMed]
53. Montefiori, D.C. Evaluating neutralizing antibodies against HIV, SIV, and SHIV in luciferase reporter gene assays. *Curr. Protoc. Immunol.* **2005**. [CrossRef]
54. Dezzutti, C.S.; Rohan, L.C.; Wang, L.; Uranker, K.; Shetler, C.; Cost, M.; Lynam, J.D.; Friend, D. Reformulated tenofovir gel for use as a dual compartment microbicide. *J. Antimicrob. Chemother.* **2012**, *67*, 2139–2142. [CrossRef]
55. Chou, T.C. Theoretical basis, experimental design, and computerized simulation of synergism and antagonism in drug combination studies. *Pharmacol. Rev.* **2006**, *58*, 621–681. [CrossRef]
56. Zhang, Y.; Huo, M.; Zhou, J.; Zou, A.; Li, W.; Yao, C.; Xie, S. DDSolver: An add-in program for modeling and comparison of drug dissolution profiles. *AAPS J.* **2010**, *12*, 263–271. [CrossRef]
57. Bilati, U.; Allemann, E.; Doelker, E. Poly(D,L-lactide-co-glycolide) protein-loaded nanoparticles prepared by the double emulsion method–processing and formulation issues for enhanced entrapment efficiency. *J. Microencapsul.* **2005**, *22*, 205–214. [CrossRef]
58. Xu, B.; Dou, H.; Tao, K.; Sun, K.; Lu, R.; Shi, W. Influence of experimental parameters and the copolymer structure on the size control of nanospheres in double emulsion method. *J. Polym. Res.* **2011**, *18*, 131–137. [CrossRef]

59. Liu, J.; Qiu, Z.; Wang, S.; Zhou, L.; Zhang, S. A modified double-emulsion method for the preparation of daunorubicin-loaded polymeric nanoparticle with enhanced in vitro anti-tumor activity. *Biomed. Mater.* **2010**, *5*, 065002. [CrossRef] [PubMed]
60. Marques, M.R.C.; Loebenberg, R.; Almukainzi, M. Simulated Biological Fluids with Possible Application in Dissolution Testing. *Dissolution Technol.* **2011**, *18*, 15–28. [CrossRef]
61. Destache, C.J.; Belgum, T.; Goede, M.; Shibata, A.; Belshan, M.A. Antiretroviral release from poly(DL-lactide-co-glycolide) nanoparticles in mice. *J. Antimicrob. Chemother.* **2010**, *65*, 2183–2187. [CrossRef] [PubMed]
62. Friend, D.R.; Kiser, P.F. Assessment of topical microbicides to prevent HIV-1 transmission: concepts, testing, lessons learned. *Antiviral. Res.* **2013**, *99*, 391–400. [CrossRef] [PubMed]
63. Abdool Karim, Q.; Abdool Karim, S.S.; Frohlich, J.A.; Grobler, A.C.; Baxter, C.; Mansoor, L.E.; Kharsany, A.B.; Sibeko, S.; Mlisana, K.P.; Omar, Z.; et al. Effectiveness and safety of tenofovir gel, an antiretroviral microbicide, for the prevention of HIV infection in women. *Science* **2010**, *329*, 1168–1174. [CrossRef] [PubMed]
64. Dolgin, E. Long-acting HIV drugs advanced to overcome adherence challenge. *Nat. Med.* **2014**, *20*, 323–324. [CrossRef]
65. Andrews, C.D.; Spreen, W.R.; Mohri, H.; Moss, L.; Ford, S.; Gettie, A.; Russell-Lodrigue, K.; Bohm, R.P.; Cheng-Mayer, C.; Hong, Z.; et al. Long-acting integrase inhibitor protects macaques from intrarectal simian/human immunodeficiency virus. *Science* **2014**, *343*, 1151–1154. [CrossRef] [PubMed]
66. Kramzer, L. Preformulation and Biological Evaluations for the Intravaginal Delivery of Griffithsin for HIV Prevention. Ph.D. Thesis, University of Pittsburgh, Pittsburgh, PA, USA, 2015.
67. Zolnik, B.S.; Burgess, D.J. Effect of acidic pH on PLGA microsphere degradation and release. *J. Control. Release* **2007**, *122*, 338–344. [CrossRef] [PubMed]
68. Ding, A.G.; Schwendeman, S.P. Acidic microclimate pH distribution in PLGA microspheres monitored by confocal laser scanning microscopy. *Pharm. Res.* **2008**, *25*, 2041–2052. [CrossRef] [PubMed]
69. Smith, P. Estrogens and the urogenital tract. Studies on steroid hormone receptors and a clinical study on a new estradiol-releasing vaginal ring. *Acta Obstet. Gynecol. Scand.* **1993**, *72*, 5–26. [CrossRef]
70. Boskey, E.R.; Cone, R.A.; Whaley, K.J.; Moench, T.R. Origins of vaginal acidity: high D/L lactate ratio is consistent with bacteria being the primary source. *Hum. Reprod.* **2001**, *16*, 1809–1813. [CrossRef]
71. Fredenberg, S.; Wahlgren, M.; Reslow, M.; Axelsson, A. The mechanisms of drug release in poly(lactic-co-glycolic acid)-based drug delivery systems—A review. *Int. J. Pharm.* **2011**, *415*, 34–52. [CrossRef]
72. Secchi, M.; Xu, Q.; Lusso, P.; Vangelista, L. The superior folding of a RANTES analogue expressed in lactobacilli as compared to mammalian cells reveals a promising system to screen new RANTES mutants. *Protein Expr. Purif.* **2009**, *68*, 34–41. [CrossRef]
73. Chou, T.C. Drug combination studies and their synergy quantification using the Chou-Talalay method. *Cancer Res.* **2010**, *70*, 440–446. [CrossRef]

pharmaceutics

MDPI

Review

Reverse Transcriptase Inhibitors Nanosystems Designed for Drug Stability and Controlled Delivery

Fedora Grande *, Giuseppina Ioele *, Maria Antonietta Occhiuzzi, Michele De Luca, Elisabetta Mazzotta, Gaetano Ragno, Antonio Garofalo and Rita Muzzalupo

Department of Pharmacy, Health and Nutritional Sciences, University of Calabria, Via P. Bucci, 87036 Rende (CS), Italy; mariaantonietta.occhiuzzi@unical.it (M.A.O.); michele.deluca@unical.it (M.D.L.); mazzotta-elisabetta@libero.it (E.M.); gaetano.ragno@unical.it (G.R.); antonio.garofalo@unical.it (A.G.); rita.muzzalupo@unical.it (R.M.)

* Correspondence: fedora.grande@unical.it (F.G.); giuseppina.ioele@unical.it (G.I.); Tel.: +39-098-449-3019 (F.G.); +39-098-449-3268 (G.I.)

Received: 20 March 2019; Accepted: 22 April 2019; Published: 27 April 2019

Abstract: An in-depth analysis of nanotechnology applications for the improvement of solubility, distribution, bioavailability and stability of reverse transcriptase inhibitors is reported. Current clinically used nucleoside and non-nucleoside agents, included in combination therapies, were examined in the present survey, as drugs belonging to these classes are the major component of highly active antiretroviral treatments. The inclusion of such agents into supramolecular vesicular systems, such as liposomes, niosomes and lipid solid NPs, overcomes several drawbacks related to the action of these drugs, including drug instability and unfavorable pharmacokinetics. Overall results reported in the literature show that the performances of these drugs could be significantly improved by inclusion into nanosystems.

Keywords: HIV; antiretrovirals; nanoformulations; drug degradation; drug protection

1. Introduction

The human immunodeficiency virus (HIV), belonging to the lentivirus genus of the large family of retroviridae, is the etiological agent of AIDS. The infection causes severe consequences to the immune system including a loss of CD4+T lymphocytes that leads to an increased susceptibility to even fatal opportunistic infections. The identification of various antiretroviral drugs allowed defining efficacious therapeutic regimens for the prevention and treatment of the disease by the combined administration of two, three or more different drugs acting on crucial steps of viral replication. In particular, the targets of conventional drugs are proteins involved in the viral entry or specific enzymes necessary for the virus replication such as protease (PR), reverse transcriptase (RT) and integrase (IN). This approach known as HAART (highly active antiretroviral therapy) nowadays represents the most useful therapeutic treatment, even though it is affected by many drawbacks such as lifetime administration with a consequent reduced patients' compliance, severe side effects, and quick viral outbreak after drug resistance emergence. HAART was demonstrated to be particularly effective in cutting down the overall number of viral particles, but is unable to completely eradicate infection in sanctuary sites, such as the brain, liver and lymphatic system [1–5]. Moreover, HAART always includes one or more nucleoside and non-nucleoside reverse transcriptase inhibitors (NRTI and NNRTI, respectively), which, despite a high antiviral efficacy, unavoidably show important clinical drawbacks. Relevant information on common RTI is summarized in Table 1 [6,7].

Table 1. Relevant information of currently used RTI. *Year of FDA approval; LS = Lipid Solubility; OB = Oral Bioavailability; t/2 = Plasma Half-life.

Drug Class	Name (Acronym)	Year*	LS	OB (%)	t/2 (hours)	Side Effects
NRTIs	Stavudine (STV)	1996	low	86	1.3–1.4	Peripheral neuropathy, pancreatitis, asymptomatic acidosis, lipoatrophy, hepatic steatosis
	Zidovudine (AZT)	1986	low	60	0.5–3	Neutropenia, anemia, nausea, vomiting, asthenia, headache, insomnia, skin hyperpigmentation, acidosis, hepatic steatosis
	Lamivudine (3TC)	1995	low	86	5–7	Cough, diarrhea, fatigue, headache, malaise, nasal symptoms, lactic acidosis, hepatic steatosis
	Abacavir (ABV)	1998	low	83	0.8–1.5	Systemic respiratory hypersensitivity, gastrointestinal symptoms, fever, tiredness, sore throat
	Emtricitabine (FTC)	2006	low	93	8–10	Headache, nausea, upset stomach, diarrhea, trouble sleeping, dizziness, skin rash, strange dreams, cough, runny nose
	Zalcitabine (ZCT)	1992	low	85		Peripheral neuropathy, stomatitis, esophageal ulcerations, acidosis, hepatic steatosis
	Didanosine (DDN)	1991	low	30	2	Gastrointestinal intolerance, peripheral neuropathy, pancreatitis, asymptomatic acidosis, lipoatrophy, hepatic steatosis
	Tenofovir (TDF)	2001	low	25-39	12–15	Nausea, depression, confusion, headache, hitching, weakness, kidneys problems
NNRTIs	Nevirapine (NVR)	1996	moderate	92	25–30	Rash, Stevens-Johnson syndrome, elevated transaminases blood level, hepatitis, severe hypersensitivity reaction
	Efavirenz (EFV)	1998	high	50	40–55	Rash, Stevens-Johnson syndrome, sleep disturbances, dizziness, vertigo, depression, euphoria, difficulty concentrating, hallucination.
	Etravirine (ETV)	2008	high	–	30–40	Rash, Stevens-Johnson syndrome, toxic epidermal necrosis and multiform erythema, hypersensitivity reactions, hepatic failure
	Rilpivirine (RPV)	2011	high	50	19	Rash, depression, liver problems, mood changes

The above therapeutic strategy is not even capable of stimulating a lasting immune response of memory cells necessary to antagonize the infective agent [8]. Even after taking into account all these considerations, innovative therapeutic approaches based on both the identification of alternative drugs and innovative pharmaceutic formulations still have demanding requirements [9–11]. Nanotechnologies could help to reach this latter crucial point in order to increase cellular uptake, enhance drug distribution, prolong half-life and reduce side effects depending on the lower drug dosage in the nanosystem. In particular, application of nanoformulations consisting of a given drug and a supramolecular matrix such as niosomes, liposomes and solid lipid nanoparticles (SLN), already led to some improvements of pharmacokinetic and pharmacodynamic parameters. Very interesting results have been recorded in the anti-cancer research field where an altered microenvironment of cancer cells facilitates a selective drug delivery [12]. A similar approach could be adopted in the case of cells infected by HIV [13].

In the light of these findings, the incorporation of new or customary anti-HIV drugs into supramolecular carriers could be particularly effective in suppressing viral replication. This strategy is corroborated by the possibility of encapsulating drugs or genes to not only be delivered next to the infected cells but also to target reservoir tissues to eradicate latent HIV [14]. This innovative strategy is suitable for improving the distribution of both hydrophilic and hydrophobic small molecules, as well as macromolecular drugs, which can be driven toward specific tissues thanks to the reduced size of the nanoscale delivery systems. Antiretroviral drugs can be carried as nanoparticles for their potential to better reach macrophages, CD4+T cells and latent reservoirs organs, such as brain and lymph nodes, that are particularly responsible of viral survival. [15–19]. Drug delivery systems (DDS) for RTI, developed in the last few years, are described in this survey and depicted in Figure 1, while their main properties are summarized in Table 2.

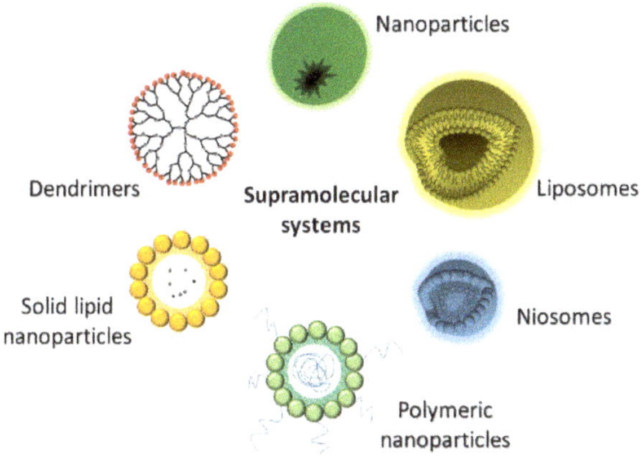

Figure 1. Drug delivery systems for RTI nanoformulations.

Table 2. Targets of DDS designed for anti-HIV therapy.

DDS		TARGET
Matrix	**Surface**	
Liposome NPs	Mannose	Liver, spleen, lung, brain, macrophages
Liposomes	Galactose	Liver
Chitosan NPs	Glycyrrhizin	Liver
NPs	Transferrin	Brain, endothelial cells
NPs	Serum albumin	Brain, liver, spleen
SLN	Phenylalanine	Blood brain barrier
Polymeric micelles	Anti-GP2 antibody	M-cell of gut-associated lymphoid tissue
Dendrimers	Tuftsin	Macrophages, monocytes, polymorph nuclear leukocytes

Once the supramolecular system reaches the site of action, a controlled drug release provides high local concentration and longer residence time, resulting in an improved antiviral effect (Figure 2). Furthermore, some nanomaterials possess themselves favorable biological effects. The nanotechnology approach has been advanced for many aspects dealing with HIV infection, namely theranostic, vaccine prophylaxis and gene therapy. This survey however focuses on studies describing NRTI and NNRTI based nanoformulations for prevention or treatment of HIV infection. Although these systems allow a remarkable improvement of the pharmacokinetics and pharmacodynamics of RTI, several drawbacks still need to be overcame. The major advantages and limitations of known nanosystems are listed in Table 3.

Table 3. Advantages and limitations of anti-HIV DDS.

DDS	Advantages	Limitations
Liposomes	Co-delivery of hydrophilic and lipophilic drug Selective uptake by mononuclear phagocytes Surface modification with target moiety of virus reservoir	Low drug loading capacity Physical and chemical instability Drug leakage Difficulty in sterilization Short half-life Poor scale up
Niosomes	Chemical stability Protection of drug from degradation Large uptake by mononuclear phagocytes and localization in virus reservoir organs Less expensive respect liposomes Functionalization with target ligand	Physical instability during the storage Difficulty in sterilization Difficulty in large-scale production
Polymeric NPs	High drug loading capacity Co-delivery of different drug for anti-HIV combination therapy Selective uptake by lymphoid organ Prolonged circulation time Surface functionalization with target moiety	Fast burst release Limited safe correlated to polymer toxicity High cost production
SLN	Higher stability and biological compatibility than liposomes and polymeric NPs Increase the bioavailability of poorly water soluble drug Avoidance of organic solvent Slow uptake by the RES Feasible-large scale production and sterilization Less expensive than polymeric and surfactant carriers	Low drug solubility in lipid matrix and loading capacity Drug leakage Particle growth Unpredictable gelation tendency
Dendrimers	Uniform particle size Large surface functional groups for the conjugation with target moieties	Toxicity problems

Several of these clinically used drugs show an undesirable stability profile when exposed to stressing conditions either in solution or in solid form [20,21]. Similarly, stress tests have been performed on supramolecular systems in order to confirm any improved stability, under different conditions [22]. In particular, studies published in the last few years on the RT inhibitors included in combined therapy have been taken into consideration [15,16,23].

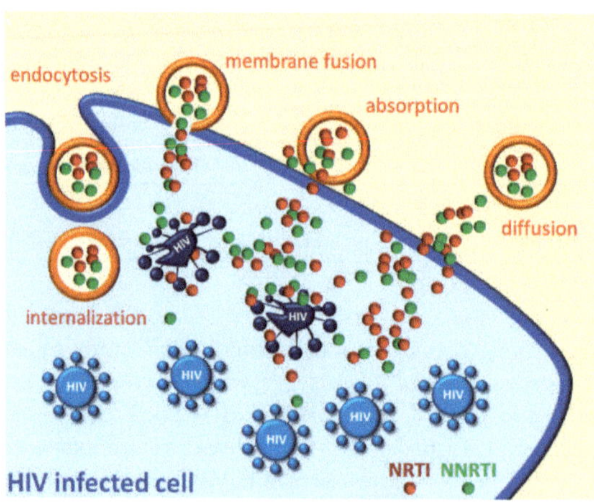

Figure 2. Representative delivery modalities for NRTI/NNRTI nanosystems to HIV reservoirs.

2. Drug Protection Nanosystems

Most of the protocols adopted for studying the drug stability are suggested by the ICH (International Conference on Harmonization) guidelines [24]. According to such rules, clearly defined drug storage conditions are required to prevent the degradation effects related to pH, temperature, light, air and humidity for either solution/suspension or commercial formulations/packaging [25]. The protection of sensitive drugs could often be assured by shielding with an adequate packaging [26]. When this simple precaution showed not to be sufficient, the stability of the drug, exposed to different environmental conditions, could be improved by suitable delivery devices, such as vesicular matrices (i.e., liposomes and niosomes), nanoparticles (NPs), and solid lipid nanoparticles (SLN).

Niosomal and liposomal vesicles consist of amphiphilic molecules and an aqueous compartment and differ for their structural chemical units. Both systems are very versatile. The hydrophilic drugs can be entrapped in their aqueous core while the lipophilic drugs can be partitioned into the bilayer domains. Liposomes are made of natural phospholipids, resulting in a greater stability, a low production cost and reduced toxicity [27–32], while niosomes are prepared by means of synthetic, non-ionic surfactants, as alkyl ethers, alkyl esters and pluronics copolymers, or fatty acid and amino acid compounds [31–33]. The preparation of these vesicles requires a simple procedure based on a gentle agitation or sonication of an aqueous solution of phospholipid/surfactant and drug mixtures taken from an ultracentrifugation or low-pressure gel filtration chromatography to purify the formed systems [34].

More recent SLN have been proposed as alternative formulations for both hydrophilic and hydrophobic drugs. These systems are colloidal carriers based on a solid phase lipid and a surfactant and are characterized by a spherical shape in which the lipid portion is always solid and the surfactant acts as a stabilizing factor [22,35–40]. Fatty acids, monoglycerides, diglycerides, triglycerides, waxes and steroids can be applied in the preparation of SLN in the absence of organic solvent [35]. Low cost, good physical stability, large scale production, no toxicity and high biodegradability represent the greatest advantages in the use of these formulations with respect to the liposomes matrices [41].

Nanoparticles systems are known promising carriers for the improvement of solubility and pharmacokinetics of drugs as well as vaccines, nucleic acids and therapeutic proteins. These delivery devices can influence therapeutic efficiency of a drug, enhance its protection from degradation and reduce dose-limiting side effects. A variety of hydrophobic or hydrophilic active molecules can be dissolved, encapsulated, absorbed or conjugated to polymeric nanoparticles following different techniques [42]. Several natural and biodegradable materials like chitosan have been proposed for the realization of anti-HIV drug nanosystems. An alternative approach, based on the formation of crystalline complex with a fixed range size, was attempted by inclusion of the pure drug into a hydrophobic synthetic polymer [43]. Polymeric nanoparticles based on poly(lactic acid) (PLA) or poly(lactide-*co*-glycolide) (PLGA) are reported as ideal delivery systems, showing an improved therapeutic efficacy with lower incidence of side effects [42,44–49].

A higher local concentration of active molecules is often reached by integration of classic antiretroviral drugs in different NPs of metals [50]. According to a relative inertness and low toxicity, silver or gold NPs have been explored in biomedicine as multifunctional scaffolds. In particular, the application of gold NPs has been employed to conjugate biomolecules on the outer surface. Alternative inorganic multifunctional materials, such as silver NPs coated with poly(vinyl)pyrrolidone, have also been exploited as drug carriers. [51].

Synthetic well-defined nanopolymers with a three-dimensional architecture, known as dendrimers, have been proposed for the vehiculation of several drugs. Generally, a dendrimer is a symmetric and hyper-branched macromolecule characterized by the presence of reactive groups in the central core, repeated branching units in the interior layers of the core and functional groups spanning from the outer surface. The drug molecule can be either entrapped inside the structure or linked to the external functional groups. This approach was proposed for the carrying of several anti HIV agents [52–55].

Some of these matrices are also able to improve the stability of drugs. Several studies described the use of niosomes, metal based and polymeric NPs to prevent the degradation effects caused by stressing conditions [33,50,56].

3. Nucleoside Reverse Transcriptase Inhibitors Nanosystems

The therapy based on the administration of nucleoside antiviral derivatives such as stavudine, zidovudine, lamivudine and emtricitabine represented a first and effective approach adopted for the management of HIV infection (Figure 3).

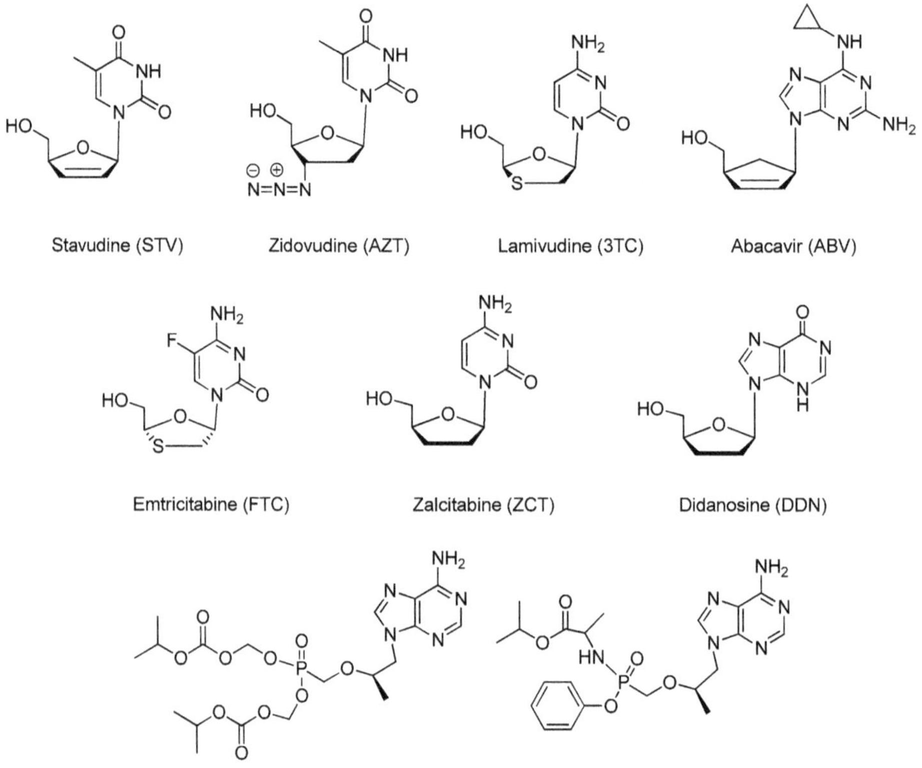

Figure 3. Chemical structures of NRTIs.

Due to the structural similarity to purine or pyrimidine nucleosides, the mode of action of these drugs consists of the competition for the incorporation into viral DNA, catalyzed by RT, so causing chain termination (Figure 4, panel b). Nowadays the most common nucleoside derivatives used in therapy are lamivudine, abacavir, emtricitabine, tenofovir and tenofovir alafenamide [57]. Stavudine and zidovudine, not yet recommended in first-line therapy, were early examples of NRTIs included in nanoformulations designed for limiting their severe side effects and improving pharmacokinetics.

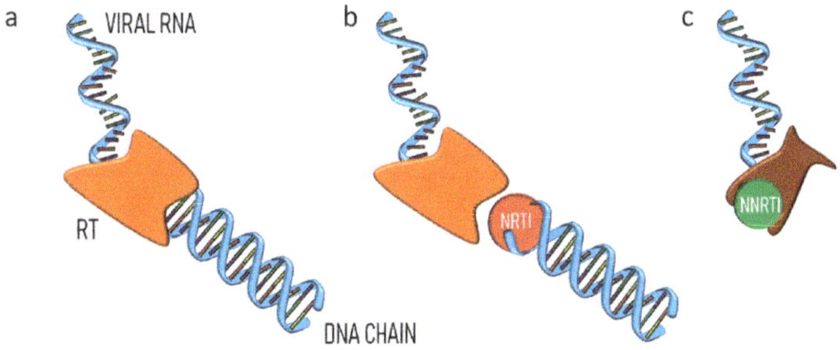

Figure 4. (**a**) RT catalyzes the conversion of viral RNA into pro-viral DNA before its incorporation into the target cell genome; (**b**) NRTIs are incorporated into the DNA causing chain termination; (**c**) NNRTIs bind the enzyme inhibiting its function.

3.1. *Stavudine*

Stavudine (1-((2R,5S)-5-(hydroxymethyl)-2,5-dihydrofuran-2-yl)-5-methylpyrimidine-2,4(1H, 3H)-dione, STV) is one of the most commonly used forms of NRT approved by FDA in 1996 and its use was recommended in association with other antiretroviral agents. The drug showed major side effects, such as high blood lactate, pancreatitis and hepatomegaly. STV was characterized by a serum half-life of 1h only, while that of its phosphorylated active metabolite was calculated as being 3.5 h [27]. Thus, STV loaded formulations able to concomitantly increase cellular uptake and sustain release should reduce unwanted effects. Accordingly, galactosyl or mannosyl coated liposomes loaded with STV were described. These formulations reached the desired results showing an increased *in vitro* anti-HIV activity together with a remarkable decrease of side effects. The efficacy was tested in a mononuclear phagocyte system, a major reservoir of HIV, proving advantages in terms of bio-stability, site-specific and ligand-mediated delivery, compared to free drug and uncoated liposomes [27,28]. More recently, STV-containing nanoformulations were proposed for the dual utilization to control the residual viremia as well as to target the reservoir sites. To achieve this aim, gelatin nanoformulations containing very low dosage of the drug were prepared through a simple desolvation process and loaded into soya lecithin based liposomes [29]. A study on STV degradation under different stress conditions (hydrolysis, oxidation, photolysis and thermal stress) was initially reported. A stability-indicating reversed-phase HPLC assay method showed the hydrolysis of the drug to thymine in acidic, neutral, alkaline and under oxidative stress conditions [20]. In order to improve the stability of this drug, STV-loaded SLN for intravenous injection were produced by high-pressure homogenization of drug lipid melt dispersed in hot surfactant solution [22]. This SLN formulation was also studied for its active delivery to lymphatic tissues by ex vivo cellular uptake evaluation in macrophages. Reported experiments confirmed an improved cellular uptake together with a prolonged activity next to the delivery site of the formulation compared to the simple drug solution. This could account for an efficient and safe therapeutic profile of the drug-carrier system [58].

3.2. *Zidovudine*

Zidovudine, also known as azidothymidine (1-((2R,4S,5S)-4-azido-5-(hydroxymethyl)tetra hydrofuran-2-yl)-5-methylpyrimidine-2,4(1H,3H)-dione, AZT), the first antiretroviral medication proposed to prevent and treat HIV/AIDS, has been approved in 1986. An extensive first pass metabolism often requires an in vein administration. This feature and a long list of severe side effects limit the use of this drug, which is however still present in many therapeutic anti-HIV regimens. Its incorporation into supramolecular matrices was extensively exploited in order to increase bioavailability and to reduce dose-dependent unwanted effects.

Positively and negatively charged liposomes based on stearylamine and diacetyl phosphate were used as AZT carriers. In order to enhance localization to lymph nodes and spleen, these systems were even coated with a site-specific mannose-terminated stearylamine ligand. Fluorescent microscopy images showed an enhanced uptake and localization of these liposomes in the target tissues [59]. In an early paper, a dispersed system comprising polyoxypropylene, polyoxyethylene, oleic acid, water and cetyl alcohol as surfactant, was described as a potential DDS. The release profile experimental analysis showed that the delivery of AZT could be controlled this way, in accordance with a mathematical theoretical approach [60]. This system has been proposed as a carrier, which potentially could overcome the main drawbacks of conventional pharmaceutical formulations [61].

AZT loaded in polymeric NPs based on PLA and poly(L-lactide)poly(ethyleneglycol) (PLA/PEG) were prepared by double emulsion solvent evaporation and thoroughly investigated *in vitro* for uptake into polymorphonuclear leucocytes of rat peritoneal exudate. The cells activation by NPs was assessed by a chemiluminescence assay suggesting a more favorable behavior of PLA vs. PLA/PEG complexes [62]. On the other hand, the drug release increased proportionally to the PEG amount in the blend [63]. AZT was encapsulated in alginate-glutamic acid amide based NPs obtained by an emulsion solvent evaporation method. The polymeric NPs were coated with pluronic F-68 to favor cellular internalization through the endocytosis mechanism. As a result, the antiviral drug loaded in these nanosystems was released in a prolonged manner. Intracellular uptake and cell viability assays also confirmed an efficient uptake of AZT in glioma cell lines [64]. Solid lipid NPs based on modified stearic acid and *Aloe vera* extract were described as an alternative drug delivery carrier for controlled release and targeting of AZT. The plant extract was used because of its high content of polysaccharides that showed synergistic antiretroviral activity with AZT. The described nanocarriers did not interact with plasma proteins and showed high drug loading and entrapment efficiency. Moreover, fluorescent microscopy images suggested that the natural gel facilitated the cellular uptake of AZT in brain cells [36]. The drug proved to decompose when exposed to light or under hydrolytic conditions, while it was more stable toward oxidation agents and thermal stress [65]. In particular, the acid degradation induced the formation of a pyrimidine derivative endowed with higher toxicity when compared to AZT, as demonstrated by a mutagenicity and an aerobic biodegradability assay [21]. Recently, three novel prodrugs of AZT, obtained by functionalization with dicarboxylic acids, were designed in order to enhance pharmacokinetics, chemical stability and affinity for human serum albumin [66].

3.3. *Lamivudine*

The oral agent lamivudine, (4-amino-1-((2S,5R)-2-(hydroxymethyl)-1,3-oxathiolan-5-yl)pyrimidin-2(1H)-one, 3TC), an analog of nucleoside cytidine, was approved by FDA for the combined therapy with AZT in 1995 and for monotherapy in 2002. However, the emergence of drug resistance, associated to the gene mutation of RT, limited its clinical application [67,68].

Several nanocarriers were prepared by mixing biodegradable networks (i.e. PEG, pluronic-polyethyleneimmine (PEI), glycyrrhizin conjugated chitosan, mannosylated-PLGA) or dendritic networks (i.e. starPEG-PEI, poly(amidoamine)dendrimer-PEI-PEG) or nanogels with AZT or didanosine triphosphates under a freeze-drying method, to specifically deliver the antiviral agent next to macrophages in the CNS.

All nano-NRTIs demonstrated high efficacy in inhibiting HIV at low μM drug concentration. The major cause of NRTI neurotoxicity, consisting in the mitochondrial DNA depletion, was also reduced 3-fold compared to uncoated NRTIs [69]. Acid or alkaline conditions as well as an oxidative environment caused the degradation of the drug into five different products. On the other hand, light exposure or thermal stress did not affect drug stability [70,71].

More recently, a mass spectrometry study evidenced the formation of an additional degradation product when the solid drug was exposed to oxidative conditions [72]. 3TC was incorporated into polymethacrylic acid NPs in different drug/polymer ratio by nanoprecipitation method in order to overcome some drug limitations, such as accumulation during multi dose therapy and poor patient compliance. These polymers offer several advantages, including high stability and simple preparation route compared to remaining colloidal carriers. Moreover, these nanocarriers were shown to increase drug bioavailability and optimize the release time to the target site without significant chemical interactions between the drug and matrix [73]. An alternative encapsulation for 3TC was exploited by using PLGA NPs coated with bovine serum albumin through a double emulsion procedure. It was then demonstrated that PLGA NPs were rapidly internalized into the human liver cells after oral administration, at different drug concentrations, confirming their high potential as ideal 3TC delivery systems [74]. The PLGA/3TC system was also investigated for the formulation of a thermosensitive vaginal gel. The system was obtained in the form of NPs by the formation of an amide bond between the biodegradable polymer and the free amine group of the drug. An analogue formulation was prepared using emtricitabine as an alternative NRTI. The NPs were finally incorporated into a thermosensitive gel for vaginal administration. The nanoformulations showed to be non-toxic in HeLa cells assay up to a 100 μg/mL concentration. Similar preparations containing fluorescent NPs were found to be active for up to 5 days, suggesting a potential long-lasting application in therapy [75]. A similar approach was adopted for the achievement of transdermal formulation of PLGA/3TC complex. NPs obtained resulted in an ideal spherical shape and an external smooth shell. The high drug entrapment rate resulted in an improved physical stability of the drug together with an efficient delivery after skin permeation. This last property was enhanced after microneedles skin pre-treatment [76]. Mannosylated-PLGA NPs were prepared to ensure an efficient delivery of 3TC into brain macrophages. The experimental data confirmed the increased drug release from nanocarriers and this effect may be due to the presence of sugar receptors on the luminal surface of blood-brain barrier (BBB) [77]. 3TC was also entrapped into PLA/chitosan (CS) NPs by an emulsion technique. The drug was efficaciously entrapped and protected at low values of pH, while it was rapidly released at higher pH values, thus allowing the drug to be selectively absorbed in the intestinal tract. These NPs were also proven to be non-toxic in a mouse fibroblasts model. Efficient biomedical applications could be accordingly envisaged for such an inclusion system [78].

Similar results were obtained for 3TC-CS NPs prepared by ionic gelation of CS with tripolyphosphate anions. These formulations offer several advantages with respect to conventional dosage forms of the drug, particularly in terms of bioavailability [79]. The CS functionalization with glycyrrhizin was realized for a liver targeting and a 3TC controlled release. In fact, the results of this research confirmed a lower drug release and an augmented level of 3TC in hepatocyte tissues, if compared with CS NPs or the free-drug solution [80]. Successively, 3TC was loaded into poly(ε-caprolactone) through a double emulsion spray-drying method giving rise to NPs with spherical morphology. This system also proved to be effective in improving drug bioavailability and reducing side effects [81]. Multiple drugs combined in a single nanosystem showed significant advantages over therapy based on a single drug. Accordingly, an example of polymeric NPs based on methyl methacrylate or ε-caprolactone was designed for the release of four different anti HIV drugs, AZT, 3TC, nevirapine and the IN inhibitor raltegravir [82]. In a recent paper, the incorporation of 3TC into CS with sodium alginate/calcium chloride by the above gelation method, in different experimental conditions, was described and showed an impressive drug release rate lasting up to 24 h. The method proved to furnish highly homogenous particles capable of improve bioavailability together with a constant drug release, following a first order mechanism, with diffusion of the drug after swelling of the polymer [83].

Protein-based NPs were prepared using lactoferrin for the controlled release of 3TC combined with AZT and EFV, after its application for a single DDS. The lactoferrin possesses itself antiviral activity and therefore acts synergistically with the entrapped drugs. The assessment of pharmacokinetic profile for each entrapped drug and *in vitro* data suggested that these NPs are able to release drugs intracellularly in a controlled and sustained manner [84,85].

Alternative nanotechnologies used for anti-HIV drugs release included the use of inorganic components such as iron or silica. Preliminary results suggested a potential application for these new formulations. In particular, chemical-physical characterization of SiO2 NPs coated with magnetic

Fe_2O_3 loaded with 3TC were investigated for their pharmacokinetic and cytotoxic profiles showing more favorable features compared to the free drug [86].

A similar preparation was attempted for 3TC and zalcitabine both in form of triphosphosphates. Preliminary results showed that these nanosystems of dideoxynucleoside triphosphates/SiO2 NPs were useful transport systems for delivering these drugs to target cells with increased antiviral efficiency [87].

Moreover, the co-encapsulation of 3TC and AZT, both in form of triphosphates, into iron carboxylate mesoporous NPs gave biocompatible systems endowed with peculiar delivery properties. In particular, the drugs were released with different kinetics: 3TC showed accelerated release, while AZT was released more slowly. This vector protected drug from degradation, conferring at the same time improved *in vitro* anti-HIV activity. In fact, these formulations contribute to stabilizing the drugs since no alterations were detected after two-month storage and freeze-drying reconstitution [50]. High drug bioavailability and patient compliance were recorded after the administration of the of 3TC encapsulated into a new gum odina based biopolymer obtained by a multiple water-in-oil-in-water emulsion approach. The long-term stability study showed the improvement of the stability of the emulsions after a 90-day storage compared to a similar emulsion comprising Tween 80 as a stabilizer [88]. 3TC was also loaded in nanovesicles based on phospholipids or non-ionic surfactants (niosomes and liposomes). The best components and reparation methods able to produce formulations with suitable size, improved drug encapsulation efficiency and release profile were formulated for these systems [89,90].

3.4. Abacavir

Abacavir ((1S,4R)-4-(2-amino-6-(cyclopropylamino)-9H-purin-9-yl)cyclopent-2-en-1-yl)methanol, ABV) introduced in 1998, represented an alternative nucleoside derivative administered orally in solid or solution form for the prevention and treatment of HIV infection. Similarly to other nucleoside analogs, its use is recommended in combination therapy because of its severe side effects like hypersensitivity, liver damage and lactic acidosis, which all preclude monotherapy. ABV and its congener 3TC, after transformation into the corresponding thiol ending ester derivatives, were conjugated to glucose-coated gold NPs, which were investigated for their pH dependent drug release performances. This drug-delivery system was in turn studied for new multifunctional devices since such gold NPs, themselves endowed with microbicide properties, proved useful for the loading of more than one active agent differently targeting the viral replication cycle, and therefore representing a multivalent therapeutic approach [51]. Albumin NPs loaded with ABV sulfate were prepared by solvation method and studied for their mechanism of drug release. Results obtained revealed a remarkable drug loading capacity together with a sustained and controlled release within 24 h in HIV reservoir organs [91]. A myristoylated ABV prodrug entrapped into poloxamers was evaluated for the pharmacokinetic properties after injection in mice. Comparison of such nanoformulation with the free drug was performed on human monocyte-derived macrophages by proton nuclear magnetic resonance studies in terms of anti-HIV activity.

As a result, an efficacy comparable to that of the native drug was detected for the encased polymer, which showed a two-week lasting release [92]. A detailed study described the formation of innovative nanocarriers named ProTide (PROdrug and nucleoTIDE) obtained by the loading of L-alanine and L-phenylalanine ester phosphoramidates of ABV into PLGA and poloxamer NPs. Such formulations showed sustained retention and antiretroviral activities for up to one month [93].

3.5. Emtricitabine

Emtricitabine (4-amino-5-fluoro-1-((2S,5R)-2-(hydroxymethyl)-1,3-oxathiolan-5-yl)pyrimidin-2(1H)-one, FTC), is a deoxycytidine nucleoside analog approved in 2006 for anti-HIV therapy. Even if it showed reduced side effects when compared to other NRTIs, FTC is largely used in triple or quadruple drug combinations.

A customary PLGA nanoformulation of this water soluble drug was achieved through the water-in-oil-in-water emulsion method and showed a sustained release profile in rats, with adequate drug concentration up to two weeks [94,95]. The large volume distribution, beside a short plasma half-life, suggests the use of FTC in alternative formulations, such as PLGA NPs. This particular administration form proved to be able in enhancing drug stability and intracellular retention time, as demonstrated by an ex vivo endosomal assay. A once-biweekly dosing for HIV infection prevention or treatment was accordingly hypothesized [96]. In addition, eight degradation products were separated and characterized by LC–MS/MS from ABV sulfate when subjected to forced degradation under hydrolysis, oxidation, photolysis and thermal stressing conditions [97]. More recently, a solution state study showed the formation of eleven degradation products [98].

The thermal decomposition of FTC was well investigated by applying different methods. FTC largely decomposed to small molecules and insoluble substances. A small amount decomposed to 5-fluorocytosine due to an oxidation reaction [99]. When the drug was exposed to the action of acids or bases as well as oxidative stress conditions, an additional degradation product was detected [100].

3.6. *Tenofovir*

Tenofovir disoproxil fumarate ((R)-(((((1-(6-amino-9H-purin-9-yl)propan-2-yl)oxy)methyl) phosphoryl)bis(oxy))bis(methylene) diisopropyl dicarbonate, TDF) is a more recently approved NRTI (2001) used in the treatment of chronic hepatitis B and in the prevention and treatment of HIV infection. Successively its prodrug tenofovir alafenamide fumarate ((isopropyl 2-((((((R)-1-(6-amino-9H-purin-9-yl)propan-2-yl)oxy)methyl)(phenoxy)phosphoryl)amino)propanoate, TAF) was laun-ched in the market due to its more favorable properties after oral administration. TAF has greater antiviral activity and better distribution into lymphoid tissues than TDF. Both drugs are recommended in combination therapy along with other antiretroviral agents.

3.6.1. Tenofovir SLN

Lipid NPs loaded with NRTI and NNRTI agents including TDF or TAF were extensively studied for the improvement of bioavailability and long lasting drug release. Several lipid matrices were designed and showed a very promising behavior under different experimental conditions [37,101,102]. Toxic effects of TDF loaded in nanoemulsions on liver and kidney were assessed using an animal model. Although any behavioral toxicity and mortality were not detected, moderate alterations were however observed on both organs [103]. Extensive chemical-physical studies were performed on hybrid inclusion complexes obtained by encasement of TDF into lipid and polymer matrices by engineered melt emulsification-probe sonication technique. The carrier obtained by combining TDF, lauric acid and pemulen polymer was shown to promote a noteworthy increase of TDF trans-nasal flux, so potentially useful for nasal administration [104]. Nanocarriers based on a hydrogel-core and a lipid-shell were designed for the controlled loading and topical vaginal release of TDF and maraviroc, a virus entry inhibitor. These nanolipogels proved to be efficient systems and robust carriers for the encapsulation and the prolonged *in vivo* release of antiretroviral drugs, showing solubility concerns that are useful during the prevention and treatment of HIV infection [105].

3.6.2. Tenofovir/Dendrimers Complexes

A drug combination including TDF into dendrimers was designed for the evaluation in an *in vitro* model of semen-enhanced viral infection. The results obtained suggested that this therapeutic strategy could bypass the detrimental effects of amyloid fibrils, present in semen, which seem responsible of the failure of topical vaginal gels action [54]. An approach to the treatment of neuro-AIDS was based on the use of co-encapsulated drugs into ultra-small iron oxide NPs with the addition of dextran sulfate. The inclusion complex of TDF and vorinostat, a latency-breaking agent, was assembled by magnetically guided layer-by-layer method and a noteworthy blood–brain barrier transmigration of drugs was then observed. This strategy, aimed to the activation of latent virus and its simultaneous killing, would result in a high efficacy to eradicate completely the infection from the CNS [106]. Nanosystems such as carbosilane dendrimers seem themselves able to inhibit HIV replication with a potential as local antiviral agents. Nevertheless, the concomitant administration of specific antiretroviral agents led to a potent synergistic activity. TDF, along with AZT and EFV or with maraviroc was encapsulated into anionic carbosilane dendrimers, bringing sulfated and naphthyl sulfonated groups to generate potential microbicides to prevent the sexual transmission of HIV [55,107–110]. An innovative therapeutic strategy could be based on the TDF prolonged release from NPs obtained by hyaluronic acid (HA) cross-linked with adipic acid dihydrazide. This nanosystem did not show detectable toxicity under the control of hyaluronidase enzyme. Comparative experiments with a simple TDF/HA gel suggested an essential role of the enzyme during the HA degradation and TDF release. The potential of these formulations for topical delivery of antiviral agents for the prevention of sexually transmitted diseases was accordingly hypothesized [111].

3.6.3. Chitosan based TDF Nanoparticles

TDF was also used as a model drug in a CS based nanopreparation coated with sodium acetate, an aggregation-preventing agent, realized by the freeze-drying method. The NPs cytotoxic profile on macrophages was assessed by neutral red, resazurin, nitrite oxide and cytokines assays. Satisfactory encapsulation rate together with a good stability of the colloidal dispersions was observed for the formulation. Moreover, a sustained drug release beside a lack of cytotoxicity and a pro-inflammatory effect was recorded [112]. Further improvements in terms of mucoadhesive performance were obtained by a formulation based on TDF-loaded CS NPs dispersed in vaginal thermogels [113]. CS based oral NPs loaded with TDF were prepared by the ionic gelation technique and studied for their potential in preventing esterase metabolism and facilitate active transport uptake. Both processes were affected as confirmed by *in vitro* experiments. Moreover, data obtained suggested that a clathrin-mediated mechanism is involved in the enhancement of drug oral absorption [114]. A triple combination of TDF, FTC and bictegravir, an integrase inhibitor, was loaded into trimethyl CS to generate a nanoconjugate with improved cellular uptake. The efficiency of the nanocarrier was determined by spectrophotometry while XTT and ELISA tests were used to determine cytotoxicity and anti-retroviral efficiency, respectively. As a result, this formulation proved to inhibit viral replication at lower concentrations than the free drugs combination, without a significant cytotoxicity, therefore resulting in a lower drug resistance [115]. Colloids based on polyelectrolyte complexes of CS and chondroitin sulfate were loaded with TDF and examined for the stability at physiological conditions. This property was assured by the use of Zn(II) throughout the formulation procedure. *In vitro* studies did not reveal toxicity of such NPs on human peripheral blood mononuclear cells, while a remarkable dose-dependent antiretroviral activity was detected [116]. TDF was loaded into thiolated CS core/shell nanofibers in order to investigate the rate of drug loading, mucoadhesion properties and *in vivo* safety. The formulation was fabricated by assembling poly(ethylene oxide) with the CS component and PLA by a coaxial electrospinning technique. An enhanced drug loading together with a prolonged drug release and an increased mucoadhesion were assessed by in vitro studies, whereas a significant toxicity was not detected in neither *in vitro* nor *in vivo* experimental models. These new formulations could be therefore considered promising tools for the local delivery of microbicide agents [117].

3.6.4. Alternative Polymeric TDF NPs

An original formulation to be used for vaginal administration was fabricated by oil-in-water emulsification of the inclusion product of TFV into PLGA and sodium deoxycholate as an ion-pairing agent and a thermosensitive gel. Sustained release properties in humanized BLT mice were shown for these nanoformulations when instilled locally [118]. Similar results were obtained by loading TFV into PLGA/stearylamine and incorporating such NPs into a hydroxypropyl methylcellulose/PVA-based film [119,120]. TAF and FTC entrapped NPs were prepared for subcutaneous administration during pre-exposure prophylaxis. Drugs were included into the PLGA/PVA system and investigated for their long-acting potency detectable even after 14 days by a humanized mice model [121,122]. A similar approach was exploited for the incorporation of TAF and elvitegravir, an integrase inhibitor, during the fabrication of devices to be used during vaginal prevention [123,124]. The drug absorption following oral administration were also positively affected by the use of TAF/PGLA loaded NPs, as highlighted by a statistical model study [125]. Formulations containing mono- or by-layered films of PVA and pectin were coupled with Eudragit NPs loaded with TDF/FTC, by nano spray-drying technique. These systems were designed for vaginal use with a better patient compliance. The time of disintegration and drug release was evaluated in a simulated vaginal fluid, showing favorable results. The by-layered films equipped with NPs loaded with drugs showed the best performances in terms of drug release delay. Moreover, this topic formulation was shown particularly safe by MTT and lactate dehydrogenase assays using different cervical cell lines [126]. Multifunctional magneto-plasmonic liposomes charged with TDF were obtained with the aim to study guided systems for enhancing efficiency of antiviral treatment. The distribution of such a hybrid system can be monitored by image technique and activated magnetically into the brain. The gold shell of such nanocomplexes can be followed by computed tomography. This way, these particular systems proved to be efficient against HIV in infected microglia cells after adequately crossing the BBB [127]. A nanosuspension of drug combination particles consisting of TDF, ritonavir and lopinavir, two protease inhibitors, and lipids were prepared for the development of innovative topical formulations. This system was highly efficient in targeting lymphocytes during anti-HIV therapy with a long-lasting action after a single subcutaneous administration [128]. Similar results were described after the addition of 3TC to the previously combination to give a four-drug components nanosuspension [129]. Similar results were obtained for alternative combinations of TDF and other RTI. In all cases, a persistent drug concentration was detected after single subcutaneous injection in different HIV reservoir cells [130,131]. A stability study was performed on TAF and compared with the stress degradation behavior of TDF. Gastrointestinal stability studies were conducted on both drugs, showing the formation of six degradation products. These studies revealed a higher stability of TAF, except for with the acid condition, where the drug was extensively degraded [132].

4. Non-Nucleoside Reverse Transcriptase Inhibitors Nanosystems

An alternative approach to anti-HIV treatment with RTIs is represented by the combined therapy using both NRTIs and NNRTIs, exploiting the synergism of the two distinct classes of drugs. Accordingly, the multi-therapy is nowadays the most widely adopted strategy for the treatment of HIV infection in the clinic. NNRTI act directly by binding the enzyme, so preventing its DNA polymerase function. In fact, their heterogeneous structures do not resemble those of nucleobases, the natural substrate of RT (Figure 4, panel c).

After a first generation NNRTI drugs introduced in the 90s (i.e. nevirapine, delavirdine, efavirenz) approved for anti-HIV therapy, recently some new and effective compounds entered the market (i.e. etravirine, rilpivirine). Some more interesting compounds are currently under clinical investigation. The structures of representative NNTRI are reported in Figure 5.

4.1. Nevirapine

Nevirapine (11-cyclopropyl-4-methyl-5H-dipyrido[3,2-b:2′,3′-e][1,4]diazepin-6(11H)-one, NVR) was the first NNRTI approved by FDA in 1996 for the treatment of HIV infection. In order to improve the pharmacokinetics of this hydrophobic drug, NVR was loaded into liposomes prepared by thin film hydration and extrusion method to give uniform spherical vesicles. The matrix, obtained from egg phospholipid and cholesterol, proved to release the drug during 22 h at physiological pH values. The presence of proteins into the medium or the exposition of the system to ultrasounds greatly impair the delivery mode of the drug. However, this encapsulation method could optimize the efficacy of NVR in terms of drug stability and controlled release to the target tissues [30]. Transferrin grafted PLGA NPs have been designed in order to facilitate NVR in crossing vascular endothelial cells of the human brain. This particular nanosystem allowed a favorable drug loading with a desired controlled release, so proving to act as an efficient carrier to promote vascular diffusion of the compound [133]. Nanoparticles of PLA/PEG, the surface of which was modified with serum albumin, were prepared to improve release of the drug to the target tissues. This favorable feature was measured after i.v. injection in rats, showing an improved bioavailability, cellular uptake and drug accumulation in the brain, liver and spleen, compared to pure drug solution or uncoated nanoformulations. Moreover, no additional cytotoxicity was recorded. The capability to cross the BBB makes these formulations potentially useful for the treatment of AIDS related dementia [134,135]. The stability of NVP was well investigated by a stability-indicating ultra-high performance liquid chromatography (UHPLC) method. Drug product efficacy, safety and quality were verified in different degradation conditions by using acids, bases, water, metal ions, heat, light and oxidation agents. The tests were applied on the pure compound and on its tablet formulation leading to the formation of five degradation products [136]. A physically stable formulation of NVP was prepared by forming a crystalline inclusion complex with biodegradable and hydrophobic poly(ε-caprolactone). Compared to pure NVP crystals, the formulation assured a sustained drug release at physiological conditions in PBS solution up to 6 weeks, due to the reduction of drug solubility [43].

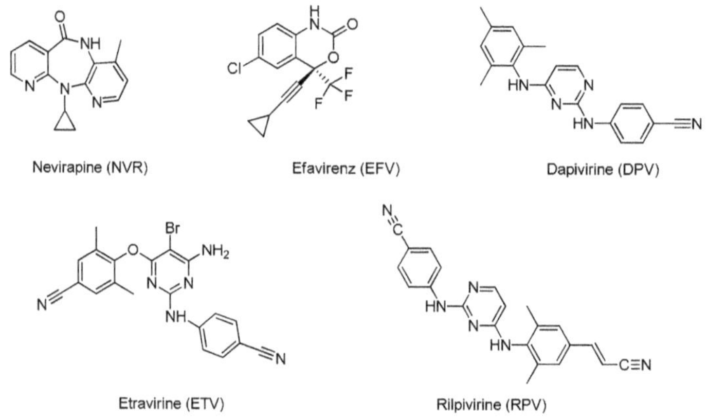

Figure 5. Chemical structures of NNRTI.

4.2. *Efavirenz*

Efavirenz, ((S)-6-chloro-4-(cyclopropylethynyl)-4-(trifluoromethyl)-1H-benzo[d][1,3]oxazin-2(4H)-one, EFV) was approved in 1998 and is largely utilized with other drugs for anti-HIV association therapies. However, its potential was limited by a low bioavailability due to its high lipophilicity and other drawbacks related to irritant effects on mucosae [137,138]. A strategy devoted to the improvement of EFV pharmacokinetics resides in its incorporation into SLN, which should drive the delivery of the drug next to the lymphoid system and brain [38]. Phenylalanine anchored SLN (PA-SLN) were used to encapsulate EFV and the resulting nanocomplex was tested for its potential to cross BBB. Phenylalanine was chosen in order to exploit the aromatic amino acid transporter active within the barrier. The nanocomplex showed good entrapment efficiency and a favorable drug release, with a remarkable accumulation in brain assuring a long-lasting therapeutic effect [139].

4.2.1. Efavirenz SLN

SLN of selected lipids as matrix medium and EFV were prepared with the addition of a surfactant by high-pressure homogenization technique and evaluated *in vivo* for their enhanced bioavailability and brain targeting. In particular, such properties were assessed after an intranasal administration route, which could be useful for therapy devoted to the complete eradication of HIV [39]. As an example, EFV was loaded into SLN assembled with the use of mono- and tri-glycerides with the aid of a surfactant. This particular formulation allowed the drug to partly by-pass liver metabolism after oral administration and in doing so increasing bioavailability and accumulation into spleen [140]. A prolonged drug release together with a lower incidence of side effects, consequent to a reduced drug dosage, was achieved after EFV incorporation into SLN.

4.2.2. Polymeric EFV NPs

Efavirenz was loaded into NPs based on methacrylate polymers, which conferred an increased drug uptake in monocytes and macrophages [141,142]. Emulsion or nanoprecipitation methods allowed loading EFV into biodegradable PLGA NPs. The effects of the resulting formulations administered in combination with other free or encapsulated anti-HIV drugs were investigated. As a result, the new formulations proved more efficient compared to the free drugs and also a noteworthy synergistic effect was recorded for encapsulated EFV combined with TDF, a second NRTI [143]. EFV was dispersed with α-tocopherol polyethylene glycol succinate and PVA by an emulsion-templated freeze-drying technique to realize solid inclusion NPs. Dry monoliths charged with these NPs proved to be stable for several months. Their reconstitution in water furnished nanodispersions, which showed reduced cytotoxicity together with an improved bioavailability and pharmacokinetics [49]. A nanoformulation was obtained by combining EFV with cellulose acetate phthalate, acting as an HIV entry inhibitor, by the nanoprecipitation method. The resulting NPs were formulated into a thermosensitive gel, which resulted in an efficient nanomicrobicide for long-term HIV prophylaxis [144]. Polymeric micelles based on pluronic F127 loaded with EFV and bio-conjugated with anti-M-cell-specific antibodies were prepared in order to evaluate their preferential target delivery to gut micro fold cells of lymphoid tissue, one of the major HIV reservoirs in the body. The efficiency of such nanosystem was showed to be higher than the free drug and a possible oral application by enteric-coated capsule was accordingly hypothesized [145]. EFV was loaded into eudragit, pluronic and alginate sodium polymeric based NPs by solvent evaporation method. Cytotoxicity and antiviral characterization was investigated by syncytium inhibition assay. The polymeric nanocarrier was proven to be more effective than the pure free drug by an enhanced drug dissolution and bio-distribution, especially after BBB crossing. A reduced toxicity was also detected [146,147]. NPs to be used during intranasal administration were prepared by the encasement of EFV into CS grafted hydroxypropyl beta cyclodextrin matrices. The results obtained confirmed a better CNS access due to an improved cellular permeation consequent to both a higher drug solubility and a concomitant beneficial action of CS on cell membrane [148]. Rectal polymeric formulations were prepared by incorporation of EFV into PLGA NPs coated with PEG. This particular form assured a prolonged drug residence in the lower colon with a long lasting prophylactic action against HIV transmission [149]. A better BBB crossing together with a reduction of side effects incidence was attempted by the preparation of transferrin functionalized PLGA NPs loaded with EFV. Although a higher deposition rate of the drug to the targeted site was recorded, the formulation did not improve drug membrane permeation [150]. Lactoferrin NPs were fabricated for oral or vaginal administration giving raise to significant results in terms of drug release and bioavailability. A combination of EFV and curcumin was also attempted in order to reach synergism for the two-microbicide agents. The results obtained confirmed an improved pharmacokinetic profile for the drug combination nanoformulation with respect to free drugs [151,152].

4.2.3. Efavirenz/Dendrimer Complexes

EFV was incorporated into t-Boc–glycine and mannose conjugated dendrimer-based poly(propyleneimine) (PPI). These branched three-dimensional polymers were shown to be less toxic than free PPI and could allow the drug to reach monocytes and macrophages, both reservoirs of HIV in the body. In fact, it is known that inhibitors targeting only lymphocytes are ineffective in completely eradicating the infection. These formulations demonstrated to be particularly effective since they are able to promote a significant increase of EFV cellular uptake [153]. The same authors loaded EFV into dendrimer-based PPI complexed with tuftsin (Tu). This latter is a tetrapeptide coming from the cleavage of immunoglobulin G and showed to be capable of selectively target and activate macrophages, monocytes and polymorph nuclear leukocytes. This strategy would help EFV in avoiding unwanted effects acting in a synergistic fashion with the peptide. Such an option was suggested by the fact that the TuPPI complex showed no toxicity and prolonged EFV cellular uptake in HIV infected macrophages with respect to uninfected cells [154].

4.2.4. Alternative Supramolecular EFV NPs

Vesicular systems consisting of gold NPs entrapped inside the aqueous core were loaded with EFV in the bilayer membrane. The niosomes so formed were dispersed in carrageenan/hyaluronic acid/poloxamer based thermogel after coating with an opportunely functionalized mannose, in order to combine the action of the protein and the sugar for an effective prophylactic vaginal application. A remarkable inhibition of viral transmission as well as a drastic reduction of side effects was recorded [31]. A soya lecithin/cholesterol and PEG liposomal DDS was designed in order to overcome the limited solubility, dissolution rate and bioavailability of EFV [32]. Similar results were achieved by the entrapment of EFV into boron nitride and carbon nanotubes as delivery vehicles. Both the systems showed a favorable behavior. Between the two distinct formulations, carbon nanotubes assured a higher drug adsorption [155]. Enhanced solubility and dissolution of the drug were also obtained by inclusion complex with hydroxypropyl-β-cyclodextrin. The solubility of the complex was further increased by the addition of l-Arginine [156]. Degradation behavior of the drug in water solution was assessed by HPLC analyses, detecting a total twelve degradation products under acidic conditions. Alkaline stress resulted in the formation of only two degradation products, whereas oxidative and photolytic conditions did not promote any degradation of the drug. All the degradation derivatives showed remarkable toxicity, including carcinogenicity, mutagenicity and skin irritation, as confirmed by *in silico* experiments [72]. The thermal behavior of glass EFV was evaluated in different experimental conditions, showing a good stability only at room temperature [157].

4.3. *Dapivirine*

Dapivirine (4-((4-(mesitylamino)pyrimidin-2-yl)amino)benzonitrile, DPV) is a new, sparingly soluble NNRTI, under investigation for vaginal application during the prevention of HIV sexual transmission. This compound was shown to be a noncompetitive inhibitor of RT that is particularly useful for topic treatments. In order to improve efficacy, a microbicide film for vaginal delivery was formulated by loading DPV into PLGA NPs and the nanosystems were in turn coupled to a PVA in a cellulose based platform film using solvent casting technique. TDF was also added to the formulation to achieve a synergist antiviral effect. A rapid release of active principles was accordingly recorded suggesting an ideal behavior of such a formulation as prophylactic microbicide [158,159]. Some advantages were also obtained by alternative formulations of this drug. In particular, surface-engineered poly(ε-caprolactone) NPs were manufactured and evaluated in pig vaginal and rectal mucosa, resulting in favorable drug release properties [160]. Physical–chemical properties of these nanocarriers were evaluated upon one-year storage to a variable temperature range. Colloidal instability affected the *in vitro* drug release of the NPs, although no detectable degradation was observed for the entrapped drug [56].

4.4. *Etravirine*

Etravirine (4-((6-amino-5-bromo-2-((4-cyanophenyl)amino)pyrimidin-4-yl)oxy)-3,5-dimethylbenzonitrile, ETV) is a NNRTI approved in 2008 for monotherapy against HIV, also clinically used in combination with other antiviral agents in antiretroviral treatment-experienced adult patients with onset of resistance. A combination of ETV, maraviroc and raltegravir was loaded into PLGA by emulsion-solvent evaporation technique. The antiviral potency of the resulting NPs was compared to the free triple drug combination in an *in vitro* cells assay and on a macaque cervicovaginal explant model *in vivo*. The nanoformulation provided a prolonged release by extension of the half-life of drugs, leading to an enhanced synergistic antiviral action [161]. An innovative approach for the design of NPs overcoming the drawback, consisting in low drug retention and massive leakage, was undertaken by the preparation of NP-releasing nanofiber vaginal devices. Mucoadhesive PVA fibers coupled to PEGylated NPs should ensure an adequate retention together with a rapid mucus diffusion. These composite nanoformulations proved to assure a sustained ETV release for up to seven days [162].

4.5. *Rilpivirine*

Rilpivirine ((E)-4-((4-((4-(2-cyanovinyl)-2,6-dimethylphenyl)amino)pyrimidin-2-yl)amino) benzonitrile, RPV) is a second-generation NNRTI, approved in 2011, possessing increased potency, longer half-life and lesser side-effects with respect to former non-nucleoside agents, today mainly used in combination with other anti-HIV drugs [163]. RPV met therapeutic success either in combination with other anti-HIV agents or in long-acting injectable nanoformulations during maintenance therapy. Its use could also be advantageous for the prophylactic treatment of high-risk uninfected individuals [164,165]. In two separate papers, the potential use of long acting RPV NPs associated with cabotegravir, an HIV integrase inhibitor, was described and an innovative monthly dosing therapeutic regimen was accordingly proposed. Recently, the result of a phase IIb study was reported on the effectiveness of this combination nanoformulation in maintaining adequate drug concentration in plasma or vaginal mucus for up to one month [166–168]. RPV-loaded PLGA NPs were fabricated by emulsion-solvent evaporation method. A sustained release in plasma as well as faster clearance were observed in animal models, suggesting a prophylactic use after the preparation of thermosensitive gels as well as long acting injectable nanosuspensions [169].

Biodistribution of tri-modal theranostic NPs was studied by single-photon emission computed tomography, magnetic resonance imaging and fluorescence techniques. These devices were prepared by the incorporation of Indium radiolabeled, europium doped cobalt-ferrite particles and RPV loaded into a poly(ε-caprolactone) matrix included into a lipid shell. A sustained drug release and antiretroviral activity were observed in HIV infected macrophages. These multi-functional NPs represent a platform for the monitoring and optimization of the antiretroviral drug pharmacokinetic profile [170].

5. Conclusions

There is a high level of interest in nanotechnologies for their relevant roles in the design and development of innovative anti-HIV formulations either for oral or topical administration. In the first case, a more specific targeted delivery together with a sustained release represents the major goal in the field. Mucoadhesive performance, prolonged retention time and improved patients' compliance were desired for vaginal or rectal application of microbicide nanoformulated antiretroviral agents. A large number of diversely assembled nanocarriers loaded with different classes of antiviral drugs were then developed and deeply investigated to improve both pharmacokinetic characteristics and stability behavior. Overall results demonstrated a potential success of such an approach for monotherapy, even though more profitable applications were recorded for the combination therapy, the most diffused method nowadays. In fact, anti-HIV therapeutic regimens today comprise multiple daily doses of more drugs acting at different stages of viral replication, which lead to poor patient compliance. Although the use of protease, integrase and viral entry inhibitors has become customary, NRTIs and NNRTIs remain pivotal tools during the infection management. In the light of these findings, this overview focuses mainly on the application of nanotechnology applied to RTI as vehicles for challenging virus eradication from depot organs and/or as a platform for the design of modern prophylactic devices for stable or stressing conditions.

Author Contributions: Conceptualization: F.G., G.I.; Supervision: R.M., A.G.; Writing-original draft preparation: M.A.O., E.M., M.D.L., F.G.; Writing-review and editing: A.G., G.R., F.G., G.I., R.M. All authors agree with the final version of the manuscript. The authors declare that the content of this paper has not been published or submitted for publication elsewhere.

Funding: This research received no external funding.

Conflicts of Interest: The authors declare no conflict of interest.

References

1. Vandamme, A.M.; Van Vaerenbergh, K.; De Clercq, E. Anti-human immunodeficiency virus drug combination strategies. *Antivir. Chem. Chemoth.* **1998**, *9*, 187–203. [CrossRef] [PubMed]
2. Lisziewicz, J.; Toke, E.R. Nanomedicine applications towards the cure of HIV. *Nanomed-Nanotechnol.* **2013**, *9*, 28–38. [CrossRef] [PubMed]
3. Bangsberg, D.R.; Hecht, F.M.; Charlebois, E.D.; Zolopa, A.R.; Holodniy, M.; Sheiner, L.; Bamberger, J.D.; Chesney, M.A.; Moss, A. Adherence to protease inhibitors, HIV-1 viral load, and development of drug resistance in an indigent population. *Aids* **2000**, *14*, 357–366. [CrossRef] [PubMed]
4. Schrager, L.K.; D'Souza, M.P. Cellular and anatomical reservoirs of HIV-1 in patients receiving potent antiretroviral combination therapy. *JAMA* **1998**, *280*, 67–71. [CrossRef] [PubMed]
5. Richman, D.D. HIV chemotherapy. *Nature* **2001**, *410*, 995–1001. [CrossRef] [PubMed]
6. Yilmaz, A.; Price, R.W.; Gisslen, M. Antiretroviral drug treatment of CNS HIV-1 infection. *J. Antimicrob. Chemother.* **2012**, *67*, 299–311. [CrossRef]
7. Guidelines for the Use of Antiretroviral Agents in Adults and Adolescents with HIV. Available online: http://aidsinfo.nih.gov/guidelines/html/1/adult-and-adolescent-treatment-guidelines/0 (accessed on 10 April 2019).
8. Rohit, S.; Ramesh, J.; Karan, G.; Raman, K.; Anil, K.S. Nanotechnological interventions in HIV drug delivery and therapeutics. *Biointerface Res. Appl. Chem.* **2014**, *4*, 820–831.
9. Harrigan, P.R.; Hogg, R.S.; Dong, W.W.; Yip, B.; Wynhoven, B.; Woodward, J.; Brumme, C.J.; Brumme, Z.L.; Mo, T.; et al. Predictors of HIV drug-resistance mutations in a large antiretroviral-naive cohort initiating triple antiretroviral therapy. *J. Infect. Dis.* **2005**, *191*, 339–347. [CrossRef] [PubMed]
10. Chun, T.W.; Davey, R.T., Jr.; Engel, D.; Lane, H.C.; Fauci, A.S. Re-emergence of HIV after stopping therapy. *Nature* **1999**, *401*, 874–875. [CrossRef] [PubMed]
11. Marsden, M.D.; Zack, J.A. Eradication of HIV: current challenges and new directions. *J. Antimicrob. Chemother.* **2009**, *63*, 7–10. [CrossRef] [PubMed]
12. Nie, S.; Xing, Y.; Kim, G.J.; Simons, J.W. Nanotechnology applications in cancer. *Annu. Rev. Biomed. Eng.* **2007**, *9*, 257–288. [CrossRef] [PubMed]
13. Mazzuca, P.; Caruso, A.; Caccuri, F. HIV-1 infection, microenvironment and endothelial cell dysfunction. *New. Microbiol.* **2016**, *39*, 163–173. [PubMed]
14. Mahajan, S.D.; Aalinkeel, R.; Law, W.C.; Reynolds, J.L.; Nair, B.B.; Sykes, D.E.; Yong, K.T.; Roy, I.; Prasad, P.N.; Schwartz, S.A. Anti-HIV-1 nanotherapeutics: promises and challenges for the future. *Int. J. Nanomed.* **2012**, *7*, 5301–5314. [CrossRef] [PubMed]
15. Vyas, T.K.; Shah, L.; Amiji, M.M. Nanoparticulate drug carriers for delivery of HIV/AIDS therapy to viral reservoir sites. *Expert Opin. Drug Deliv.* **2006**, *3*, 613–628. [CrossRef]
16. Amiji, M.M.; Vyas, T.K.; Shah, L.K. Role of nanotechnology in HIV/AIDS treatment: potential to overcome the viral reservoir challenge. *Discov. Med.* **2006**, *6*, 157–162.
17. Ferrari, M. Cancer nanotechnology: opportunities and challenges. *Nat. Rev. Cancer* **2005**, *5*, 161–171. [CrossRef] [PubMed]
18. Farokhzad, O.C. Nanotechnology for drug delivery: the perfect partnership. *Expert Opin. Drug Deliv.* **2008**, *5*, 927–929. [CrossRef] [PubMed]
19. Farokhzad, O.C.; Langer, R. Impact of nanotechnology on drug delivery. *ACS Nano* **2009**, *3*, 16–20. [CrossRef] [PubMed]
20. Dunge, A.; Sharda, N.; Singh, B.; Singh, S. Establishment of inherent stability of stavudine and development of a validated stability-indicating HPLC assay method. *J. Pharm. Biomed. Anal.* **2005**, *37*, 1115–1119. [CrossRef] [PubMed]
21. Devrukhakar, P.S.; Shiva Shankar, M.; Shankar, G.; Srinivas, R. A stability-indicating LC-MS/MS method for zidovudine: Identification, characterization and toxicity prediction of two major acid degradation products. *J. Pharm. Biomed. Anal.* **2017**, *7*, 231–236. [CrossRef]
22. Shegokar, R.; Singh, K.K.; Muller, R.H. Production & stability of stavudine solid lipid nanoparticles–from lab to industrial scale. *Int. J. Pharm.* **2011**, *416*, 461–470. [PubMed]

23. Nowacek, A.; Gendelman, H.E. NanoART, neuroAIDS and CNS drug delivery. *Nanomedicine (Lond)* **2009**, *4*, 557–574. [CrossRef] [PubMed]
24. *ICH Q1A(R2)- Stability Testing of New Drug Substances and Products*; International Council for Harmonisation of Technical Requirements for Pharmaceuticals for Human Use (ICH): Geneve, Switzerland, 2003.
25. Tonnesen, H.H. *Photostability of Drugs and Drug Formulations*; CRC Press: New York, NY, USA, 2004.
26. De Luca, M.; Ioele, G.; Spatari, C.; Ragno, G. Photostabilization studies of antihypertensive 1,4-dihydropyridines using polymeric containers. *Int. J. Pharm.* **2016**, *505*, 376–382. [CrossRef] [PubMed]
27. Garg, M.; Asthana, A.; Agashe, H.B.; Agrawal, G.P.; Jain, N.K. Stavudine-loaded mannosylated liposomes: *in-vitro* anti-HIV-I activity, tissue distribution and pharmacokinetics. *J. Pharm. Pharmacol.* **2006**, *58*, 605–616. [CrossRef] [PubMed]
28. Garg, M.; Dutta, T.; Jain, N.K. Reduced hepatic toxicity, enhanced cellular uptake and altered pharmacokinetics of stavudine loaded galactosylated liposomes. *Eur. J. Pharm. Biopharm.* **2007**, *67*, 76–85. [CrossRef]
29. Nayak, D.; Boxi, A.; Ashe, S.; Thathapudi, N.C.; Nayak, B. Stavudine loaded gelatin liposomes for HIV therapy: Preparation, characterization and *in vitro* cytotoxic evaluation. *Mater. Sci. Eng. C Mater. Biol. Appl.* **2017**, *73*, 406–416. [CrossRef] [PubMed]
30. Ramana, L.N.; Sethuraman, S.; Ranga, U.; Krishnan, U.M. Development of a liposomal nanodelivery system for nevirapine. *Int. J. Biomed. Sci.* **2010**, *17*, 57. [CrossRef] [PubMed]
31. Malik, T.; Chauhan, G.; Rath, G.; Kesarkar, R.N.; Chowdhary, A.S.; Goyal, A.K. Efaverinz and nano-gold-loaded mannosylated niosomes: a host cell-targeted topical HIV-1 prophylaxis via thermogel system. *Artif. Cells Nanomed. Biotechnol.* **2018**, *46*, 79–90. [CrossRef] [PubMed]
32. Rao, M.R.P.; Babrekar, L.S. Liposomal Drug Delivery for Solubility and Bioavailability Enhancement of Efavirenz. *Indian J. Pharm. Sci.* **2018**, *80*, 1115–1124.
33. Abdelkader, H.; Alani, A.W.; Alany, R.G. Recent advances in non-ionic surfactant vesicles (niosomes): self-assembly, fabrication, characterization, drug delivery applications and limitations. *Drug Deliv.* **2014**, *21*, 87–100. [CrossRef] [PubMed]
34. Khan, R.; Irchhaiya, R. Niosomes: a potential tool for novel drug delivery. *J. Pharm. Investig.* **2016**, *46*, 195–204. [CrossRef]
35. Naseri, N.; Valizadeh, H.; Zakeri-Milani, P. Solid Lipid Nanoparticles and Nanostructured Lipid Carriers: Structure, Preparation and Application. *Adv. Pharm. Bull.* **2015**, *5*, 305–313. [CrossRef]
36. Joshy, K.S.; Chandra, P.S.; Nandakumar, K.; Sandeep, K.; Sabu, T.; Laly, A.P. Evaluation of *in-vitro* cytotoxicity and cellular uptake efficiency of zidovudine-loaded solid lipid nanoparticles modified with Aloe Vera in glioma cells. *Mater. Sci. Eng. C Mater. Biol. Appl.* **2016**, *66*, 40–50.
37. Freeling, J.P.; Koehn, J.; Shu, C.; Sun, J.; Ho, R.J. Anti-HIV drug-combination nanoparticles enhance plasma drug exposure duration as well as triple-drug combination levels in cells within lymph nodes and blood in primates. *AIDS Res. Hum. Retroviruses* **2015**, *31*, 107–114. [CrossRef] [PubMed]
38. Mishra, N.; Mishra, M.; Padh, H. Formulation Development and Optimization of Efavirenz Loaded SLNs and NLCs using Plackett- Burman Design and its Statistical Elucidation. *Int. J. Pharm. Res. Health Sci.* **2018**, *6*, 2379–2388.
39. Gupta, S.; Kesarla, R.; Chotai, N.; Misra, A.; Omri, A. Systematic Approach for the Formulation and Optimization of Solid Lipid Nanoparticles of Efavirenz by High Pressure Homogenization Using Design of Experiments for Brain Targeting and Enhanced Bioavailability. *Biomed. Res. Int.* **2017**, *2017*, 5984014. [CrossRef] [PubMed]
40. Madhusudhan, A.; Bhagavanth, R.G.; Venkatesham, M.; Veerabhadram, G. Design and Evaluation of Efavirenz loaded Solid Lipid Nanoparticles to Improve the Oral Bioavailability. *Int. J. Pharm. Pharm. Sci.* **2012**, *2*, 84–89.
41. Pardeshi, C.; Rajput, P.; Belgamwar, V.; Tekade, A.; Patil, G.; Chaudhary, K.; Sonje, A. Solid lipid based nanocarriers: an overview. *Acta Pharm.* **2012**, *62*, 433–472. [CrossRef]
42. Date, A.A.; Destache, C.J. A review of nanotechnological approaches for the prophylaxis of HIV/AIDS. *Biomaterials* **2013**, *34*, 6202–6228. [CrossRef]
43. Yang, X.; Yu, B.; Zhong, Z.; Guo, B.H.; Huang, Y. Nevirapine-polycaprolactone crystalline inclusion complex as a potential long-acting injectable solid form. *Int. J. Pharm.* **2018**, *543*, 121–129. [CrossRef] [PubMed]
44. Cao, S.; Woodrow, K.A. Nanotechnology approaches to eradicating HIV reservoirs. *Eur. J. Pharm. Biopharm.* **2018**, *138*, 48–63. [CrossRef] [PubMed]

45. Singh, L.; Kruger, H.G.; Maguire, G.E.M.; Govender, T.; Parboosing, R. The role of nanotechnology in the treatment of viral infections. *Therap. Adv. Infect. Dis.* **2017**, *4*, 105–131. [CrossRef] [PubMed]
46. Parboosing, R.; Maguire, G.E.; Govender, P.; Kruger, H.G. Nanotechnology and the treatment of HIV infection. *Viruses* **2012**, *4*, 488–520. [CrossRef] [PubMed]
47. das Neves, J.; Amiji, M.M.; Bahia, M.F.; Sarmento, B. Nanotechnology-based systems for the treatment and prevention of HIV/AIDS. *Adv. Drug Deliv. Rev.* **2010**, *62*, 458–477. [CrossRef] [PubMed]
48. Patel, B.K.; Parikh, R.H.; Patel, N. Targeted delivery of mannosylated-PLGA nanoparticles of antiretroviral drug to brain. *Int. J. Nanomed.* **2018**, *13*, 97–100. [CrossRef] [PubMed]
49. McDonald, T.O.; Giardiello, M.; Martin, P.; Siccardi, M.; Liptrott, N.J.; Smith, D.; Roberts, P.; Curley, P.; Schipani, A.; Khoo, S.H.; et al. Antiretroviral solid drug nanoparticles with enhanced oral bioavailability: production, characterization, and *in vitro-in vivo* correlation. *Adv. Healthc. Mater.* **2014**, *3*, 400–411. [CrossRef] [PubMed]
50. Marcos-Almaraz, M.T.; Gref, R.; Agostoni, V.; Kreuz, C.; Clayette, P.; Serre, C.; Couvreurb, P.; Horcajada, P. Towards improved HIV-microbicide activity through the co-encapsulation of NRTI drugs in biocompatible metal organic framework nanocarriers. *J. Mater. Chem. B* **2017**, *5*, 8563–8569. [CrossRef]
51. Chiodo, F.; Marradi, M.; Calvo, J.; Yuste, E.; Penades, S. Glycosystems in nanotechnology: Gold glyconanoparticles as carrier for anti-HIV prodrugs. *Beilstein J. Org. Chem.* **2014**, *10*, 1339–1346. [CrossRef]
52. Peng, J.; Wu, Z.; Qi, X.; Chen, Y.; Li, X. Dendrimers as potential therapeutic tools in HIV inhibition. *Molecules* **2013**, *18*, 7912–7929. [CrossRef]
53. Ionov, M.; Ciepluch, K.; Klajnert, B.; Glinska, S.; Gomez-Ramirez, R.; de la Mata, F.J.; Munoz-Fernandez, M.A.; Bryszewska, M. Complexation of HIV derived peptides with carbosilane dendrimers. *Colloids Surf. B Biointerfaces* **2013**, *101*, 236–242. [CrossRef]
54. Garcia-Broncano, P.; Cena-Diez, R.; de la Mata, F.J.; Gomez, R.; Resino, S.; Munoz-Fernandez, M.A. Efficacy of carbosilane dendrimers with an antiretroviral combination against HIV-1 in the presence of semen-derived enhancer of viral infection. *Eur. J. Pharmacol.* **2017**, *811*, 155–163. [CrossRef] [PubMed]
55. Cena-Diez, R.; Vacas-Cordoba, E.; Garcia-Broncano, P.; de la Mata, F.J.; Gomez, R.; Maly, M.; Munoz-Fernandez, M.A. Prevention of vaginal and rectal herpes simplex virus type 2 transmission in mice: mechanism of antiviral action. *Int. J. Nanomed.* **2016**, *11*, 2147–2162. [PubMed]
56. das Neves, J.; Amiji, M.; Bahia, M.F.; Sarmento, B. Assessing the physical-chemical properties and stability of dapivirine-loaded polymeric nanoparticles. *Int. J. Pharm.* **2013**, *456*, 307–314. [CrossRef] [PubMed]
57. Nucleoside Reverse Transcriptase Inhibitors (NRTIs or nukes). Available online: https://www.hiv.va.gov/patient/treat/nrtis.asp (accessed on 10 April 2019).
58. Shegokar, R.; Singh, K.K. Stavudine entrapped lipid nanoparticles for targeting lymphatic HIV reservoirs. *Pharmazie* **2011**, *66*, 264–271. [PubMed]
59. Kaur, C.D.; Nahar, M.; Jain, N.K. Lymphatic targeting of zidovudine using surface-engineered liposomes. *J. Drug Target* **2008**, *16*, 798–805. [CrossRef] [PubMed]
60. Weibull, W. A statistical distribution function of wide applicability. *J. Appl. Mech.* **1951**, *18*, 293–297.
61. Carvalho, F.C.; Sarmento, V.H.; Chiavacci, L.A.; Barbi, M.S.; Gremiao, M.P. Development and *in vitro* evaluation of surfactant systems for controlled release of zidovudine. *J. Appl. Mech.* **2010**, *99*, 2367–2374. [CrossRef]
62. Mainardes, R.M.; Gremiao, M.P.; Brunetti, I.L.; da Fonseca, L.M.; Khalil, N.M. Zidovudine-loaded PLA and PLA-PEG blend nanoparticles: influence of polymer type on phagocytic uptake by polymorphonuclear cells. *J. Pharm. Sci.* **2009**, *98*, 257–267. [CrossRef] [PubMed]
63. Mainardes, R.M.; Gremiao, M.P. Nanoencapsulation and characterization of zidovudine on poly(L-lactide) and poly(L -lactide)-poly(ethylene glycol)-blend nanoparticles. *J. Nanosci. Nanotechnol.* **2012**, *12*, 8513–8521. [CrossRef] [PubMed]
64. Joshy, K.S.; Susan, M.A.; Snigdha, S.; Nandakumar, K.; Laly, A.P.; Sabu, T. Encapsulation of zidovudine in PF-68 coated alginate conjugate nanoparticles for anti-HIV drug delivery. *Int. J. Biol. Macromol.* **2018**, *107*, 929–937. [CrossRef] [PubMed]
65. Dunge, A.; Sharda, N.; Singh, B.; Singh, S. Validated specific HPLC method for determination of zidovudine during stability studies. *J. Pharm. Biomed. Anal.* **2005**, *37*, 1109–1114. [CrossRef]

66. Schenfeld, E.M.; Ribone, S.R.; Quevedo, M.A. Stability and plasmatic protein binding of novel zidovudine prodrugs: Targeting site ii of human serum albumin. *Eur. J. Pharm. Sci.* **2018**, *115*, 109–118. [CrossRef] [PubMed]
67. Perry, C.M.; Faulds, D. Lamivudine. A review of its antiviral activity, pharmacokinetic properties and therapeutic efficacy in the management of HIV infection. *Drugs* **1997**, *53*, 657–680. [CrossRef]
68. Vogenthaler, N.S. Lamivudine and second-line antiretroviral regimens. *Clin. Infect. Dis.* **2007**, *44*, 1387. [CrossRef] [PubMed]
69. Vinogradov, S.V.; Poluektova, L.Y.; Makarov, E.; Gerson, T.; Senanayake, M.T. Nano-NRTIs: efficient inhibitors of HIV type-1 in macrophages with a reduced mitochondrial toxicity. *Antivir. Chem. Chemother.* **2010**, *21*, 1–14. [CrossRef] [PubMed]
70. Bedse, G.; Kumar, V.; Singh, S. Study of forced decomposition behavior of lamivudine using LC, LC-MS/TOF and MS(n). *J. Pharm. Biomed. Anal.* **2009**, *49*, 55–63. [CrossRef]
71. Konari, N.S.; Jacob, J.T. Stability indicating validated UPLC technique for the simultaneous analysis of raltegravir and lamivudine in pharmaceutical dosage forms. *HIV AIDS Rev.* **2016**, *15*, 161–169. [CrossRef]
72. Kurmi, M.; Sahu, A.; Singh, D.K.; Singh, I.P.; Singh, S. Stability behaviour of antiretroviral drugs and their combinations. 8: Characterization and in-silico toxicity prediction of degradation products of efavirenz. *J. Pharm. Biomed. Anal.* **2018**, *148*, 170–181. [CrossRef] [PubMed]
73. Tamizhrasi, S.; Shukla, A.; Shivkumar, T.; Rathi, V.; Rathi, J.C. Formulation and evaluation of lamivudine loaded polymethacrylic acid nanoparticles. *Int. J. Pharmtech Res.* **2009**, *1*, 411–415.
74. Bing, W.; Guan Qun, C.; Zheng Wei, M.; Yu Ying, Z.; Da Hai, Y.; Chang You, G. Preparation and cellular uptake of PLGA particles loaded with lamivudine. *Chin. Sci. Bull.* **2012**, *57*, 3985–3993.
75. Rajca, A.; Li, Q.; Date, A.; Belshan, M.; Destache, C. Thermosensitive Vaginal Gel Containing PLGA-NRTI conjugated nanoparticles for HIV prophylaxis. *NSTI-Nanotech* **2013**, *3*, 293–296.
76. Ramadan, E. Transdermal microneedle–mediated delivery of polymeric lamivudine loaded nanoparticles. *J. Pharm. Technol. Drug Res.* **2016**, 5. [CrossRef]
77. Hatami, E.; Mu, Y.; Shields, D.N.; Chauhan, S.C.; Kumar, S.; Cory, T.J.; Yallapu, M.M. Mannose-decorated hybrid nanoparticles for enhanced macrophage targeting. *Biochem. Biophys. Rep.* **2019**, *17*, 197–207. [CrossRef]
78. Dev, A.; Binulal, N.S.; Anitha, A.; Nair, S.V.; Furuike, T.; Tamura, H.; Jayakumar, R. Preparation of poly(lactic acid)/chitosan nanoparticles for anti-HIV drug delivery applications. *Carbohydr. Polym.* **2010**, *80*, 833–838. [CrossRef]
79. Nesalin, J.A.J.; Smith, A.A. Stability study of chitosan nanoparticles containing some antiretroviral drugs. *Res. J. Pharm. Biol. Chem. Sci.* **2014**, *5*, 193–203.
80. Deepak, M.; Nivrati, J.; Vaibhav, R.; Ashish, K.J. Glycyrrhizin conjugated chitosan nanoparticles for hepatocyte-targeted delivery of lamivudine. *J. Pharm. Pharmacol.* **2014**, *66*, 1082–1093.
81. Tshweu, L.; Katata, L.; Kalombo, L.; Swai, H. Nanoencapsulation of water-soluble drug, lamivudine, using a double emulsion spray-drying technique for improving HIV treatment. *J. Nanopart. Res.* **2013**, *15*. [CrossRef]
82. Oluwaseun, O.; Namita, K.; Kahli, A.S.; Oleg, B.; Simeon, A.; Ayele, G.; Winston, A.A.; Sergei, N.; Emmanuel, O.A. Antiretroviral Drugs-Loaded Nanoparticles Fabricated by Dispersion Polymerization with Potential for HIV/AIDS Treatment. *Inf. Dis. Res. Treat.* **2016**, *9*, 21–32.
83. Pravalika, R.P.; Madhava, R.B.; Prakash, D.J.; Latha, K.; Prashanthi, D. Formulation and characterisation of chitosan based lamivudine nanoparticles. *Eur. J. Pharm. Med. Res.* **2017**, *4*, 377–383.
84. Kumar, P.; Lakshmi, Y.S.; C, B.; Golla, K.; Kondapi, A.K. Improved Safety, Bioavailability and Pharmacokinetics of Zidovudine through Lactoferrin Nanoparticles during Oral Administration in Rats. *PLoS ONE* **2015**, *10*, e0140399. [CrossRef]
85. Kumar, P.; Lakshmi, Y.S.; Kondapi, A.K. Triple Drug Combination of Zidovudine, Efavirenz and Lamivudine Loaded Lactoferrin Nanoparticles: an Effective Nano First-Line Regimen for HIV Therapy. *Pharm. Res.* **2017**, *34*, 257–268. [CrossRef] [PubMed]
86. Shahabadia, N.; Khorshidia, A.; Zhalehc, H.; Kashaniand, S. Synthesis, characterization, cytotoxicity and DNA binding studies of Fe3O4@ SiO2 nanoparticles coated by an antiviral drug lamivudine. *J. Drug Deliv. Sci. Technol.* **2018**, *46*, 55–65. [CrossRef]

87. Vasilyeva, S.V.; Shtil, A.A.; Petrova, A.S.; Balakhnin, S.M.; Achigecheva, P.Y.; Stetsenko, D.A.; Silnikov, V.N. Conjugates of phosphorylated zalcitabine and lamivudine with SiO2 nanoparticles: Synthesis by CuAAC click chemistry and preliminary assessment of anti-HIV and antiproliferative activity. *Bioorg. Med. Chem.* **2017**, *25*, 1696–1702. [CrossRef] [PubMed]
88. Kumar, J.A.; Kumar, A.N.; Arnab, D.; Debmalya, M.; Amalesh, S. Development of lamivudine containing multiple emulsions stabilizedby gum odina. *Future J. Pharm. Sci.* **2018**, *4*, 71–79.
89. Suma, U.S.; Parthiban, S.; Senthil, K.G.P.; Tamiz, M.T. Effect of span-80 in the formulation lamivudine niosomal gel. *Asian J. Res. Biol. Pharm. Sci.* **2015**, *4*, 35–45.
90. Godbole, M.D.; Mathur, V. Selection of phospholipid and method of formulation for optimum entrapment and release of lamivudine from liposome. *J. Drug Deliv. Ther.* **2018**, *8*, 175–183. [CrossRef]
91. Wilson, B.; Paladugu, L.; Priyadarshini, S.R.; Jenita, J.J. Development of albumin-based nanoparticles for the delivery of abacavir. *Int. J. Biol. Macromol.* **2015**, *81*, 763–767. [CrossRef] [PubMed]
92. Corbo, C.; Molinaro, R.; Parodi, A.; Toledano Furman, N.E.; Salvatore, F.; Tasciotti, E. The impact of nanoparticle protein corona on cytotoxicity, immunotoxicity and target drug delivery. *Nanomedicine (Lond.)* **2016**, *11*, 81–100. [CrossRef]
93. Lin, Z.; Gautam, N.; Alnouti, Y.; McMillan, J.; Bade, A.N.; Gendelman, H.E.; Edagwa, B. ProTide generated long-acting abacavir nanoformulations. *Chem. Commun.* **2018**, *54*, 8371–8374. [CrossRef]
94. Singh, G.; Pai, R.S. Pharmacokinetics and *in vivo* biodistribution of optimized PLGA nanoparticulate drug delivery system for controlled release of emtricitabine. *Drug Deliv.* **2014**, *21*, 627–635. [CrossRef] [PubMed]
95. Singh, G.; Pai, R.S. Optimization (central composite design) and validation of HPLC method for investigation of emtricitabine loaded poly(lactic-*co*-glycolic acid) nanoparticles: *in vitro* drug release and *in vivo* pharmacokinetic studies. *Sci. World J.* **2014**, *2014*, 583090. [CrossRef] [PubMed]
96. Mandal, S.; Belshan, M.; Holec, A.; Zhou, Y.; Destache, C.J. An Enhanced Emtricitabine-Loaded Long-Acting Nanoformulation for Prevention or Treatment of HIV Infection. *Antimicrob. Agents Chemother.* **2017**, 61. [CrossRef]
97. Rao, R.N.; Vali, R.M.; Ramachandra, B.; Raju, S.S. Separation and characterization of forced degradation products of abacavir sulphate by LC-MS/MS. *J. Pharm. Biomed. Anal.* **2011**, *54*, 279–285. [CrossRef] [PubMed]
98. Kurmi, M.; Sahu, A.; Singh, S. Stability behaviour of antiretroviral drugs and their combinations. 5: Characterization of novel degradation products of abacavir sulfate by mass and nuclear magnetic resonance spectrometry. *J. Pharm. Biomed. Anal.* **2017**, *134*, 372–384. [CrossRef] [PubMed]
99. Wang, X.J.; Youa, J.; Yub, F. Study on the thermal decomposition of emtricitabine. *J. Anal. Appl. Pyrolysis* **2015**, *115*, 344–352. [CrossRef]
100. Kurmi, M.; Singh, S. Stability behavior of antiretroviral drugs and their combinations. 7: Comparative degradation pathways of lamivudine and emtricitabine and explanation to their differential degradation behavior by density functional theory. *J.Pharm. Biomed. Anal.* **2017**, *142*, 155–161. [CrossRef]
101. Duan, J.; Freeling, J.P.; Koehn, J.; Shu, C.; Ho, R.J. Evaluation of atazanavir and darunavir interactions with lipids for developing pH-responsive anti-HIV drug combination nanoparticles. *J. Pharm. Sci.* **2014**, *103*, 2520–2529. [CrossRef] [PubMed]
102. Freeling, J.P.; Koehn, J.; Shu, C.; Sun, J.; Ho, R.J. Long-acting three-drug combination anti-HIV nanoparticles enhance drug exposure in primate plasma and cells within lymph nodes and blood. *Aids* **2014**, *28*, 2625–2627. [CrossRef] [PubMed]
103. Peter, A.I.; Naidu, E.C.; Akang, E.; Ogedengbe, O.O.; Offor, U.; Rambharose, S.; Kalhapure, R.; Chuturgoon, A.; Govender, T.; Azu, O.O. Investigating Organ Toxicity Profile of Tenofovir and Tenofovir Nanoparticle on the Liver and Kidney: Experimental Animal Study. *Toxicol. Res.* **2018**, *34*, 221–229. [CrossRef] [PubMed]
104. Pokharkar, V.B.; Jolly, M.R.; Kumbhar, D.D. Engineering of a hybrid polymer-lipid nanocarrier for the nasal delivery of tenofovir disoproxil fumarate: physicochemical, molecular, microstructural, and stability evaluation. *Eur. J. Pharm. Sci.* **2015**, *71*, 99–111. [CrossRef]
105. Ramanathan, R.; Jiang, Y.; Read, B.; Golan-Paz, S.; Woodrow, K.A. Biophysical characterization of small molecule antiviral-loaded nanolipogels for HIV-1 chemoprophylaxis and topical mucosal application. *Acta Biomater.* **2016**, *36*, 122–131. [CrossRef] [PubMed]
106. Jayant, R.D.; Atluri, V.S.; Agudelo, M.; Sagar, V.; Kaushik, A.; Nair, M. Sustained-release nanoART formulation for the treatment of neuroAIDS. *Int. J. Nanomed.* **2015**, *10*, 1077–1093. [CrossRef]

107. Vacas-Cordoba, E.; Galan, M.; de la Mata, F.J.; Gomez, R.; Pion, M.; Munoz-Fernandez, M.A. Enhanced activity of carbosilane dendrimers against HIV when combined with reverse transcriptase inhibitor drugs: searching for more potent microbicides. *Int. J. Nanomed.* **2014**, *9*, 3591–3600.
108. Briz, V.; Sepulveda-Crespo, D.; Diniz, A.R.; Borrego, P.; Rodes, B.; de la Mata, F.J.; Gomez, R.; Taveira, N.; Munoz-Fernandez, M.A. Development of water-soluble polyanionic carbosilane dendrimers as novel and highly potent topical anti-HIV-2 microbicides. *Nanoscale* **2015**, *7*, 14669–14683. [CrossRef]
109. Sepulveda-Crespo, D.; Sanchez-Rodriguez, J.; Serramia, M.J.; Gomez, R.; De La Mata, F.J.; Jimenez, J.L.; Munoz-Fernandez, M.A. Triple combination of carbosilane dendrimers, tenofovir and maraviroc as potential microbicide to prevent HIV-1 sexual transmission. *Nanomedicine (Lond)* **2015**, *10*, 899–914. [CrossRef]
110. Sepulveda-Crespo, D.; Gomez, R.; De La Mata, F.J.; Jimenez, J.L.; Munoz-Fernandez, M.A. Polyanionic carbosilane dendrimer-conjugated antiviral drugs as efficient microbicides: Recent trends and developments in HIV treatment/therapy. *Nanomed.- Nanotechnol. Biol. Med.* **2015**, *11*, 1481–1498. [CrossRef] [PubMed]
111. Agrahari, V.; Zhang, C.; Zhang, T.; Li, W.; Gounev, T.K.; Oyler, N.A.; Youan, B.B. Hyaluronidase-sensitive nanoparticle templates for triggered release of HIV/AIDS microbicide *in vitro*. *AAPS J.* **2014**, *16*, 181–193. [CrossRef]
112. Ngo, A.N.; Ezoulin, M.J.; Murowchick, J.B.; Gounev, A.D.; Youan, B.B. Sodium Acetate Coated Tenofovir-Loaded Chitosan Nanoparticles for Improved Physico-Chemical Properties. *Pharm. Res.* **2016**, *33*, 367–383. [CrossRef]
113. Timur, S.S.; Sahin, A.; Aytekin, E.; Ozturk, N.; Polat, K.H.; Tezel, N.; Gursoy, R.N.; Calis, S. Design and *in vitro* evaluation of tenofovir-loaded vaginal gels for the prevention of HIV infections. *Pharm. Dev. Technol.* **2018**, *23*, 301–310. [CrossRef] [PubMed]
114. Shailender, J.; Ravi, P.R.; Reddy Sirukuri, M.; Dalvi, A.; Keerthi Priya, O. Chitosan nanoparticles for the oral delivery of tenofovir disoproxil fumarate: formulation optimization, characterization and ex vivo and *in vivo* evaluation for uptake mechanism in rats. *Drug Dev. Ind. Pharm.* **2018**, *44*, 1109–1119. [CrossRef]
115. Shohani, S.; Mondanizadeh, M.; Abdoli, A.; Khansarinejad, B.; Salimi-Asl, M.; Ardestani, M.S.; Ghanbari, M.; Haj, M.S.; Zabihollahi, R. Trimethyl Chitosan Improves Anti-HIV Effects of Atripla as a New Nanoformulated Drug. *Curr. HIV Res.* **2017**, *15*, 56–65. [CrossRef] [PubMed]
116. Wu, D.; Ensinas, A.; Verrier, B.; Primard, C.; Cuvillier, A.; Champier, G.; Paul, S.; Delair, T. Zinc-Stabilized Chitosan-Chondroitin Sulfate Nanocomplexes for HIV-1 Infection Inhibition Application. *Mol. Pharm.* **2016**, *13*, 3279–3291. [CrossRef] [PubMed]
117. Meng, J.; Agrahari, V.; Ezoulin, M.J.; Zhang, C.; Purohit, S.S.; Molteni, A.; Dim, D.; Oyler, N.A.; Youan, B.C. Tenofovir Containing Thiolated Chitosan Core/Shell Nanofibers: *In Vitro* and *in Vivo* Evaluations. *Mol. Pharm.* **2016**, *13*, 4129–4140. [CrossRef] [PubMed]
118. Destache, C.J.; Mandal, S.; Yuan, Z.; Kang, G.; Date, A.A.; Lu, W.; Shibata, A.; Pham, R.; Bruck, P.; Rezich, M.; et al. Topical Tenofovir Disoproxil Fumarate Nanoparticles Prevent HIV-1 Vaginal Transmission in a Humanized Mouse Model. *Antimicrob. Agents Chemother.* **2016**, *60*, 3633–3639. [CrossRef] [PubMed]
119. Machado, A.; Cunha-Reis, C.; Araujo, F.; Nunes, R.; Seabra, V.; Ferreira, D.; das Neves, J.; Sarmento, B. Development and *in vivo* safety assessment of tenofovir-loaded nanoparticles-in-film as a novel vaginal microbicide delivery system. *Acta Biomater.* **2016**, *44*, 332–340. [CrossRef]
120. Cunha-Reis, C.; Machado, A.; Barreiros, L.; Araujo, F.; Nunes, R.; Seabra, V.; Ferreira, D.; Segundo, M.A.; Sarmento, B.; das Neves, J. Nanoparticles-in-film for the combined vaginal delivery of anti-HIV microbicide drugs. *J. Control. Release* **2016**, *243*, 43–53. [CrossRef] [PubMed]
121. Mandal, S.; Kang, G.; Prathipati, P.K.; Fan, W.; Li, Q.; Destache, C.J. Long-acting parenteral combination antiretroviral loaded nano-drug delivery system to treat chronic HIV-1 infection: A humanized mouse model study. *Antiviral Res.* **2018**, *156*, 85–91. [CrossRef] [PubMed]
122. Mandal, S.; Kang, G.; Prathipati, P.K.; Zhou, Y.; Fan, W.; Li, Q.; Destache, C.J. Nanoencapsulation introduces long-acting phenomenon to tenofovir alafenamide and emtricitabine drug combination: A comparative pre-exposure prophylaxis efficacy study against HIV-1 vaginal transmission. *J. Control. Release* **2019**, *294*, 216–225. [CrossRef] [PubMed]
123. Mandal, S.; Prathipati, P.K.; Kang, G.; Zhou, Y.; Yuan, Z.; Fan, W.; Li, Q.; Destache, C.J. Tenofovir alafenamide and elvitegravir loaded nanoparticles for long-acting prevention of HIV-1 vaginal transmission. *Aids* **2017**, *31*, 469–476. [CrossRef]

124. Prathipati, P.K.; Mandal, S.; Pon, G.; Vivekanandan, R.; Destache, C.J. Pharmacokinetic and Tissue Distribution Profile of Long Acting Tenofovir Alafenamide and Elvitegravir Loaded Nanoparticles in Humanized Mice Model. *Pharm. Res.* **2017**, *34*, 2749–2755. [CrossRef] [PubMed]
125. Shailender, J.; Ravi, P.R.; Saha, P.; Dalvi, A.; Myneni, S. Tenofovir disoproxil fumarate loaded PLGA nanoparticles for enhanced oral absorption: Effect of experimental variables and *in vitro*, *ex vivo* and *in vivo* evaluation. *Colloids Surf. B Biointerfaces* **2017**, *158*, 610–619. [CrossRef] [PubMed]
126. Cautela, M.P.; Moshe, H.; Sosnik, A.; Sarmento, B.; das Neves, J. Composite films for vaginal delivery of tenofovir disoproxil fumarate and emtricitabine. *Eur. J. Pharm. Biopharm.* **2018**, *138*, 3–10. [CrossRef] [PubMed]
127. Tomitaka, A.; Arami, H.; Huang, Z.; Raymond, A.; Rodriguez, E.; Cai, Y.; Febo, M.; Takemura, Y.; Nair, M. Hybrid magneto-plasmonic liposomes for multimodal image-guided and brain-targeted HIV treatment. *Nanoscale* **2017**, *10*, 184–194. [CrossRef]
128. Kraft, J.C.; McConnachie, L.A.; Koehn, J.; Kinman, L.; Sun, J.; Collier, A.C.; Collins, C.; Shen, D.D.; Ho, R.J.Y. Mechanism-based pharmacokinetic (MBPK) models describe the complex plasma kinetics of three antiretrovirals delivered by a long-acting anti-HIV drug combination nanoparticle formulation. *J. Control. Release* **2018**, *275*, 229–241. [CrossRef]
129. McConnachie, L.A.; Kinman, L.M.; Koehn, J.; Kraft, J.C.; Lane, S.; Lee, W.; Collier, A.C.; Ho, R.J.Y. Long-Acting Profile of 4 Drugs in 1 Anti-HIV Nanosuspension in Nonhuman Primates for 5 Weeks After a Single Subcutaneous Injection. *J. Pharm. Sci.* **2018**, *107*, 1787–1790. [CrossRef]
130. Koehn, J.; Iwamoto, J.F.; Kraft, J.C.; McConnachie, L.A.; Collier, A.C.; Ho, R.J.Y. Extended cell and plasma drug levels after one dose of a three-in-one nanosuspension containing lopinavir, efavirenz, and tenofovir in nonhuman primates. *Aids* **2018**, *32*, 2463–2467. [PubMed]
131. Perazzolo, S.; Shireman, L.M.; Koehn, J.; McConnachie, L.A.; Kraft, J.C.; Shen, D.D.; Ho, R.J.Y. Three HIV Drugs, Atazanavir, Ritonavir, and Tenofovir, Coformulated in Drug-Combination Nanoparticles Exhibit Long-Acting and Lymphocyte-Targeting Properties in Nonhuman Primates. *J. Pharm. Sci.* **2018**, *107*, 3153–3162. [CrossRef] [PubMed]
132. Golla, V.M.; Kurmi, M.; Shaik, K.; Singh, S. Stability behaviour of antiretroviral drugs and their combinations. 4: Characterization of degradation products of tenofovir alafenamide fumarate and comparison of its degradation and stability behaviour with tenofovir disoproxil fumarate. *J. Pharm. Biomed. Anal.* **2016**, *131*, 146–155. [CrossRef] [PubMed]
133. Kuo, Y.C.; Lin, P.I.; Wang, C.C. Targeting nevirapine delivery across human brain microvascular endothelial cells using transferrin-grafted poly(lactide-co-glycolide) nanoparticles. *Nanomedicine (Lond)* **2011**, *6*, 1011–1026. [CrossRef] [PubMed]
134. Shegokar, R.; Singh, K.K. Nevirapine nanosuspensions for HIV reservoir targeting. *Die Pharmazie* **2011**, *66*, 408–415.
135. Shegokar, R.; Singh, K.K. Surface modified nevirapine nanosuspensions for viral reservoir targeting: *In vitro* and *in vivo* evaluation. *Int. J. Pharm.* **2011**, *421*, 341–352. [CrossRef] [PubMed]
136. Reis, N.F.; de Assis, J.C.; Fialho, S.L.; Pianetti, G.A.; Fernandes, C. Stability-indicating UHPLC method for determination of nevirapine in its bulk form and tablets: identification of impurities and degradation kinetic study. *J. Pharm. Biomed. Anal.* **2016**, *126*, 103–108. [CrossRef]
137. Aungst, B.J.; Nguyen, N.H.; Taylor, N.J.; Bindra, D.S. Formulation and food effects on the oral absorption of a poorly water soluble, highly permeable antiretroviral agent. *J. Pharm. Sci.* **2002**, *91*, 1390–1395. [CrossRef] [PubMed]
138. Csajka, C.; Marzolini, C.; Fattinger, K.; Decosterd, L.A.; Fellay, J.; Telenti, A.; Biollaz, J.; Buclin, T. Population pharmacokinetics and effects of efavirenz in patients with human immunodeficiency virus infection. *Clin. Pharmacol. Ther.* **2003**, *73*, 20–30. [CrossRef]
139. Vyas, A.; Jain, A.; Hurkat, P.; Jain, A.; Jain, S.K. Targeting of AIDS related encephalopathy using phenylalanine anchored lipidic nanocarrier. *Colloids Surf. B Biointerfaces* **2015**, *131*, 155–161. [CrossRef]
140. Makwana, V.; Jain, R.; Patel, K.; Nivsarkar, M.; Joshi, A. Solid lipid nanoparticles (SLN) of Efavirenz as lymph targeting drug delivery system: Elucidation of mechanism of uptake using chylomicron flow blocking approach. *Int. J. Pharm.* **2015**, *495*, 439–446. [CrossRef] [PubMed]

141. Gaur, P.K.; Mishra, S.; Bajpai, M.; Mishra, A. Enhanced Oral Bioavailability of Efavirenz by Solid Lipid Nanoparticles: In Vitro Drug Release and Pharmacokinetics Studies. *BioMed Res. Int.* **2014**, *2014*, 363404. [CrossRef]
142. Vedha Hari, B.N.; Dhevendaran, K.; Narayanan, N. Development of Efavirenz nanoparticle for enhanced efficiency of anti-retroviral therapy against HIV and AIDS. In Proceedings of the First International Science Symposium on HIV and Infectious Diseases (HIV SCIENCE 2012), Chennai, India, 20–22 January 2012; Springer-Verlag GmbH: Heidelberg, Germany, 2012.
143. Chaowanachan, T.; Krogstad, E.; Ball, C.; Woodrow, K.A. Drug synergy of tenofovir and nanoparticle-based antiretrovirals for HIV prophylaxis. *PLoS ONE* **2013**, *8*, e61416. [CrossRef]
144. Date, A.A.; Shibata, A.; McMullen, E.; La Bruzzo, K.; Bruck, P.; Belshan, M.; Zhou, Y.; Destache, C.J. Thermosensitive Gel Containing Cellulose Acetate Phthalate-Efavirenz Combination Nanoparticles for Prevention of HIV-1 Infection. *J. Biomed. Nanotechnol.* **2015**, *11*, 416–427. [CrossRef]
145. Roy, U.; Ding, H.; Pilakka-Kanthikeel, S.; Raymond, A.D.; Atluri, V.; Yndart, A.; Kaftanovskaya, E.M.; Batrakova, E.; Agudelo, M.; Nair, M. Preparation and characterization of anti-HIV nanodrug targeted to microfold cell of gut-associated lymphoid tissue. *Int. J. Nanomed.* **2015**, *10*, 5819–5835. [CrossRef] [PubMed]
146. Haria, V.B.N.; Lu, C.L.; Narayananc, N.; Wang, R.R.; Zheng, Y.T. Engineered polymeric nanoparticles of Efavirenz: Dissolution enhancement through particle size reduction. *Chem. Eng. Sci.* **2016**, *155*, 366–375. [CrossRef]
147. Hari, B.N.V.; Narayanan, N.; Dhevendaran, K.; Ramyadevi, D. Engineered nanoparticles of Efavirenz using methacrylate co-polymer (Eudragit-E100) and its biological effects in-vivo. *Mater. Sci. Eng. C. Mater. Biol. Appl.* **2016**, *67*, 522–532. [CrossRef] [PubMed]
148. Belgamwar, A.; Khan, S.; Yeole, P. Intranasal chitosan-g-HPbetaCD nanoparticles of efavirenz for the CNS targeting. *Artif. Cells Nanomed. Biotechnol.* **2018**, *46*, 374–386. [CrossRef] [PubMed]
149. Nunes, R.; Araujo, F.; Barreiros, L.; Bartolo, I.; Segundo, M.A.; Taveira, N.; Sarmento, B.; das Neves, J. Noncovalent PEG Coating of Nanoparticle Drug Carriers Improves the Local Pharmacokinetics of Rectal Anti-HIV Microbicides. *ACS Appl. Mater. Interfaces* **2018**, *10*, 34942–34953. [CrossRef] [PubMed]
150. Martins, C.; Araújo, F.; Gomes, M.J.; Fernandes, C.; Nunes, R.; Li, W.; Santos, H.A.; Borges, F.; Sarmento, B. Using microfluidic platforms to develop CNS-targeted polymeric nanoparticles for HIV therapy. *Eur. J. Pharm. Biopharm.* **2018**, *138*, 111–124. [CrossRef] [PubMed]
151. Kumar, P.; Lakshmi, Y.S.; Kondapi, A.K. An oral formulation of efavirenz-loaded lactoferrin nanoparticles with improved biodistribution and pharmacokinetic profile. *HIV Med.* **2017**, *18*, 452–462. [CrossRef] [PubMed]
152. Lakshmi, Y.S.; Kumar, P.; Kishore, G.; Bhaskar, C.; Kondapi, A.K. Triple combination MPT vaginal microbicide using curcumin and efavirenz loaded lactoferrin nanoparticles. *Sci. Rep.* **2016**, *6*, 25479. [CrossRef]
153. Dutta, T.; Agashe, H.B.; Garg, M.; Balakrishnan, P.; Kabra, M.; Jain, N.K. Poly (propyleneimine) dendrimer based nanocontainers for targeting of efavirenz to human monocytes/macrophages *in vitro*. *J. Drug Target.* **2007**, *15*, 89–98. [CrossRef] [PubMed]
154. Dutta, T.; Garg, M.; Jain, N.K. Targeting of efavirenz loaded tuftsin conjugated poly(propyleneimine) dendrimers to HIV infected macrophages *in vitro*. *Eur. J. Pharm. Sci.* **2008**, *34*, 181–189. [CrossRef]
155. Hong, X.; Long, L.; Guohong, F.; Xiangfeng, C. DFT study of nanotubes as the drug delivery vehicles of Efavirenz. *Comput. Theor. Chem.* **2018**, *1131*, 57–68.
156. Suvarna, V.; Thorat, S.; Nayak, U.; Sherje, A.; Murahari, M. Host-guest interaction study of Efavirenz with hydroxypropyl-β-cyclodextrin and L-arginine by computational simulation studies: Preparation and characterization of supramolecular complexes. *J. Mol. Liq.* **2018**, *259*, 55–64. [CrossRef]
157. Moura Ramos, J.J.; Piedade, M.F.M.; Diogo, H.P.; Viciosa, M.T. Thermal Behavior and Slow Relaxation Dynamics in Amorphous Efavirenz: A Study by DSC, XRPD, TSDC, and DRS. *J. Pharm. Sci.* **2019**, *108*, 1254–1263. [CrossRef] [PubMed]
158. das Neves, J.; Sarmento, B.; Amiji, M.M.; Bahia, M.F. Development and validation of a rapid reversed-phase HPLC method for the determination of the non-nucleoside reverse transcriptase inhibitor dapivirine from polymeric nanoparticles. *J. Pharm. Biomed. Anal.* **2010**, *52*, 167–172. [CrossRef] [PubMed]
159. Akil, A.; Devlin, B.; Cost, M.; Rohan, L.C. Increased Dapivirine tissue accumulation through vaginal film codelivery of dapivirine and Tenofovir. *Mol. Pharm.* **2014**, *11*, 1533–1541. [CrossRef] [PubMed]

160. das Neves, J.; Araújo, F.; Andrade, F.; Michiels, J.; Ariën, K.K.; Vanham, G.; Amiji, M.; Bahia, M.F.; Sarmento, B. *In Vitro* and *Ex Vivo* Evaluation of Polymeric Nanoparticles for Vaginal and Rectal Delivery of the Anti-HIV Drug Dapivirine. *Mol. Pharm.* **2013**, *10*, 2793–2807. [CrossRef] [PubMed]
161. Jiang, Y.; Cao, S.; Bright, D.K.; Bever, A.M.; Blakney, A.K.; Suydam, I.T.; Woodrow, K.A. Nanoparticle-Based ARV Drug Combinations for Synergistic Inhibition of Cell-Free and Cell-Cell HIV Transmission. *Mol. Pharm.* **2015**, *12*, 4363–4374. [CrossRef] [PubMed]
162. Krogstad, E.A.; Ramanathan, R.; Nhan, C.; Kraft, J.C.; Blakney, A.K.; Cao, S.; Ho, R.J.Y.; Woodrow, K.A. Nanoparticle-releasing nanofiber composites for enhanced *in vivo* vaginal retention. *Biomaterials* **2017**, *144*, 1–16. [CrossRef]
163. Goebel, F.; Yakovlev, A.; Pozniak, A.; Vinogradova, E.; Boogaerts, G.; Hoetelmans, R.; de Béthune, M.P.; Peeters, M.; Woodfall, B. Short-term antiviral activity of TMC278 – a novel NNRTI – in treatment-naive HIV-1-infected subjects. *Aids* **2006**, *20*, 1721–1726. [CrossRef] [PubMed]
164. Jackson, A.; McGowan, I. Long-acting rilpivirine for HIV prevention. *Curr. Opin. HIV AIDS* **2015**, *10*, 253–257. [CrossRef] [PubMed]
165. Viciana, P. Rilpivirine: The Key for Long-term Success. *AIDS Rev.* **2017**, *19*, 156–166. [CrossRef]
166. Spreen, W.R.; Margolis, D.A.; Pottage, J.C. Long-acting injectable antiretrovirals for HIV treatment and prevention. *Curr. Opin. HIV AIDS* **2013**, *8*, 565–571. [CrossRef] [PubMed]
167. Margolis, D.A.; Boffito, M. Long-acting antiviral agents for HIV treatment. *Curr. Opin. HIV AIDS* **2015**, *10*, 246–252. [CrossRef] [PubMed]
168. Ferretti, F.; Boffito, M. Rilpivirine long-acting for the prevention and treatment of HIV infection. *Curr. Opin. HIV AIDS* **2018**, *13*, 300–307. [CrossRef] [PubMed]
169. Kovarova, M.; Council, O.D.; Date, A.A.; Long, J.M.; Nochi, T.; Belshan, M.; Shibata, A.; Vincent, H.; Baker, C.E.; Thayer, W.O.; et al. Nanoformulations of Rilpivirine for Topical Pericoital and Systemic Coitus-Independent Administration Efficiently Prevent HIV Transmission. *PLoS Pathog.* **2015**, *11*, e1005075.
170. Ottemann, B.M.; Helmink, A.J.; Zhang, W.; Mukadam, I.; Woldstad, C.; Hilaire, J.R.; Liu, Y.; McMillan, J.M.; Edagwa, B.J.; Mosley, R.L.; et al. Bioimaging predictors of rilpivirine biodistribution and antiretroviral activities. *Biomaterials* **2018**, *185*, 174–193. [CrossRef] [PubMed]

MDPI
St. Alban-Anlage 66
4052 Basel
Switzerland
Tel. +41 61 683 77 34
Fax +41 61 302 89 18
www.mdpi.com

Pharmaceutics Editorial Office
E-mail: pharmaceutics@mdpi.com
www.mdpi.com/journal/pharmaceutics

www.ingramcontent.com/pod-product-compliance
Lightning Source LLC
LaVergne TN
LVHW071946080526
838202LV00064B/6684